Exploration, Religion, and Empire in the Sixteenth-century Ibero-Atlantic World

# Exploration, Religion, and Empire in the Sixteenth-century Ibero-Atlantic World

*A New Perspective on the History of Modern Science*

*Mauricio Nieto Olarte*

Routledge
Taylor & Francis Group
LONDON AND NEW YORK

Originally published as: *Las máquinas del imperio y el reino de Dios. Reflexiones sobre ciencia, tecnología y religión en el Atlántico del siglo XVI*, Ediciones Uniandes 2013

Translation: Jimmy Weiskopf

First published in 2022 by Amsterdam University Press Ltd.

Published 2025 by Routledge
4 Park Square, Milton Park, Abingdon, Oxon OX14 4RN
605 Third Avenue, New York, NY 10158

*Routledge is an imprint of the Taylor & Francis Group, an informa business*



ISBN: 9789463725316 (hbk)
ISBN: 9781041179092 (pbk)
ISBN: 9781003695189 (ebk)
NUR 685

Cover illustration: Frans Huys (engraver) and Pietr Brueghel the Elder (designer), *Armed three-masted ship off a coast, with the Fall of Daedalus and Icarus*, c. 1561–1562. Engraving and etching; three states known. CC BY-NC © Fondation Custodia, Collection Frits Lugt, Paris

Cover design: Coördesign, Leiden

DOI 10.5117/9789463725316

*'Muchos viajarán y el conocimiento aumentará'.*[1]

'[...] *España descubrió el Nuevo Mundo para que todas las naciones estuvieran bajo una sola ley'.*[2]

1 *Multi pertransibunt & augebitur cientia* ('Many shall run to and for, and knowledge shall be increased'). Prophecy from the Book of Daniel 12:4, on the frontispiece of Francis Bacon's *Instauratio Magna*, first published in London in 1620.

2 Campanella, Tommaso, ('Spain discovered the New World so that all nations would be under a single law') 'La Imaginaria ciudad del sol, idea de una República Filosófica', in: Ímaz Eugenio, *Utopías del Renacimiento*, 17th ed., Fondo de Cultura Económica, Mexico, 2009, page 30.

*Emma, your life resembles that of the protagonists of this book: it has been full of questions, discoveries, adventures, and new horizons. Like the explorers of the Atlantic, you do not like limits: you are independent and stubborn, and I hope that you will be like that forever.*

# Table of Contents


**List of illustrations** 11

**Acknowledgments** 15

**Introduction** 17
- The New World and the problem of Eurocentrism 17
- Science and empire 27
- Summary of the chapters in this book 32

**1. The Iberian Peninsula and the Atlantic** 37
- Portugal and Spain 37
- Winds, currents, and sailing ships in the Atlantic 44
- Gold, silver, slaves, souls, and a thousand kinds of trees 50

**2. The imperial bureaucracy and the appropriation of the New World** 55
- Seville and the *Casa de Contratación* 58
- The universal monarchy 64

**3. The *piloto mayor*: cosmography and the art of navigation** 75
- The post of *piloto mayor*: seamanship and cartography 76
- The navigation manuals 80
- Manuals for the Empire 81
- Publications, dissemination, and secrecy 89
- Humanism and the classics 92
- Experience and authority 105
- Man against the sea: shipwrecks and meteorology 109
- Routes and chorographic descriptions: The New World within the new global order 116

**4. Machines of the empire** 119
- The ships 125
- Shipbuilding 132
- War and artillery 141
- Navigational instruments 143
  - The astrolabe 145
  - The cross-staff 151
  - The mariner's compass 158

Time and clocks 166
The sounding/plumb line 172
The navigation charts 172
Astronomical tables 175
Instruments, measurements, precision, and standardization 182
The crew 185
The captain/admiral 191
The pilot 191
The shipmaster (*maestre*) and quartermaster (*contramaestre*) 195
The boatswain (*guardián*) 195
The ordinary seamen (*marineros*) 196
Midshipmen (*grumetes*) and cabin boys (*pajes*) 196
The carpenter, steward, cooper, and cook 197
The scribe, master-at-arms, and overseer 198
The cannoneer 198
The barber/surgeon 199
The priest 199
Life on board 199
The argot of the sailors 200
Overcrowding 202
Food and health 204
Men of the sea and men of God 207

**5. The Master Map (*Padrón Real*) and the cartography of the New World** 217
Nautical charts and how they were made 223
The making of a chart 225
The charts of *tierra firme*: the earliest maps of the New World 229
Three early maps of the New World 234
Juan de la Cosa (1500) 234
Waldseemüller (1507) 236
Diego Ribero (1520) 240

**6. The creatures of God never seen before: natural history** 245
Nature in the New World 245
The classics and the Bible 252
Monsters in paradise 258
To describe, classify, and name 263
Medicine, botany, and the knowledge of the natives 271
The Empire and natural history 281

7. **The New World, global science, and Eurocentrism** 285
*Plus ultra* 285
Experience and authority 292
The Empire and the challenge of standardization 300
Eurocentrism 303

**Bibliography** 307

**About the Author** 319

**Index** 321

# List of illustrations


| | | |
|---|---|---|
| *Illustration 1.1.* | On the winds, their quality and names, and how one should sail in them | 47 |
| *Illustration 1.2.* | The planetary winds | 48 |
| *Illustration 1.3.* | The main ocean currents on the Earth | 48 |
| *Illustration 1.4.* | The voyages of Columbus, 1492–1503 | 49 |
| *Illustration 2.1.* | Panoramic view of Seville in the 17th century | 60 |
| *Illustration 3.1.* | *Suma de Geographia*, by Martín Fernández de Enciso | 83 |
| *Illustration 3.2.* | *Regimiento de Navegación*, by Pedro de Medina | 84 |
| *Illustration 3.3.* | Cartagena and Punta de los Icacos | 93 |
| *Illustration 3.4.* | *Arte de Navegar*, by Pedro de Medina | 96 |
| *Illustration 3.5.* | The sphere and the four elements | 97 |
| *Illustration 3.6.* | Ptolemaic map | 99 |
| *Illustration 3.7.* | The four elements: earth, water, air and fire | 100 |
| *Illustration 3.8.* | The celestial sphere, with the Earth in the center, around which the Sun, planets, and stars revolve | 101 |
| *Illustration 3.9.* | Poles of the world and the basic coordinates of the heavenly sphere | 102 |
| *Illustration 3.10.* | Height of the Sun above the equinoctial line | 103 |
| *Illustration 3.11.* | The seven planets | 104 |
| *Illustration 3.12.* | Rules of the tides | 111 |
| *Illustration 3.13.* | Navigational instructions XIIII: 'On when the ship is grasped by the force of the weather and is in danger of being lost: what one should do to correct it' | 112 |
| *Illustration 3.14.* | Navigational instructions XI: 'On the measures [remedy] which should be taken when the vessel is sailing and ships a lot of water' | 114 |
| *Illustration 4.1.* | *Columbus the first discoverer*, by Theodor de Bry | 122 |
| *Illustration 4.2.* | Navigational instructions II: 'How the pilot should know the ship in which he has to sail and be familiar with the peculiarities it has' | 126 |
| *Illustration 4.3.* | Displays of a ship that sails against the wind and into the wind | 128 |
| *Illustration 4.4.* | *Sailing ship with the Fall of Icarus*, by Peter Brueghel | 132 |

| | | |
|---|---|---|
| *Illustration 4.5.* | 'On the calculations and other things which pertain to the measurements of the Hull of any Ship' | 138 |
| *Illustration 4.6.* | 'On the calculations and other things which pertain to the measurements of the Hull of any Ship' | 138 |
| *Illustration 4.7.* | On the mainsail and foresail | 140 |
| *Illustration 4.8.* | Maritime astrolabe | 147 |
| *Illustration 4.9.* | Explanation of the position the sailor must follow when he is sailing on the sea and wishes to know the height of the Sun | 150 |
| *Illustration 4.10.* | Nautical astrolabe | 152 |
| *Illustration 4.11.* | Explanation of the astrolabe for taking the height of the Sun and its use | 153 |
| *Illustration 4.12.* | Explanation of the quadrant for taking the height of the Sun and the stars, and its use | 154 |
| *Illustration 4.13.* | Drawing of the use of the astronomical staff or cross-staff | 155 |
| *Illustration 4.14.* | On the height of the north | 156 |
| *Illustration 4.15.* | Use of the cross-staff | 157 |
| *Illustration 4.16.* | Order and rule of the Southern Cross in order to take its height, both on sea and on land | 159 |
| *Illustration 4.17.* | Explanation of the mariner's compass | 160 |
| *Illustration 4.18.* | Compass needle | 161 |
| *Illustration 4.19.* | Compass rose | 162 |
| *Illustration 4.20.* | *Carta Pisana*, c. 1275 | 163 |
| *Illustration 4.21.* | On the mariner's compasses | 165 |
| *Illustration 4.22.* | Sixteenth-century mariner's sandglass | 169 |
| *Illustration 4.23.* | Night chart to determine the hours with the North Star | 170 |
| *Illustration 4.24.* | On sailing in the western seas and lands | 171 |
| *Illustration 4.25.* | Table of the Sun's declination | 176 |
| *Illustration 4.26.* | On the height of the Sun and the rules for navigating by it | 179 |
| *Illustration 4.27.* | Perpetual table to find out lunar conjunctions | 181 |
| *Illustration 4.28.* | Phases of the Moon | 182 |
| *Illustration 4.29.* | Tables to find out at what hours the tides come in | 183 |
| *Illustration 4.30.* | Calendar | 214 |
| *Illustration 5.1.* | Map of the recently discovered New World | 220 |
| *Illustration 5.2.* | Map of the New World | 221 |

*Illustration 5.3.* Maps of different parts of America 222
*Illustration 5.4.* Navigational chart containing the navigation for most of Europe, Africa, the Indies or New World, and the altitude and degrees that each thing has 225
*Illustration 5.5.* The leagues calculated by degree in each course of the voyage 233
*Illustration 5.6.* Juan de la Cosa's Map, 1500 235
*Illustration 5.7.* Compass rose with a picture of the Virgin Mary and Jesus at the center of the map 236
*Illustration 5.8.* In the middle of the New World the figure of Saint Christopher crosses the seas, suggesting the idea of a passage to the Orient 237
*Illustration 5.9.* *Universalis Cosmographie descriptio in plano*, by Martin Waldseemüller 238
*Illustration 5.10.* Universal map containing all of the world that has been discovered up to now, by Diego Ribero 241
*Illustration 5.11.* Quadrant 242
*Illustration 5.12.* Toponymy of the New World 243
*Illustration 6.1.* Iguana 259
*Illustration 6.2.* Pineapple 267
*Illustration 6.3.* *Tlazolteozacatl, Tlayapaloni, Axocotl, Chicomacatl* 276
*Illustration 6.4.* *Tezonpahtli, Huitzquilitl, Tecuammaitl* 278
*Illustration 6.5.* *Atzitzicaztli* 279
*Illustration 6.6.* Museum of Ferrante Imperato in Naples, 1599 283
*Illustration 7.1.* Frontispiece of Francis Bacon's *Instauratio Magna* 287
*Illustration 7.2.* *Historia general y natural de las Indias*, by Gonzalo Fernández de Oviedo, 1535. 290
*Illustration 7.3.* *Regimiento de Navegación*, by Andrés García de Céspedes, Folio I. 291

# Acknowledgments

I must clarify that the contents of this book and its proposals of a theoretical nature are obviously indebted to specific books and authors. My aim here is not to present an exhaustive account of the state of art in this field but to acknowledge those studies that motivated and shaped the questions and arguments in this book. For example, John Law's works on Portuguese navigation have been an obvious inspiration for the general framework of this investigation.

There is a long list of authors who have dealt with the sixteenth-century Iberian science and I hope that I have given them their due credit in the footnotes of this book. However, I must single out the pioneering studies of José María López Piñero, as well as the more recent editorial emendation of Antonio Barrera, Jorge Cañizares, María Portuondo, and Alison Sandman, who have conclusively shown that there is an urgent need to study these aspects of the history of European modernity.

This book would not have been written without the support of the Universidad de los Andes. The first stage of the project was aided by the Centro de Estudios Sociales (Center of Social Studies – CESO) of the School of Social Sciences. Subsequently, the second stage was financed by Colciencias (the Administrative Department of Science, Technology and Innovation, a Colombian government agency that supports fundamental and applied research); Lastly the final writing of the book would not have been possible without the collaboration of the History Department of Social Science of the Univesity de los Andes, which granted me a sabbatical in the first semester of 2011.

Throughout the whole process of researching and writing this book, I had the support of graduate students. The help of Rafael Acevedo was a decisive factor in the initial stages of research and the assembling and organization of information When it came to editing the manuscript, the help of Daniela Samur was equally crucial, as she put the chaos of quotes and references of the first draft into order. In addition, Lina Rocío Medina undertook a general revision of the text. The contributions and suggestions of all these people were very opportune. The comments made by Camilo Quintero and Alexis De Greiff after reading the manuscript helped me to spot flaws, which I have tried to correct.

# Introduction

**Abstract**
The book's introduction exposes its historiographical challenges and theoretical framework. This includes a discussion of the limitations of the traditional forms of considering the relationship between science and empire, and proposes original ways of dealing with old problems related to both Eurocentrism and modern science.

**Key words:** Discovery, Modern Science, Voyages of exploration, Eurocentrism

## The New World and the problem of Eurocentrism

'The greatest thing after the creation of the world, except for the incarnation and the death of he who gave birth to it, is the discovery of the Indies'.[1] Francisco López de Gómara wrote those words in the Introduction to his *Historia General de las Indias* (General History of the Indies), published in 1552. Far from being an extravagant idea at the time, the providential notion that God wanted Spain to conquer the New World for the spread and final triumph of Christianity[2] was widely shared by the chroniclers and cosmographers of the Spanish Crown in the 16$^{th}$ century.[3] As the Spaniards saw it, 1492 was the year that split the history of the world in two; five centuries later, we are still trying to understand what happened and the world has not finished adjusting itself to the changes that began in the Atlantic.

1 López de Gómara, Francisco, *Historia General de las Indias*, [1552], Pilar Guibelalde and Emiliano Aguilera (eds.), Barcelona, Iberia, 1965, 'To Don Carlos, Emperor of the Romans, King of Spain, Lord of the Indies and the New World', page 5.
2 Ibid., 'God wished the Indies to be discovered in your age and by your vassals, so that they would be converted to his holy law, as many wise and Christian men say', page 6.
3 López Piñero, José María, *El arte de navegar en la España del Renacimiento*, Barcelona, Editorial Labor, 1986, Chapter VIII.

Nieto Olarte, M., *Exploration, Religion, and Empire in the Sixteenth-century Ibero-Atlantic World: A New Perspective on the History of Modern Science*. Taylor & Francis Group, 2022
DOI 10.5117/9789463725316_INTRO

In the abovementioned introduction to his book, López de Gómara wrote that 'every history, even if is not well written, delights',[4] this may be a questionable statement but it is a convenient way to raise your spirits when you set out to write an incredible history of the European Renaissance. Here, is yet another attempt. This time, the account will deal with personages and regions which are little seen in the historiography of modern science: the Iberian Catholics in the Atlantic world.

Many have called the 16th century the century of discoveries. Equating European expansion with the 'age of discoveries' forms part of a view of history that is centered on Europe and that is only possible when the great explorers who preceded Columbus are forgotten. One might mention, for example, the Polynesians' maritime explorations of the Pacific, the Nordic navigators in the Atlantic, and, naturally, the Chinese who, led by admiral Zheng, crossed the Indian Ocean.[5] However, it is undeniable that at the end of the 15th century and throughout the 16th century the Christian navigators, and the Iberians in particular, opened routes, sailed long distances, and entered into contact with parts of the globe and cultures that were then completely unknown. This is not the place to write a full account of the maritime expansion of Europe in the 16th century, but it is nevertheless worth recalling some names in order to form an idea of the scale of the endeavor of Christian expansion in that period. Admiral Christopher Columbus sought an alternative trade route to the Orient and managed to cross the Atlantic four times; and, while he never reached Cipango, he found a continent that was new to Europeans along the way. Vasco de Gama reached India from Lisbon and returned to the Portuguese capital after slightly more than two years circumnavigating the continent of Africa: a voyage in which less than half of his men survived. After a voyage of more than 40 days, Pedro Álvares Cabral reached the coasts of Brazil, while Amerigo Vespucci, Juan de la Cosa and Alonso de Ojeda, among others, explored a large part of the eastern coast of America. Ferdinand Magellan, following Columbus' idea of finding a passage to the east by sailing west, set out on an expedition that circled the world. In August 1519, Magellan departed from Seville with 234 men on board and five ships: after a voyage of slightly more than two years, only eighteen survivors returned in a single ship commanded by Juan Sebastián Elcano. Basque navigators and cosmographers such as Andrés Urdaneta and Miguel López de Legaspi conquered the Pacific and

4 López de Gómara, Francisco. *Historia General de las Indias*, op. cit, 'To the readers', page 3.

5 See: Fernández-Armesto, Felipe, *Pathfinders: A Global History of Exploration*, New York, W. W. Norton, 2006.

took possession of the Philippine islands. All of this happened in just a few decades and, for the first time in the history of the world, one people—the Christian monarchies—thought it would be possible to conquer the whole world. *Plus Ultra* was their motto and Christian monarchs were regarded as universal rulers.

It would be difficult to determine the precise moment at which what is known as the 'process of globalization' began, nor is there is a birthdate for things such as world trade or universal science, but there is no doubt that the 16th century witnessed unprecedented changes on a global scale. The Iberian exploration of the Atlantic in the first half of the 16th century consolidated two enormous commercial axes and two great monopolies: that which was forged between Portugal and India, specifically, between Lisbon and Goa, and that which was forged between Spain and America, which linked Seville with different ports in the Caribbean and the Gulf of Mexico.[6] These axes were the main commercial networks of Europe outside of the Mediterranean and also the foundation upon which the new world order was built, when Christian Europe proclaimed its hegemony over a large part of the planet. A number of similarities can be seen between the two empires: among them, their Iberian and Catholic origin in the context of a fierce 'holy war' against any non-Christian nation and the urgent need to establish more efficient trade routes beyond the circuits of the Mediterranean. However, there are also important differences. The Spanish vessels that sailed toward the Indies carried passengers, casks of wine, barrels of flour, pitchers of oil, tools, agricultural implements, seeds, domestic animals, and, of course, a powerful armaments. By contrast, the Portuguese vessels, which were also strongly armed and fitted out, sailed to the east loaded with ballast and passengers who would work in their trading posts (*factorías*).[7] On the return voyage, their cargoes were also different, since the ships from India arrived in Europe with spices—pepper, cloves, cinnamon, and nutmeg—and manufactured goods such as silks, porcelains, and other oriental works of art, comparatively the most valuable cargoes brought back by the Spanish ships were gold and silver. The East Indies and West Indies —as sixteenth-century Europeans referred to those parts of the world— were two different worlds at that time and therefore, established singular commercial and political relations with the Christian empires.

6 Braudel, Fernand, *El Mediterráneo y el mundo mediterráneo en la época de Felipe II*, 4th ed., Mexico, Fondo de Cultura Económica, 1997, page 399.

7 Martínez, José Luis, Pasajeros de Indias. *Viajes transatlánticos en el siglo XVI*, Mexico, Fondo de Cultura Económica, 1999, page 155.

Nevertheless, despite these differences, both the Spaniards and Portuguese had to face the same challenge: long-distance control. The conquest of remote places, the control of trade or the establishment of an imperial system is, in the end, a problem of communication: to proclaim their rule, they needed a safe means for the circulation of information, goods and people.

The objects and products that this book deals with are not traditional merchandises: they do not take up space nor are they very heavy cargoes. For the most part, they are printed products, visual records, chronicles, maps, texts, and collections of data that constantly circulated between the New and the Old World. The focus of study in this investigation is the knowledge of those techno-scientific practices, which, as they interacted with other political, religious and geographical factors, enabled Christian Europe to dream of controlling the whole of the Earth.

From the earliest chroniclers of the 16th century to today, the idea of 'discovery' has been a dominant concept in the historiography of the modern world, a key notion in our Eurocentric conception of history and one that is central to the exalted idea of progress in the West. 'To discover' implies an individual feat or achievement in which someone, at a specific time, sees or encounters something that no one had seen before; it also implies that the discovered object existed as such, in itself, before and independently of its discoverer. 'Discoveries' have also been presented as forms of appropriation in which, in a more or less natural way, the discoverers proclaim their right to possess and rule over the discovered places and objects, so that the narratives of 'discoveries' are celebrations of the European's power over nature, commemorative acts that have helped to idealize the scientific practices through which Western culture proclaims its control and power over the world. The idea of discovery has also been key in the creation of the idea of modern science, which, in turn, has been definitive for the notion of Western Europe as the center and driving force of history. Thus, it is often assumed that the history of modern science and, by extension, the success of the expansion of the European world and man's conquest of nature can be reduced to a series of crucial 'discoveries' and individual feats. In the field of geographical exploration, specific remarks, such as 'Christopher Columbus discovered America in October, 1492' or 'Vasco Núñez de Balboa discovered the Pacific Ocean', are common and generally accepted, but the narratives of the discovery of universal laws, principles, or truths[8] are also frequent and applauded.

8 For a critical look at the idea of discovery, see: Brannigan, Augustine, *The Social Basis of Scientific Discoveries*, Cambridge, Cambridge University Press, 1981.

Philosophical subtleties are not needed to acknowledge the absurdity of statements such as 'Christopher Columbus discovered America on October 12, 1492'. The 'discovery' of America was not a unique event, restricted to the voyages of Christopher Columbus at the end of the 15$^{th}$ century. Instead, it should be understood as process that stretches from before 1492 to, if you like, our time. We are speaking of an inhabited continent and, therefore, one known by human beings and probably visited by Europeans and Asians before Columbus. More important still, the statement is meaningless and turns out to be anachronistic since neither in 1492, nor in the moment when he died after four transatlantic voyages, did something like our idea of an American continent exist in Columbus' mind. The statement only became possible and acquired meaning after the cartographers, politicians, kings, popes, chroniclers, and political elites acknowledged and agreed on the new borders, characteristics, and reality of the New World. Likewise, this notion of discovery presupposes a one-directional and asymmetric process in which America and its native population are reduced to an object whose reality depends on the feats of Europeans.

Other concepts have been used to explain the appearance of America in world history. As an alternative to this one-directional perception of the discovery of America, and in the hope of offering a symmetrical view in which the voices of both Europeans and the natives of America are acknowledged, some have preferred to speak of an 'encounter' of the two worlds. It is an attractive idea and a subject of growing interest for cultural history. As Peter Burke points out, there are no genuinely pure and isolated cultures, and cultural frontiers are always diffuse and mobile.[9] In particular, the history of the empires of the Iberian Atlantic is the history of encounters and cultural interaction, and the idea of hybrid or *mestizo* cultures is a reality in the history of the Atlantic world. Recent historiographical debates try to fight against the isolation of national or regional histories and increasingly defend the notion of 'connected histories', histories of the Atlantic world, or histories of the world as a whole.[10]

However, the idea of an 'encounter' implies the possibility of a symmetrical narration in which the different cultures that are involved in the phenomenon are comparable in a balanced manner. There have numerous

9 Burke, Peter, *Formas de Historia Cultural*, 2$^{nd}$ ed., Madrid, Alianza, 2006.

10 For the particular subject that is dealt with in this book, see: Cañizares-Esguerra, Jorge and Seeman, Erik R., *The Atlantic in Global History*, 1500–2000, Upper Saddle River, NJ, Pearson Prentice Hall, 2007; or Delbourgo, James and Dew, Nicholas (eds.) *Science and Empire in the Atlantic World*, Routledge, London, 2008.

efforts by anthropologists and historians to reconstruct the viewpoint of the 'other' or the 'voice of the defeated' in order to make us see, for example, the way in which the Caribs thought of Columbus, the Aztecs to the armies of Cortés, or the Incas of Pizarro, and one of the greatest challenges for the social sciences has been to explain strange cosmologies and other forms of knowledge. This longed-for symmetry presents serious methodological and historiographical difficulties since, in most cases, the native voices and points of views, can only be reconstructed through the narrations or interpretations—or, alternatively, deconstructions—found in the European narratives. These well-intentioned efforts to make the 'subaltern' visible are often marred by ingenuousness or new ways of concealing the other that one wishes to speak for. From the sixteenth-century chroniclers to the ethnography of the 21st century, the modern or postmodern historian or ethnographer's pretense to be the legitimate spokesmen for the native inhabitants of American continent is full of insuperable difficulties.[11] Along that line of thought, it is important to stress that this books seeks to explore possible paths for a better understanding of European rule and thus it is not interested in either celebrating not denying Eurocentrism. In any case, the historical consequences of the process that this book wishes to investigate are far from balanced and the purpose of this investigation is to contribute possible explanations of a world order with an obvious asymmetry, the center of which was Christian Europe.

Such attempts to be neutral and symmetrical may not seem very realistic and for some it would be more correct to try for a narrative which unambiguously denounces the horror of the conquest. From Fray Bartolomé de las Casas to some twenty-first-century historians, the objective of historical narrative has been to show the brutality of the European conquest of America. Many have preferred terms like the 'European invasion'[12] to denounce the violent character of the European's entrance into the American continent and abandon any hint of a heroic celebration of the conquest. Tzvetan Todorov baldly denounces the conquest of America as 'the biggest genocide

11 On the difficulties of writing a non-Eurocentric history and the problem of acknowledging the voices which some call 'subaltern', see, for example: Spivak, Gayatri Chakravorty, 'Can the Subaltern Speak?', in: Nelson, Cary and Grossberg, Larry (eds.), *Marxism and the Interpretation of Culture*, Chicago, University of Illinois Press, 1988, pages 271–313; and Dirlik, Arif, 'History without a center? Reflections on Eurocentrism', in: Fuchs, Eckhardt and Stuchtey, Benedikt (eds.), *Across Cultural Borders: Historiography in Global Perspective*, Boston, Lanham, Rowman and Littlefield, 2002, pages 247–284

12 Carmagnani, Marcello, *El otro occidente. América Latina desde la invasión europea hasta la globalización*, Jaime Riera Rehren (trans.), Mexico, Fondo de Cultura Económica, 2004, page 35.

in human history' and has no problem in accepting and defending, with sound reasons, his preference for a moralist history instead of a neutral one.[13] There is abundant evidence to justify such a denouncement, but a simple accusation does not seem to be sufficient. A one-sided description of a crime may be a lesson for the future and is also necessary for a healthy reconstruction of the memory of the native peoples of America and the history of their conquerors, but it fails to explain the 'success' of Christian Europe and teaches us little about the cultural practices that made European sovereignty possible, both in America and in a large part of the world.

Other authors have preferred to speak of the 'social construction' or 'invention' of America.[14] This kind of analysis allows one to overcome some of the deficiencies of the traditional view of the discovery of ahistorical objects, but leads to another problem: reducing the reality of America to a mere 'social construction' and a European creation, where there is little room to interpret the role of geography, nature, and the peoples of America in the history of the Atlantic world. As will be shown below, the natural setting, population, and geography of the New World form an active and definitive part of that history. In addition, the concepts of 'invention' and 'construction' do not offer an alternative to the traditional dichotomies between Europe and the 'others', between culture and nature and between subject—the European subject in this case—and object, the American one. While such dichotomies are fundamental for our modern notion of science, they cannot be the starting point for historical analysis. The same dichotomies need to be historically explained. We are dealing with a lengthy process of change, during which the New World was incorporated into European culture at the same time in which Europe transformed and rebuilt its own identity.

In the face of this complex panorama it will be argued that the idea of 'comprehension' may be useful, not simply as an alternative that replaces all previous ideas, but also as a concept that may have advantages and that might help us to understand the conquest of the New World from an epistemological point of view in which scientific practices play an essential role. Both in the conquest and domestication of the new and in the construction of a new subject with pretensions to global rule, science was a central actor.

13 Todorov, Tzvetan, *La conquista de América. El problema del Otro*, Flora Botton Burlá (trans.), Mexico, Siglo XXI, 1989, page 14.

14 See, for example: O'Gorman, Edmundo, *La invención de América*, Mexico, Fondo de Cultura Económica, 1995; and Rabasa, José, *Inventing America: Spanish Historiography and the Formation of Eurocentrism*, Norman, University of Oklahoma Press, 1993.

According to the dictionary of the Royal Academy of the Spanish Language, the root of 'to comprehend' is the Latin verb *comprehendere*, from meaning *cum*, 'with', and meaning *prehendere*, 'to take hold of'. This original meaning of the term is key to understanding the practices associated with the discovery and conquest of the New World—cartography, natural history, or moral history—as powerful ways of declaring the right to own and rule over nature and people. Thus, to comprehend implies an act of appropriation, a process of translating the unknown into something familiar, of incorporation and domestication, as well as a recognition of the alien. Nevertheless, this first definition is insufficient and maintains the one-directional sense in which Europe is the *subject* and America the *object* of that understanding, and once more seems to lead to the limited perception of a passive New World that is appropriated by Europe. We should make it clear that when one refers to the 'understanding of the New World', a reflective act is implicit that cannot be limited to the apprehension or comprehension of something that is external but is, rather, one in which both the *subject* that comprehends and the *objects* of that comprehension[15] actively participate and are transformed. It is a process in which the agents and objects of the appropriation are constructed at the same time. From this point of view, 1492 is a date that should recall the 'discovery' or 'construction' of both America and Europe.

As José Rabasa has suggested, this means that the term 'New World' should not be limited to that geographical space, one that is different from Europe,

15 To answer this question, it might be useful to go back to some ideas found in hermeneutic philosophy. The problem of comprehension has been the subject of important philosophical studies in the 20th century, some of which are worth having a brief look at. For the argument we wish to develop, it is interesting to recall the existential meaning that Martin Heidegger gives to the idea of 'comprehension'. To comprehend is not the act of knowing or possessing something that lies 'before one's eyes', but it is a constitutive act of 'being there'. For Heidegger, to comprehend has the sense of a projection in which 'being in the world' arises. To understand is always a form of self-understanding, since understanding is only possible to the extent that one acknowledges the coexistence of the person who comprehends and others and the world. The ideas H.G. Gadamer draws from hermeneutic philosophy are also relevant, as when he states that 'insofar as it is a hermeneutic task, comprehending always includes a reflective dimension. At bottom, comprehending is always comprehending oneself, but not in the manner of a previous self-possessing or one that is already attained. Because this self-understanding is realized in the understanding of something'. Hence, it is through the strange or alien that human beings come close to themselves. The motive or the incitement of such understanding is alterity: it begins when something strikes one's attention and it thus implies movement, a coming and going between the strange and the familiar. Heidegger, Martin, *El ser y el tiempo*, Jose Gaos (trans.) Mexico, Fondo de Cultura Económica, 1983, page 163; Gadamer, Hans-Georg, *Verdad y Método. Fundamentos de una hermenéutica filosófica*, Salamanca, Sígueme, 1992, page 121.

that was the object of European exploration and exploitation from the 16th century onwards[16] but should instead refer to the conception of the world that arose from the European conquest of a large part of the terrestrial globe. The construction or invention of the New World is thus inseparable from the invention of Europe. Hence, the voyages of exploration and the zeal to create catalogues and inventories of nature; the creation of maps and bodies of knowledge about geography, seamanship, and cosmography; the establishment of a colonial trade and administration; the improvement of mining techniques; and the study of medical, botanical, zoological, ethnographical and climatic matters, or the customs and history of other cultures, were all fundamental practices in the consolidation of a new world in which Christian Europeans gradually emerged as spokesmen for the rest It is not only the representation or the construction of America that can be recognized in these scientific practices: it is at the same time, the construction of the Old World. That new *order* that comprehends and creates links between the familiar and the new was conceived in a culture that defined itself by its determination to domesticate the New World, as well as its ambition to rescue the rest of the world from barbarism and paganism.

Studying the 16th century means that you have to face big historiographical problems, both in political history and in the field of the history of science and technology. The aim of this book is to contribute to the historical comprehension of a notable change in the balance of world powers. How did the building of big European empires in the 16th century become possible? What actions and practices made it possible for them to proclaim their control over a world separated by enormous distances of land and sea? A definitive answer to such broad and complex questions is beyond the scope of this study; instead, its aim is to explore paths and link different fields of investigation such as political history and the history of science and technology.

In his book *Empires of the Atlantic World*, John H. Eliot suggests that Europe's domination of America must be explained on three different and complementary levels: the symbolic taking of possession, the material occupation, and the settling or resettling of lands.[17] There is no doubt that these are crucial aspects for understanding the conquest of America, but, like most of the analyses of imperial history, this approach ignores the

16 Rabasa, José, 'Inventing America: Spanish Historiography [...]', op. cit. While Rabassa deals with the 16th and 17th centuries, this idea still holds for the period covered by this book.

17 Elliot, John Huxtable. *Empires of the Atlantic World: Britain and Spain in America, 1492–1830*, New Haven/London, Yale University Press, 2006, page 64.

epistemological shift that made the appropriation of the alien possible and neglects the importance of an intense scientific activity that made it possible for the cultural centers of Europe to act on and control the New World from a distance. To attain a better understanding of the history of the Iberian world in the Atlantic, we need a careful study of the techno-scientific practices that allowed for both the European appropriation of the New World and the building of a new world order.[18] In the course of this book we will show how the grand Iberian empires of the 16th century were a great techno-scientific endeavor and how this relation between science and empire has important consequences for the history of science in the West.

The use of the term 'Eurocentrism' may be problematical if it is taken to mean that Europe is a discrete and homogenous actor, since it is obvious that Europe, as it is usually understood, shows a notable cultural diversity and has geographical, cultural, and economic centers and peripheries. However, the point is precisely to learn about the great Christian endeavor of conquering the whole world. As Tomaso Campanella said: 'thus Spain discovered the New World so that all nations would be under a single law'.[19] This investigation is therefore oriented toward a better understanding of the process of constructing the idea of a homogenous Europe or the consolidation of the West as a geographical and cultural entity, which is only possible insofar as one confronts something that is different and creates the referent of an 'other' that is common to the whole Christian world. Hence, it is a process that is clearly related to the exploration of the rest of the world and means that, despite their enormous differences, even within their own nations, Portugal, Spain, and Italy, —and, later, England, France, Holland, and Germany—, can be seen as a community with collective interests and familiar traits that were emphasized as the New World was gradually conquered.

Both 'the West' and 'Europe' are categories that should be historically explained.[20] An interesting standpoint on this problem is the one set forth by Arif Dirlik, who says that the true power of a Eurocentric view does not lie in the exclusion of 'the others' but, on the contrary, their inclusion: in the

18 On the role of technology in modern imperial history, see: Cipolla, Carlo M., *Las máquinas del tiempo y de la guerra. Estudios sobre la génesis del capitalismo*, Barcelona, Crítica, 1999; or Chaunu, Pierre, *European Expansion in the Later Middle Ages*, Oxford, North Holland Pub. Co., 1979.

19 Campanella, Tommaso, 'La imaginaria ciudad del sol [...]', op. cit., page 30.

20 On this aspect, see: Dussel, Enrique, 'Europa, modernidad y eurocentrismo', in: Lander, Edgardo (comp.), *La colonialidad del saber. Eurocentrismo y ciencias sociales. Perspectivas latinoamericanas*, Buenos Aires, Clacso, 2000, pages 41–53.

inscription of the whole world in a single order and system.[21] This idea of 'comprehension' as a process of inclusion and therefore of self-understanding will be central in this book. The 'scientific' practices discussed herein may be understood as forms of inclusion, processes through which links are created and the alien is changed into the familiar. It is in that sense that the idea of the 'comprehension of the New World' enables us to attain a better understanding of the science and political history of the Iberian Atlantic.

The Christian Europe of the Renaissance succeeded in incorporating the unknown into its own familiar frameworks and thus it proclaimed its dominion over strange seas, islands, continents, peoples, animals, and plants. In order to answer the question of how this became possible it will be useful to make use of some theoretical proposals found in current social studies of science.

## Science and empire

Whatever their object of study may be it seems that historians are forced to take sides on the nature of historical causes and superpose one cause on, or subordinate it to, others. Some opt for economic, political, ideological, or religious explanations, while others have sought to make technology itself the explanation of history. In the introduction to his book on technology and empire, *The Tools of Empire*, Daniel R. Headrick faces the question of whether it is imperialism that shapes technology or, on the contrary, technology that shapes empires .[22] Before addressing this problem, Headrick reminds us of the limitation that one of the central axioms of modern historiography entails: the explanation of the past as the result of the interaction of human decisions,[23] or, what is the same thing, the search for causes that, in one way or another, might be called social causes.[24] Such causes of a 'social' nature

21 Dirlik, Arif, 'History without a center? [...]', op. cit., page 252.

22 Headrick, Daniel R., *The Tools of Empire: Technology and European Imperialism in the Nineteenth Century*, Oxford, Oxford University Press, 1981, page 4. These questions take us back to an old debate between historians of science about the validity of an 'externalist' or 'internalist' historiography of science. See, for example: Shapin, Steven, 'Discipline and Bounding: The History and Sociology of Science as Seen through the Externalism–Internalism Debate'. In, *History of Science*, Vol. XXX, 1992, pages 333–369.

23 Headrick, Daniel R, 'The Tools of Empire: Technology [...]', op. cit., page 9.

24 This problem has been discussed by authors like Lagdon Winner in his essay 'Tienen política los artefactos?', in: *La Ballena y el Reactor. Una búsqueda de los límites en la era de la alta tecnología*, Barcelona, Gedisa, 1987; and Latour, Bruno, *La esperanza de Pandora. Ensayos sobre la realidad de los estudios de la ciencia*, Tomás Fernández Aúz (trans.), Barcelona, Gedisa, 2001.

are generally related to political, economic, and religious interests, which are being finally understood as the result of the interests and agency of certain social groups or persons who hold and exert power over the rest. Therefore, it is not completely surprising that reality, artifacts and nature are often explained as 'social constructions'.

Science and technology, on the other hand, may also be presented as the causes of major historical changes or as tools at the service of powerful human beings. For Carlo M. Cipolla, there is a reciprocal and complex interaction between the economy and technological developments, so that his central thesis tries to show that 'in Europe, the nations which were the best at inventing cannons and sails attained a supremacy over the rest. The era of human energy was over, and the era of machinery began to approach'.[25] His main argument explains this even more clearly: 'religion provided the pretext and gold, the motive. The technological progress which Atlantic Europe went through during the 14th and 15th centuries provided the means'.[26] This book does not aim to define pretexts, motives, and means, since technology should not be regarded as a resource or instrument at the service of a power that precedes it; in this case, the 'imperial power'. On the contrary, technological practices are part of the empire and the power, not its tool. Neither will we argue that religion can be reduced to a mere pretext, since it forms part of the social order, and even less that imperial expansion can be explained as the simple result of a commercial thrust.[27] In the history that we set out to narrate here, we do not look for a first and unique cause, and it will become evident that artifacts, nature, the human, and the divine form part of networks and practices that afford us a better understanding of the exercise of power and the shaping of a new world order.

As though the world of the inanimate belonged to a special dimension —a dimension whose explanation would presuppose frameworks of reference different to that which is dealt with by the human— those political history rarely give artefacts the attention they deserve. Incorporating natural causes or agents into historical explanations is equally strange for the social sciences and it would be even more surprising for them to find historical causes of a divine nature. Even though the religious ambit was inseparable from the knowledge of the world of the Renaissance chroniclers and naturalists of

25 Cipolla, Carlo M., 'Las máquinas del tiempo y [...]', op. cit., page 145.

26 Ibid., page 179.

27 These reductions to a first cause and prime mover recall the metaphysics of Aristotle and his idea of 'an unmoved mover', whose movement does not need to be explained but remains the cause of all movement.

the 16th century, including explanations of a divine or spiritual nature in the analysis of history today seems unwise or absurd nowadays. As Ulrich Beck notes, '[we historians and sociologists] bear the language of secularism in our blood',[28] and the spiritual dimension and religious experience, which were so definitive in the Catholic world of the 16th century, are excluded from historical analyses.

To sum up, in one way or another, the social sciences have always wished to find a 'social' explanation for historical processes. Following the proposals of the Actor–Network Theory,[29] this study will seek to keep these apparently heterogeneous elements together—political, religious, technical, and natural factors—without letting one of these levels explain or define the role of the others. Instead, it is a matter of understanding power as the result of a conjunction of agents that are both social—human—and technical, natural, or even divine, and thus, this work will argue that the triumph of Christians in a large part of the world is explained by the mobilization and combination of these elements. The monarchs, traders, and their human interests obviously form part of history, but not because they embody a power that precedes the practices of dominion and control. In the history of the Empire, the ships, sails, cannons, winds, and current should be made visible, along with the actions of the sailors and astronomers, demons and saints, live creatures, plants, and animals.

The sailing ships of the 16th century will be the leading actors in this history although, strictly speaking, neither the ships nor sailors sailed on their own: it would be more accurate to say that the feat of crossing the Atlantic was fundamentally an achievement of the *Casa de Contratación* in Seville (House of Trade, a Crown agency with broad responsibilities for Spain's overseas interests). Only an explanation of the conjunction of the interests of the monarchy, the Christian mission to spread its dogma, the merchants of the Mediterranean, the pilots, sailors, instruments, complex ships, and all of these in alliance with the winds of the Atlantic, can offer a more complete picture of the nature of imperial power. It is therefore necessary to show the interaction and linkage among dissimilar and apparently isolated factors.

28 Beck, Ulrich, *El Dios personal. La individualización de la religión y el 'espíritu' del cosmopolitismo*, Rosa S. Carbo (trans.), Barcelona, Paidós, 2009, page 11.

29 Latour, Bruno, *Reassembling the Social: An Introduction to Actor-Network-Theory*, Oxford, Oxford University Press, 2005; Law, John, 'On Methods of Long-Distance Control: Vessels, Navigation, and the Portuguese Route to India' in: *Power, Action, and Belief: A New Sociology of Knowledge?*, London, Routledge & Kegan Paul, 1986, pages 234–263; Law, John (ed.), *A Sociology of Monsters: Essays on Power, Technology, and Domination*, London/New York, Routledge, 1991; Law, John and Hassard, John (eds.) *Actor Network Theory and After*, Oxford, Blackwell, 1999

Only in that way, through the description of complex networks, is it possible to understand how a few people proclaimed their control over enormous stretches of land, natural resources, goods, and people in remote places.

The idea of incorporating natural, geographical, or technical factors into a historical analysis is not novel, Fernand Braudel and many others have reclaimed the importance both of geography and of ships and devices in imperial history. What may be novel about the proposal of the Actor–Network Theory (ANT) lies in the fact that, according to that theory, artefacts form part of the systems of long-distance control and are not simply means or tools that serve social interests. What is possibly a particularly theoretical novelty of this investigation, even when compared with the studies of those who created the ANT, is its attempt to incorporate the agency of the divine into secular history.

The best expression of the inseparability of these social, natural, and religious factors is found in the very narratives of the 16$^{th}$ century itself. The chroniclers, cosmographers, and naturalists described the New World without any need to distinguish human from natural or divine aspects: a history in which it was obvious that the most powerful actor, even above the great monarchs, was the Creator, by virtue of whom all of history becomes meaningful.

The problems that this books wishes to explore have less to do with the question of when science emerged in Western Europe or who invented it than they do with identifying those practices that gave rise to a form of knowledge that enabled some to proclaim their control of the world, and to come to a better understanding of this process of expansion, the relations between the technical, economic, and the political, as well as the natural and divine need to be made visible.

Considering that the subject of this investigation is the knowledge of seamanship—the norms, regulations, and instruments produced in sixteenth-century Spain—and not so much the experience of specific explorations, it will try to show how those bodies of knowledge and practice were combined in a complex web of factors of a diverse nature. Dividing this book into separate chapters is necessary in order to present the different aspects of an intricate network of actors with a certain orderliness, but it is not meant to stress their independence of one another. On the contrary, one of the central aims of this investigation is to show their interdependence. Instead of creating links between discrete elements, it sets out to describe hybrid systems or networks in which the human, the natural, the technical, and the divine are mutually assembled and form each other. It is in those terms that the 'machines of empire' will be understood. As we shall see, without a compass and magnetic needle, or without an astrolabe and sailors trained in the use of those instruments, a navigation chart would be no more than a piece of paper without any power to act on the environment. These associations are not only evident in the field of seamanship; they are also for explaining the emergence of a new natural history as something that went beyond the simple creation of the sixteenth-century naturalists. The armadillo, iguana, pineapple, and other natural wonders of the New World were not passive objects, nor were they simple 'social constructions'. In this history they should be understood as actors without whose participation the history of the New World would remain incomplete.

This introduction is not aimed at giving a 'state of the art' account of the subject: the literature on which in Spanish, English, and Portuguese is very ample, and it is obvious that this text has been built on many previous studies. The Spanish historian José María López Piñero has written about the subject[30] and his results are exemplary. However, he has been followed by a long list of authors, to whom I hope to give their due recognition in the course of this text.

The initial idea that gave rise to this book, which very clearly and evidently meshes with John Law's studies of fifteenth-century Portuguese voyages of discovery, was to write about ships and navigational techniques in the Atlantic in the 16th century. The wealth of primary sources on Spanish cosmography and seamanship in the 16th century—which, in general terms, are little studied by historians of science—and the growing body of literature on the role of the Catholic world in the history of modern science in Western Europe justified the idea of writing another book on science in the imperial context of the 16th century.

However, the central arguments of my book obliged me to present a broader panorama and include chapters that would enable one to understand the voyages of exploration as part of a general problem of political history. I also had the idea of reaching a wider public in the hope that a book on science and technology would be of interest to other fields, such as cultural history or political history. The initial chapters on the 'discovery' of America, the *Casa de Contratación*, and cartography have been written by subsequent authors who have already studied those subjects in depth, so that here the book does not pretend to offer anything novel apart from finding relationships between aspects that are often treated in an independent way. By contrast, the chapters I have written on seamanship and natural history are mostly based on primary sources. Hence, from a theoretical viewpoint, and in view

30 López Piñero, José María, 'El arte de navegar en la [...]', op. cit.

of its effort to link different fields and subjects, the book may also appeal to readers who have a more specialized interest in the history of science and its relationship with the imperial history of the Atlantic world. For those reasons, the origin and depth of the chapters differ, but their combination is useful when it comes to thinking about a major historiographical problem: the shaping of a Eurocentric world order.

## Summary of the chapters in this book

The first chapter discusses some of the main reasons why the Iberians embarked on grand projects of oceanic exploration and the impact that news of the first voyage of Columbus might have had on Spain. As an important part of this context, this section will refer to the geographical conditions of the Iberian Peninsula and the Atlantic, as well as the winds and currents that made that great ocean an ally of European expansion.

Chapter 2 discusses the institutions that were created in the framework of imperial expansion: The *Casa de Contratación* in Seville and the *Real Consejo de Indias* (the Royal Council of the Indies). Understanding the New World and controlling it from a distance required a complex bureaucracy, as well as norms and institutions whose sole aim was to ensure that there was a smooth and orderly flow of information and goods between Spain and its American possessions. Crossing the Atlantic was an endeavor linked to the land: the ship or fleet formed part of a larger project that cannot be understood without learning about its motivations and the complex institutions that were created for those purposes in Renaissance Spain.[31]

In addition to adequate vessels, voyages beyond the confines of the Mediterranean required new navigational techniques that were highly dependent on astronomy. The forms of orientation that characterized medieval navigation, that is, visual references, known winds, and the magnetic needle, were not sufficient on the great ocean. Portugal and Spain consolidated the new science of celestial navigation (*navegación de altura*), which was compiled in and spread by a set of texts and manuals that were read both in and beyond the Iberian Peninsula throughout the 16th century. These writings, most of which were printed or written in Seville, are the main source of this investigation, so that the third chapter discusses the most important manuals of navigation and cosmography that were

31 Ernesto Schäfer's study of the *Real Consejo de Indias* is an obligatory reference for this subject, which will be dealt with in Chapter 3.

produced in Spain in the 16th century, along with their importance for the history of Western science. Those books on seamanship and cosmography illustrate one of the central problems of this book: the tension between the classical authorities and the new experiences of the European explorers. The view of the New World that the first explorers of the New World had was determined by the ancient and medieval traditions of geography and natural history. The works of Aristotle, the astronomy and geography of Ptolemy, and Pliny the Younger's *Natural History*, among others, formed the frames of reference in which the natural world of America was interpreted. However, in order to transform the unknown world into something familiar, a renewed cosmography and a new natural history were required, so that the new could be incorporated into the traditions of classical learning. Learning about the New World showed the limitations of classical texts, but it was not possible to dispense with them either. In the end, the success of the European conquest of America lay in its ability to incorporate the unknown into familiar frames of reference.

As the central theme of this book, chapter 4 deals with the nautical knowledge, ships, instruments, pilots, and sailors. Any attempt at global rule presupposes the conquest of the ocean, since 70% of the Earth's surface is made of water. This allows one to argue that, in the study of the great geopolitical changes of the 16th century, sailing ships and sailors were the leading and decisive actors when it comes to understanding Europe's aspirations to world control. As we shall see, a single ship is a microcosm in which one can recognize the complex web of material and human aspects that were required to establish a stable link of control between the Old and the New World.

The knowledge and skills needed to lead a ship to its destination implied a complex division of labor and a linked sum of skills and functions. In the first place, the shipyards and the building of powerful vessels and navigational instruments—of fleets of ships equipped with costly hulls, masts, iron fittings, ropes, sails, and cannons—was already a technical challenge in itself, which entailed a combination and linkage of different resources and bodies of knowledge. Then, once at sea, the ships needed to be operated by a crew with many specific skills. Chapter 4 thus describes the ships, the activities that took place on these ships, and life aboard a transatlantic voyage, along with the difficult conditions of the crossing and the role of the religious beliefs of the mariners. Without faith and the will of God, the life and the activities of the seamen would have been meaningless.

The fifth chapter discusses one of the technological products that, in a more explicit and evident manner, represents European power over the

New World: maps. While this subject has been widely dealt with by others and, in fact, though the maps that will be analyzed in this book have been the subject of detailed studies, it is necessary to make a general reference to this subject. Navigational charts were powerful instruments of imperial expansion, and the new cartography of the world was the result of a linkage between classical traditions that spread in the Renaissance and the unprecedented experiences of voyagers. The power of cartography lies precisely in its capacity to encapsulate time and space on a flat representation of the Earth. It thus makes the world accessible: it is possible to take a Caribbean island to Madrid and put the whole world in a single place.[32] As Bruno Latour indicates, men are the lords of the world only if the world reaches them in two-dimensional representations that can be combined and are easy to handle.[33]

The sixth chapter gives an account of the main problems of the natural history of America. Once the enormous difficulty of crossing the Atlantic was overcome, the Spanish travelers faced a strange nature and the transformation of such nature into a domesticated world was the result of an arduous job of description and classification. Classical texts like the works of Aristotle, Dioscorides, or Pliny were the referent with which Europeans sought to understand nature in the New World. However, the flora and fauna of America were not to be found in the great encyclopedia of antiquity and creatures, such as the armadillo, iguana, or pineapple, among many others, defied the classical authorities and forced the naturalists of the 16th century to create a new natural history for a new world.[34] The local bodies of knowledge and tradition in America turned into sources that enabled Europeans to deal with the challenge of classifying and finding the possible uses of plants and animals that Christians had never seen before.

The final chapter goes back to this book's central questions about the role of Iberian science in the history of modern Europe. In that regard, the book sums up the work of other historians[35] who have pointed to the importance that a better understanding of the role of the Catholic and Iberian world in

32 Latour, Bruno, 'Drawing things together', in: Lynch, Michael and Woolgar, Steve (eds.), *Representation in Scientific Practice*, Cambridge, MIT Press, 1990, pages 19–68.

33 Latour, Bruno, 'La esperanza de Pandora [...]', op. cit., page 43.

34 A very complete study of this subject is found in the book by Miguel de Asúa and Roger French, *A New World of Animals: Early Modern Europeans on the Creatures of Iberian America*, Aldershot, England, Ashgate, 2005.

35 This section of book owes much to the studies and ideas of Antonio Barrera, Jorge Cañizares-Esguerra, María Portuondo, and Alison Sandman, who, in turn, draw on the studies done by a group of historians in Spain, among which José María López Piñero stands out.

the conquest of the Atlantic and New World has for the history of Western science, as well as the relation between the practices involved in that conquest and the emergence of a new science with pretensions of universality. It is not only Spain that has been ignored in the history of the birth of modern science, but also several fields of major importance for understanding science in the modern world, such as seamanship, geography, and natural history.

By way of a conclusion, we offer some reflections on the relations between the Spanish imperial endeavors of the 16$^{th}$ century and the history of Western science. The problem of knowledge, like that of control from a distance, is fundamentally a problem of communication, and by the same token, of the processes of compiling, organizing, and systematizing information. The institutions and supply centers that were created in Spain in the course of the 16$^{th}$ century, such as the *Casa de Contratación* of Seville and the *Real Consejo de Indias*, had a definitive role in the construction of a new world order and a new technical and scientific horizon.

# 1. The Iberian Peninsula and the Atlantic

**Abstract**
This chapter discusses some of the main reasons why the Iberians embarked on grand projects of oceanic exploration. The author describes the political and geographical context in which the Spanish exploration of the Atlantic occurred. As an important part of this context, the chapter refers to the geographical conditions of the Iberian Peninsula and the Atlantic, and the winds and currents that turned the great ocean into an ally of European expansion.

**Key words:** Spain, Portugal, Oceanic exploration, Navigation, Geography

## Portugal and Spain

In the 15$^{th}$ century, Western traders, particularly the Italians, needed new and better routes for the trading of products from the Orient, for which very high prices were paid in Europe. Since the days of Ancient Rome, luxury products from the Orient, which were generally known as 'spices', were a powerful incentive for the exploration of better commercial routes. Routes such as those traveled by Mediterranean merchants like Niccolò Polo and his son Marco and which had been recorded in the technical manuals of Western merchants since the start of the 14$^{th}$ century had to be explored again. The Italian merchants traveled along what was known as the 'Mongol route', from the Black Sea to Cathay, without difficulties. In the second half of the 14$^{th}$ century the silk route was consolidated, which allowed Chinese silk to be transported to Italian manufacturers at accessible prices. This large-scale traffic gave birth, in Genoa and Venice, to an industry of sumptuous textiles, brocades, and velvets, which left its mark on an epoch and formed part of a lively commercial activity. Thus, the Italians, who were adamant about profits, created a market of products that were manufactured in Italy with raw material brought from the East, as well as an ample market of condiments and aromatic substances.

Nieto Olarte, M., *Exploration, Religion, and Empire in the Sixteenth-century Ibero-Atlantic World: A New Perspective on the History of Modern Science*. Taylor & Francis Group, 2022
DOI 10.5117/9789463725316_CH01

However, the Italians lost their direct access to China and the Mongol route was affected by hostility to Christians. It became necessary to go back to ancient itineraries, with stopovers in Syria and Egypt, which were places controlled by Muslims and which entailed longer and more expensive routes, often in the hands of unstable and hostile states. Chinese silk had to reach India; from there, Arab ships carried it to the Persian Gulf and then, via caravans to cities in the desert or Egyptian ports. Pepper and other spices had to pass through the hands of intermediaries and demanded the payment of taxes and freight charges. These difficulties faced by the trade in products from India and China were one of the motivations for the project that Christopher Columbus developed in Lisbon. The merchants of Genoa, Portugal, and Andalusia would enormously benefit from a new route to the Indies: a mercantile route that would avoid traveling through the Muslim World, which was as necessary as it was difficult up to then.

One possibility, full of risks but which some thought possible, was to circumnavigate Africa, which implied a long and dangerous journey. Portuguese navigators accepted the challenge and opened this route by sea. Vasco de Gama spent two years far from Portugal and sailed for eighteen months: he lost two of his ships and 80 of the 150 men who had embarked with him. His was a route full of difficulties and, in addition, it very soon fell under the exclusive dominion of Portugal. Columbus' proposal was to sail toward the west across the Atlantic by a completely unknown route, but one which would possibly be without so many obstacles since it would be a direct route to the country of Cathay and China.

This book focuses on the Spanish conquest of the Indies, but it cannot ignore the leadership of Portugal in the grand endeavor of sixteenth-century maritime exploration. Explaining why Portugal was so powerful in the exploration of the great oceans is not so obvious. It was a small and poor nation, with a population no larger than two million persons—comparatively, something more than half of the population of England, a fourth of that of Spain, and a tenth of the population of France.[1] Despite its small size, however, that nation managed to establish a commercial system of impressive dimensions. As we shall explain below, its mercantile interests were the main stimulus for the naval vocation of Portugal, although we should not lose sight of the fact that, like Spain, it was a Christian nation that had a strong mission to spread the faith. The Iberians were freeing themselves from a long period of Muslim domination and the attacks of

1 Fernández-Armesto, Felipe, 'Pathfinders: A Global History of [...]', op. cit., page 118.

the crusaders against the enemies of Christianity had been combined with piracy, so that attacks on Muslim vessels, loaded with valuable spices, in the Mediterranean or the coasts of Africa turned into an important source of wealth for the Lusitanian monarchs. Thus, it was the war of corsairs and the robbery of merchandises from ships and ports of the enemies of the Christians that turned into the first step toward the endeavor of controlling commercial routes to Africa, the Orient, and, later, America. Therefore, it was not gratuitous that the Portuguese acquired a reputation as fearful pirates who roamed the Mediterranean, the English seas, and North Africa. To find their own route to the Orient, that is, a route controlled by Christians and without intermediaries, was therefore the clearest commercial and military aim of the Portuguese. The dream of a maritime and commercial route to India was also stimulated by the widely read accounts of Marco Polo and the supposed existence of Prester John, a powerful Christian monarch in the Orient and thus a powerful ally. However, the most obvious incentive for this commercial enterprise concerned a set of products of vegetal origin with aromatic properties—peppers, cloves, cinnamon, ginger and nutmeg, to mention only a few—that are now found in any marketplace in the world at prices accessible to most of the population, but several centuries ago were difficult to acquire, highly appreciated and, therefore commanded extraordinary prices in European markets. Pepper was perhaps the most valued spice, partly, some have argued, due to its usefulness for conserving foods. With the arrival of autumn and winter, when there was no grass for cattle, many of the animals had to be slaughtered and, as a result, pepper played a vital role in conserving meat and became a staple product.

Another factor that stimulated the maritime vocation of Portugal was, of course, its fishing industry. The lands of Portugal and its agricultural production were almost exclusively limited to the export of oil and wine. Therefore, fishing was obviously a matter of survival and a field for permanent innovation in shipbuilding and navigational techniques.[2] The economic hopes of a large part of Lusitanian society seemed to lay in the sea. For the poor and the peasants who made up most of its population, especially the men and boys, the possibility of surviving was to be found in its ports and ships and the sea. Sending boys to work on the ships or in the ports was an alternative means of subsistence, since it not only meant there was one less mouth to feed, but they could also send money home and even keep some for

2 Pestana Ramos, Fábio, *Por mares nunca dantes navegados. A aventura dos descobrimentos*, São Paulo, Contexto, 2008, page 35.

themselves. Women also found work in the ports or the ships, prostitution being a frequent way of making a living.[3]

For the abovementioned reasons, Portugal sought resources for its survival on the sea and did it so successfully that, in addition to constructing maritime routes hitherto unknown to Europeans, it discovered new territories and consolidated an unprecedented naval and commercial empire.

As we shall see below, the enterprise of exploration and maritime commerce required the linkage of a wide range of knowledge, callings, and products. A complex cluster of knowledge related to navigation was consolidated in Lisbon. The *Ribeira das Naus*, as the shipyard responsible for building the ships of the empire was known, employed more than five hundred persons of different trades at the start of the 16th century. A large number of craftsmen devoted to complementary activities gravitated toward the naval industry: woodworking, the production of tallow and the manufacture of metal pieces, and fabrics for the sails and cordage for the rigging of the ships, among others. Alongside the *Ribeira de Naus*, there was what was known as the *Armazém Real*, the warehouse for the supplies and provisions of the voyages and therefore the place where the industry and commerce of products like the sea biscuit or hardtack (*pão do mar*)[4] flourished.

In addition to the abovementioned circumstances, the geographical position of Portugal was a definitive factor. Situated at the western edge of Europe, its port cities were well placed for voyages to the Mediterranean, the North Sea, and the Atlantic. Portugal and the Iberian Peninsula as a whole have coastlines with privileged natural ports, as well as maritime currents and winds that favor navigation in different directions at different times of the year. In addition, the presence of rivers that facilitated transport and commerce to the interior of the peninsula was added to the geographical conditions that made exploration and commercial expansion possible.[5]

In the second half of the 15th century, the Portuguese began to explore the coasts of Africa. After the conquest of Ceuta in 1415, Portugal gradually seized hold of the southern part of the west coast of Africa. In 1434, Gil Eanes rounded Cape Bojador; in 1488, Bartolomé Díaz reached the Cape of Good Hope and, finally, at the end of that century, Vasco da Gama reached India, which opened a new commercial route between Europe and the East. The other alternative for finding a new route for sailing to the East

3 Ibid., page 38.

4 Ibid., pages 42–43.

5 Ibid., page 16.

was even more uncertain and difficult: sailing in a western direction until you reached the eastern coasts of the Asian continent. That was the route Christopher Columbus proposed, and the same that was initially rejected by the Portuguese who, due to their knowledge and experience of the sea, did not think it was plausible.

It was in that context that Lisbon turned into a port and commercial center for routes that allowed merchandises to be moved from distant places. Portugal focused all of its productive, nautical, and military capacity on the consolidation of a monopoly of trade with the Orient. In addition to powerful and agile vessels crewed by skilled mariners, this enterprise entailed a complex economic and technological organization.

The *Casa da India*, established in 1500, was the Portuguese organization responsible for administering its overseas maritime commerce. It was preceded by institutions with different names but similar functions,[6] among them the *Casa de Guiné*, the *Casa de Guiné e Mina*, and the *Casa da Mina*. At the beginning of the 16$^{th}$ century, most of the income of the Lusitanian State came from its overseas economic activities and the *Casa da India* maintained the control and monopoly of pepper, cinnamon, and cloves. In addition to its commercial functions, the *Casa*, just like the *Casa de Contratación* of Seville later on, also became a center for the collection of information and, therefore, a place where knowledge was produced for the regulation of practices related to seamanship, astronomy. and cartography.[7]

The general problem faced by both Portuguese and Iberians was that their projects of exploration and conquest did not fit into their domination of their domestic routes. The challenge was to design and implement a new or modified system of navigation, one that would not be exhausted by the referents of European or Mediterranean geography. Sailing toward the south coast of Africa, an endeavor already begun by Henry the Navigator in the second half of the 15$^{th}$ century, meant that Portugal had to solve a major problem: finding the location of their ships on the globe in terms of latitude. There were no maps and, with the instruments used in Mediterranean navigation, —that is the compass needle and the Portolan charts of routes and winds—, it was not possible to precisely determine the distance that was covered or the vessel's position on the globe. To solve the problem, one needs to make use of techniques of astronomy and cartography such as measuring

6 Ibid., page 42.

7 On Portuguese early modern science, see Leitão, Henrique, "Ars e ratio: A náutica e a constituição da ciência moderna", in: Vicente Maroto, María Isabel and Piñeiro, Mariano Esteban (coordinators.), *La Ciencia y el Mar*, Valladolid, Server-Cuesta Impresores, 2006, pages 183–207.

the height above the horizon of a celestial body, a technique with which one calculates degrees of latitude. Measuring latitude by observing the angle of the elevation of the Pole star was not an invention of the navigators of the 15th century, but making those measurements on land, with fixed instruments, is very different from making the same measurements on the high seas using small, portable instruments. In addition, when you sail to the south there not only appear a new sea and new lands, but also a new sky. In the southern hemisphere, the Pole star disappears, and latitude has to be measured using other astronomical referents. The Sun and its declination continued to be very important astronomical referents, but new referents in the sky were also established. The Southern Cross, near the south pole of the heavenly sphere, would have similar functions to that of the Ursa Minor in the north. The geography and cartography of the Renaissance, following the models of Ptolemy, formed part of astronomy and, as a result, the conquest of the globe was inseparable from a mastery of the celestial sphere.[8]

Portugal's refusal to sponsor Columbus' project seems reasonable since he did not belong to its courtly circles, nor was he a gentleman, nor an official of the Crown, and being a foreigner helped even less. The King refused to help him without leaving any room for an appeal since the project did not coincide with the interests of his court, which was then more concerned about putting its existing possessions in order than embarking on new and hazardous projects of conquest. The technical evaluation of the project also yielded serious objections: the experts and astrologists of the court rightly thought that Cipango (Japan) was not that close and, therefore, that it would be a longer voyage than Columbus claimed and would entail greater difficulties for the transport of provisions, staying alive, and controlling the crew.

When the Portuguese court turned him down, Columbus looked for support in Castile. He settled in La Niebla, a county in the region of Palos and Moguer near Seville where he tried to get the help of powerful princes and people who were knowledgeable about the sea still believing that he would not need royal support. The Duke of Medinacelli was interested in the project, but it was clear that it would not be possible without the participation of the Crown. Queen Isabella met Columbus and submitted his project to a council of experts she chose. On May 5, 1487, Columbus received his first funding from the royal treasury, which changed his status so that he joined the circle of court protégés. The project was studied for nearly four

8 A clear introduction to the problem of latitude and the Portuguese contributions to that field is found in: López Piñero, José María, 'El arte de navegar en la [...]', op. cit.

years and in 1490 Jerónimo Hernando de Talavera, the Queen's confessor and an expert she had consulted, rendered a negative and decidedly hostile verdict: 'nothing can justify the favor of Your Highnesses for a matter that rests on such weak foundations and which seems unrealizable in the eyes of any person who has some knowledge [of it], however sketchy it may be'.[9]

The distance between Castile and Cipango was a crucial aspect in the debate on the feasibility and relevance of the project since, in the judgment of the councilors of the Crown, it would be much longer than Columbus claimed. However, both for Columbus and the advisors of the Crown it was impossible to foresee that there were new lands to be discovered, much less imagine the existence of a fourth continent. Columbus moved to Santa Fe, a district in the central-eastern part of the Vega de Granada, in order to live near the court alongside the nobles and top officers of the army. In 1491, he managed to get his project reconsidered, but once again it was rejected, partly because of the same technical objections and partly because of his ambitious claims to rights over the new lands that were to be discovered. Influenced by Luis de Santángel, it would seem, the Queen decided to invite Columbus to present his project, which, while hardly defensible, would not be so expensive and thus there would be little to lose. Finally, the Queen and some private merchants decided to assume the risks of Columbus' project and approved a voyage of exploration with three ships under his command. The Portuguese had attained a certain dominion over the route to the Orient, by circling Africa, and that surely inclined the monarchs of Castile to listen more carefully to the proposals for an alternative route by sailing to the west. His main argument was to lessen fears about the distance and difficulties of navigation and, in other words, to guarantee a safe return. When he presented his calculations and justifications for the project, Columbus always relied on respected authors, studied the ancient writers and travel narratives that were useful to him, and strove to assemble a complete documentation, full of arguments in its favor.

The big question of distance is inseparable from the question of the size of the globe. Its division into 360 degrees was already a convention used by 15$^{th}$ century cartographers, but there was still a controversy about the length of each degree and, therefore, about the circumference of the Earth.[10] According to the calculations of Ptolemy and the Alexandrian geographers,

9 Cited in: Heers, Jacques, *Cristóbal Colón*, José Esteban Calderón and Ortiz Monasterio (trans.), Mexico, Fondo de Cultura Económica, 1992, page 141.

10 The calculations of the ancient astronomers do not differ much from current ones—21,600 miles or 40,000 kilometers.

the circumference of the Earth was 33,000 kilometers. In the 11th and 12th centuries, those figures were revised in centers such as Toledo, where the works of Arab mathematicians were known. In the days of Columbus, the figures varied between 33,000 and 44,000 kilometers. Making use of a "mistaken" interpretation of the calculations of the medieval Muslim astronomer Alfraganus —only mistaken from the point of view of a modern reader—to justify his proposal, Columbus, following the ideas of Paolo del Pozzo Toscanelli concerning sailing westwards to reach Asia, defended the idea of a smaller size of the Earth: he presented a panorama of enormous bodies of land, a navigable sea, a route marked by islands that would support him, and a trip that would guarantee a relatively safe return. Columbus argued that, sailing westward, the eastern coast of the Asian continent lay at a distance of around 750 leagues. With winds that he estimated would be navigable at an approximate rate of 35 leagues daily, the crossing would take three or four weeks of sailing.

While it was owed to the tenacity of an ambitious and obstinate navigator, Columbus' project also coincided with the most urgent needs of a Christian world and an expanding commercial and political empire. Nor was the idea so absurd, because the geographical knowledge of the 16th century revealed the possibility of conquering the whole world. Instead, it was an expression of the geographical, commercial and religious thought of his era. Without belittling the merits of Columbus' endeavor, and much less depicting it as an easy task, we should not regard that project as the achievement of one individual, nor think of it as the strange obsession of solitary visionary who fought against misunderstanding and ignorance. On the contrary, it counted on the support and interests of powerful allies. In all regards, Columbus was a man of his time, an individual who managed to link his interests with the ambitions of the most powerful people of Europe. For that reason, the 'discovery of America' was not a one-off event or an individual achievement that Christopher Columbus brought off in October 1492. It was a process that cannot be explained without examining the theoretical and practical developments of the cosmography, naval engineering, navigation, geography, and natural history of the 16th century, subjects which shall be discussed in detail below.

## Winds, currents, and sailing ships in the Atlantic

The history of the sailing ships, navigational techniques, and cosmography of the 16th century is an important chapter in the history of Western technology and one that is equally notable in what might be called the history of man's dominion over nature. This book studies the means that made the voyages of exploration and conquest possible: the knowledge of cosmography and geography, the building of the ships, the training of the sailors and the use of their instruments. All of these aspects are closely linked to a bigger problem: the tension between man and nature or between the empire of man and the empire of nature. The ultimate aim of a sailing ship, with a trained crew and suitable instruments, is human dominion over the sea. The sound use of sails turns the natural power of the wind into the capacity of movement, a force that transforms vessels into powerful vehicles that, when combined with the astronomical techniques of navigation and orientation, enabled the Christians of the 16$^{th}$ century to proclaim their rule over the sea and a large part of the world.

The treatises and manuals on cosmography and navigation that the sixteenth-century Iberians used dealt with ships and their manufacture, navigational instruments and techniques, tides, currents and winds, and even storms and tempests: ultimately, the human arts for dominating the sea. Within the complex networks that this book sets out to describe, the winds, currents, tides, and storms had an essential role: they were natural factors that might have been the biggest obstacles for European expansion, but in suitable contexts and conditions they became powerful allies of the political interests of Christian Europe.

In his book *Pathfinders*, Felipe Fernández-Armesto makes the importance of incorporating nature into the history of navigation clear; initially, because the conquest of the world or the triumph of Europe in its endeavor to dominate a large part of the planet has to do with the overwhelming fact that the surface of the globe is mostly covered by water and because the sea represents the biggest natural obstacle to human movement and expansion. To connect the world, establish commercial links, or create new forms of controlling the globe presupposes the conquest of the sea. This would not have been possible without another natural phenomenon: the permanent movement of air over the surface of the Earth due to changes due to changes in atmospheric pressure, some of these changes occur on a global scale, as may happen with the trade winds, and others, with more or less regular patterns, such as monsoons, which are continental phenomena. Many winds are local phenomena and have a variable stability and regularity, but it is true that there is a permanent activity of the air over the whole of the surface of the planet. That activity has an enormous potential as a source of energy, wind, or Eolic energy, which, thanks to the human arts, is transformed into kinetic energy. To sum up, wind is a powerful ally of this human enterprise and therefore the history of navigation and conquest are inseparable from its behavior.

The Spanish and the Portuguese faced many difficulties, but they also ran into some definitive advantages. In comparison with the Indian Ocean, the Atlantic offers an open space for navigating a large part of the planet.[11] By contrast, the navigators of the Western Pacific and the Indian Ocean did not find conditions that favored sailing long distances beyond the area of the monsoons.[12] The route across the Atlantic chosen by Columbus, which allowed sailing ships from Europe to successfully reach and return from the New World, was not a whim or a lucky guess of the first transatlantic sailors: those routes are also owed to favorable natural conditions. In the Atlantic, winds behave in accordance with a regular and predictable pattern: they blow in the same direction in a relatively stable way throughout the year. From the northwest coast of Africa, near the equator, there are winds that blow from east to west, nearly up to the Caribbean islands.

Ocean currents, mainly caused by the rotation of the Earth and planetary winds in the Atlantic, also have stable patterns that facilitate certain navigational routes, such as those of most travelers between Europe and America.

More than the existence of a fourth continent, whose discovery he could never have dreamt of, the great 'discovery' of Columbus was the transatlantic route.[13] On his first voyage, in addition to the reckless optimism caused by sailing with winds in his favor without knowing the return route, it seems that Columbus had a big stroke of luck by sailing to the west near the equator through a zone that, at certain times of the year, has big calms or variable winds, two major risks for long days at sea. In his later voyages he found a safer route, several degrees to the south, where you can take advantage of the trade winds and certain ocean currents that are favorable for the route toward the American continent.

Columbus seems to have found and defined the sailing route for the *Carrera de las Indias* (Indies Run), which underwent few changes for nearly three centuries. This is partly explained by the fact that the climatic conditions seemed to be relatively stable: for the voyage to America the ships relied on trade winds and, on the return voyage, they headed north in search of more favorable winds for reaching Europe.

11 See: Fernández-Armesto, Felipe, 'Pathfinders. A Global History of [...]', op. cit., pages 149–150.

12 A seasonal wind that blows over certain seas, particularly in the Indian Ocean: it blows in one direction during some months and in the opposite direction during other months.

13 Fernández-Armesto, Felipe, *Colón*, Barcelona, Crítica, 1992, page 92.

Delos Vientos. Fol. xvij.



LIBRO.III.DELOS VIENTOS,DE SVCA LIDAD YNOMBRES Y COMO SEHA DE NAVEGARCON ELLOS.

d

Illustration 1.1. 'On the winds, their quality, and names, and how one should sail in them'. In: *Arte de Navegar*, by Pedro de Medina. Book III, Folio XVII, 1545 © Obras clásicas de náutica y navegación / José Ignacio González-Aller Hierro (comp.) Madrid: Fundación Histórica Tavera, D.L. 1998. Copy taken from the Fundación Mapfre. The fact that Medina devoted a whole chapter to the winds shows the important role they had in navigation, and, as a consequence, the process of imperial expansion.

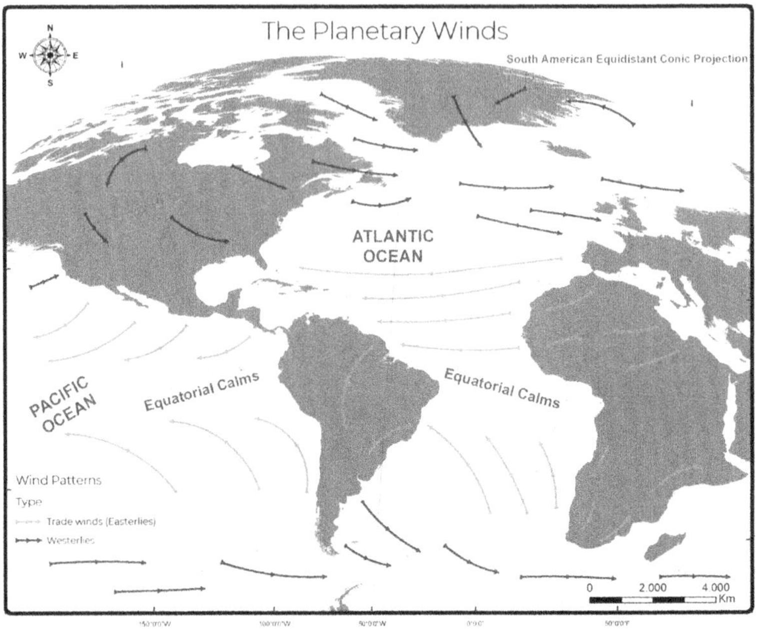


Illustration 1.2. The planetary winds. Rafael David Nieto Bello (based on the maps of Javier Eduardo Pena, Cartography studio). Universidad de los Andes.

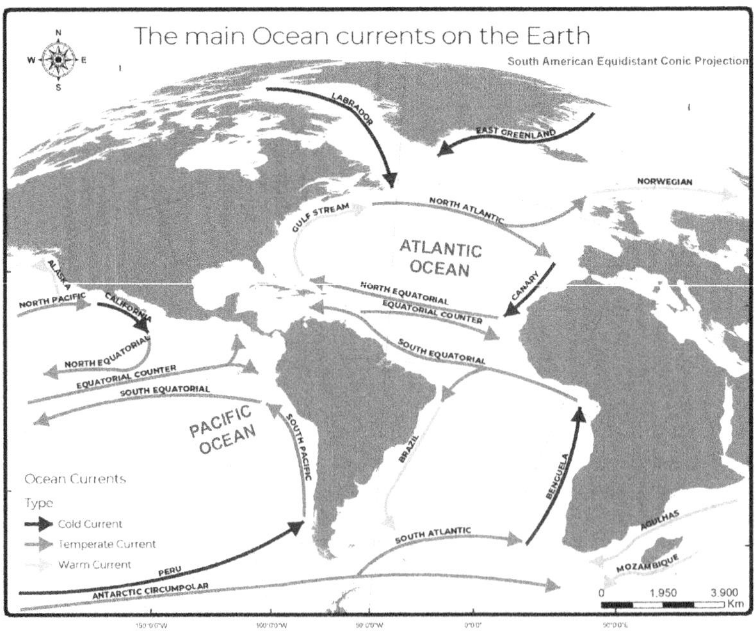


Illustration 1.3. The main Ocean currents on the Earth. Rafael David Nieto Bello (based on the maps of Javier Eduardo Pena, Cartography studio). Universidad de los Andes.

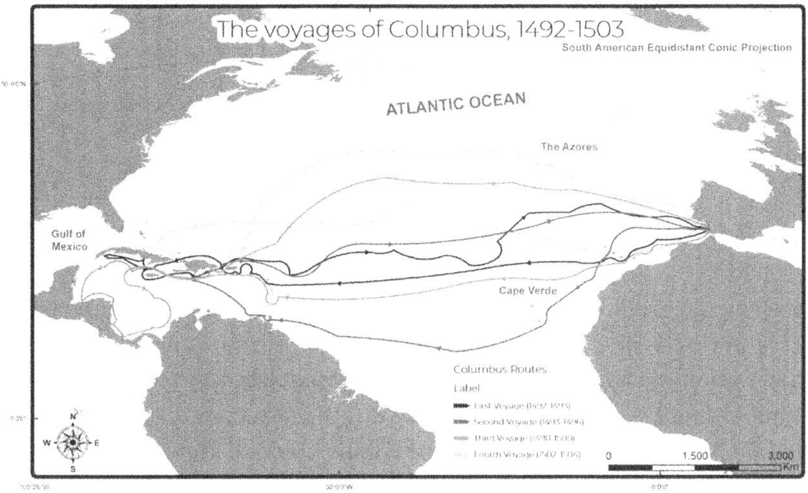


Illustration 1.4. The voyages of Columbus, 1492–1503. Rafael David Nieto Bello (based on the maps of Javier Eduardo Pena, Cartography studio). Universidad de los Andes.

Throughout the 16th century and even during the 17th century, the transatlantic itinerary was basically the same. The ships were loaded in and set sail from Seville, on the Guadalquivir River or, on some occasions, from the ports of Cádiz or Sanlúcar. They headed southwest from the Andalusian coast, toward the African coast and then turned at the level of the 38th parallel, westward toward the Canary Islands, which, at a good speed, could be reached after a week's voyage. They usually anchored at the island of Gomera, where the sailors prepared for the most difficult stretch of the voyage. From the Canaries, the ships followed a western course, going down to the 16th parallel. By taking advantage of the trade winds and without any need to change course, the ships might reach the island of La Deseada in 25 or 30 days (there was never a better name for such a longed-for rest as that 'wished for' or 'desired' island). From there, their destination varied: Havana, Santo Domingo, Cartagena, or New Spain (Mexico).

On the return trip, Santo Domingo or Havana were obligatory stops. From those two places the vessels followed a northwest course through the Bahamas channel, passing close to the Bermuda Islands before later continuing to ascend to the 38th parallel in search of the north winds that moved in an east–west direction. At the end of that run, the sailors reached the Azores where they finally made a stop. The return journey was a little slower and took between 60 and 80 days. The return crossing had to deal with two dangerous zones: The Gulf of Mexico and the passage through the

Bahamas strait and the Bermuda Islands, regions of violent hurricanes or calms that were that were also fearful for sailing ships.[14]

## Gold, silver, slaves, souls, and a thousand kinds of trees

The news that Columbus brought back on his first voyage, which was that he had found populated lands and an exuberant natural world by sailing west on the Atlantic Ocean, awakened an enormous commercial and political interest. While it was still not possible to proclaim the discovery of a fourth continent, the possibility of reaching the Orient by this transatlantic route was not only evident, but so too was the existence of new territories, abundant in natural riches, that could be exploited and a world suitable for evangelization.

The existence of large amounts of gold is one of the most interesting subjects in the diary of Columbus. The idea of ancient authors like Pliny that gold was produced in very hot places encouraged the conquistadors to believe that the tropics were abundant in that precious metal: 'For the heat which the Admiral says he suffered from, he argues that there must be much gold[15] in these Indies and the place where he traveled'. The gold of the New World is an obsessive and exaggerated subject in Columbus testimonies:

> he sent two of the Indians who were with him to bring people back from the towns that were there, near the place where the ships were, and they returned to the *nao* with a gentleman with news that in that island of Hispaniola there was a great amount of gold and that they came from other parts to buy it, and he told him that he would find as much as he wanted there. Others came who confirmed that there was much gold in it [that island] and showed him the way that you had to acquire it. The Admiral understood all that with sorrow, but he was still certain that there was a very large amount of it in those places and that, finding the place where it was dug up, it would be very cheap and even imagined it would cost nothing. And he replied by saying that he thought there must be a lot, because during the three days that they were in that port,

14 A detailed study of the routes taken by Columbus is found in: Morison, Samuel E., *El Almirante de la Mar Océano. Vida de Cristóbal Colón*, Mexico, Fondo de Cultura Económica, 1945.

15 Colón, Cristóbal, *Diario de a bordo*, Luis Arranz Márquez (ed.), Madrid, Dastin, 2000, November 21, 1492, page 145.

> there were good nuggets of gold and he doubted they were brought from another land.[16]

Just like gold, pearls also inspired legends and fantastic tales. According to Pliny, pearls are born from the dew that falls on oysters. Following Pliny once more, Columbus imagined that since dew was so abundant in the Indies there must be many pearls there.[17]

It was not only gold, pearls, and other precious metals and stones that awakened the commercial interest of Europe. A mention should also be made of the project some conquistadors, including Columbus, had to use the inhabitants of the Indies as a new source of slaves. While the initial interests of his endeavor did not lay stress on this idea, it is clear that the Admiral thought that this idea should be highlighted, even when it went against the wishes of the Catholic Monarchs. In 1495, the Admiral thought of organizing a big traffic in slaves between Spain and the Indies, and toyed with the idea of representing it as another commercial benefit of his discoveries. But, in addition to precious metals, bodies and souls, the New World was also seen as an inexhaustible source of vegetal wealth, whose uses for foodstuffs, medicines, and industrial purposes was a notable theme in sixteenth-century European science and literature. Although the earliest navigators failed in their attempt to find a new route to the Orient, they eagerly sought ways to make the Western Indies a supplier of the same products that were imported from there. Since he obviously needed to show his success, the testimonies of Columbus were always optimistic:

> He says that he presented Their Highnesses with raw, natural, unrefined copper of high purity, which weighed six *arrobas* [roughly, 25 pounds each]; indigo-dye trees; red wax; cotton; peppers; cinnamon; an infinite amount of *palo brasil* trees [the source of a red dye]; sweet gum trees; aloes; ginger; incense: all kinds of marvels, fine pearls and red pearls; and Marco Polo says that those pearls are worth more than the white ones [...] I myself have seen them and I know that you find other things in infinite amounts there; and there are also spices, which I do not want to insist on too much, so as to not go on too long.[18]

16 Ibid., December 23, 1492, page 184.

17 Heers, Jacques, 'Cristóbal Colón', op. cit., page 348; and Todorov, Tzvetan, 'La conquista de América [...]' op. cit., page 26.

18 Heers, Jacques, 'Cristóbal Colón', op. cit., page 353.

In his diary for 1492, Columbus tried to make a kind a small inventory of indigenous edible plants and fruits, along with their customs, and he wrote down some reflections on the possibility of growing European products and raising domestic animals.[19] On several occasions the Admiral made an effort to identify some trees, but there was an enormous variety of them, which led him to regret his inability to attain suitable knowledge of them:

> Here, one finds trees of a thousand classes; all bear fruit, each one in its own way, and they emanate aromas which are so strong they are really pleasurable. I regard myself as the most unfortunate man in the world for not being able to recognize them, but I think that they are all of great value. I have gathered samples of all of them, and also of the herbs [...] It is really a pity that I do not know how to distinguish the plants, which makes me very sad. I clearly see a thousand species of trees—which are always green, as in Spain in May and June—and a thousand species of herbs, all with flowers.[20]

In any case, Columbus never stopped praising the beauty of the vegetation there and swearing that, despite his ignorance, they must be useful products, with a strong commercial potential:

> The Admiral says that he has never seen such a beautiful thing, full of trees all fenced in by the river, beautiful and green and as diverse as our own, with flowers and each with its particular fruit.[21]

In another passage of the same diary, Columbus adds:

> And I still believe that [in the islands] there are many herbs and many trees which are worth a great deal in Spain for dyes and medicinal spices, but I do not know them, which fills me with sorrow.[22]

It is interesting to see the frustration the traveler felt at not being able to recognize the plants because the identification of American nature was

19 Colón, Cristóbal. 'Diario de a bordo [...]', op. cit., November 4 and November 6, 1492, pages 131–136.

20 Ibid., page 315. There is an interesting contrast between Columbus and other sixteenth century travelers,and the Linnaean naturalists of the Enlightenment, who were equipped with sophisticated systems of classification that enabled them to translate the unknown into familiar codes in an almost immediate way.

21 Ibid., October 28, 1492, page 124.

22 Ibid., October 19, 492, page 118. Also see the diary entry for October 21, 1492, page 120.

part of an enormous imperial enterprise that took the form of studies of the natural history of the New World, as shall be shown. However, America was not only a place abundant in natural riches: it was also a fruitful scenario for the expansion of the Catholic faith, an ambit inhabited by people who were willing to follow the Christians. Columbus' first impression of the natives emphasized their nudity and apparent willingness to convert to Christianity: 'Believe me, Your Highnesses, in the whole world there cannot be better people, nor more meek ones. Your Majesties must feel a great joy, because they will then be turned into Christians'.[23]

With the testimonies and discoveries of Columbus and those who followed him, the Western Indies turned into a priority for the Iberian monarchies. One proof of it is that on his second voyage Columbus had a fleet of seventeen ships, manned by 1200 people. This second expedition, expedition led by the Admiral was the start of a colossal political, religious, commercial, and scientific venture, set into motion by a royal decree in 1493.

23 Ibid., December 24, 1492, page 186.

# 2. The imperial bureaucracy and the appropriation of the New World

**Abstract**
The second chapter discusses those institutions that were created in the framework of imperial expansion: The *Casa de Contratación* in Seville and the *Real Consejo de Indias* (the Royal Council of the Indies). Understanding the New World and controlling it from a distance required a complex bureaucracy, as well as norms and institutions, whose sole aim was to ensure that there was an orderly flow of information and goods between Spain and its American possessions.

**Key words**: Seville, Council of the Indies, Spain, Casa de Contratación, American possessions

While it is not the purpose of this book to study the political organization of the empire, which others have done in some detail,[1] it may be useful to go over the juridical structure of the colonial administration that was constructed in the 16th century. A description of this legislation is opportune here because, like navigation, cartography, or natural history, political administration and commercial organization were technical challenges that shared the same aim of proclaiming dominion and control. The following section presents a general panorama of the normative and institutional development that took place in Spain for that aim.

The 'Capitulations de Santa Fe', the agreement signed between Columbus and the Catholic Monarchs in 1492, granted him the title of viceroy, but the Crown could not imagine the implication this agreement would have. After he returned, the first regulation for the administration of the Indies was

1 Schäfer, Ernesto, *El Consejo Real y Supremo de las Indias. Historia y organización del Consejo y de la Casa de la Contratación de las Indias. La labor del Consejo de Indias en la administración colonial*, 2 vols. Madrid, Junta de Castilla y Leon/Marcial Pons Historia, 2003, pages 23–29.

Nieto Olarte, M., *Exploration, Religion, and Empire in the Sixteenth-century Ibero-Atlantic World: A New Perspective on the History of Modern Science*. Taylor & Francis Group, 2022
DOI 10.5117/9789463725316_CH02

the '[i]nstruction of our lords the king and queen for don Cristóbal Colón', signed by the monarchs in Barcelona in May 1493. That 'instruction' already gave one a glimpse of the magnitude of the Spanish Crown's aims in the New World. On the one hand, it insisted on the importance of converting the indigenous people to the Catholic faith and, on the other, it was obvious that the New World was seen as a new field for commercial development.

The admiral himself was granted the powers of viceroy and governor of the Indies and, as such, was made responsible for their administration and the appointment of judges (*alcaldes*) and sheriffs (*alguaciles*) who would be in charge of civil and criminal justice. That early royal instruction ordered that the merchandise and products found in the Indies had to be sent to the Spanish Crown to strengthen the mercantile and maritime power of the Empire. It soon became evident that the scope of the project was much greater and that there was a need to establish a specific administrative entity with a staff responsible for all business dealings with the Indies.[2] To safeguard the merchandises brought from Spain or acquired in the Indies and to ensure that the royal treasury (*hacienda real*) was the direct beneficiary of any commercial activity, the Crown ordered the founding of a customs house in the Peninsula. It also stipulated that that institution would be run by the royal treasury and located in Cadiz. Very soon and in a hurried way the flow of information and goods from the New World required a centralized control and the institutions made responsible for that control were the *Casa de Contratación* (1503) and the *Real Consejo de Indias* (1524). The enormous size of the territories that had been discovered meant that the appointment of Columbus as viceroy would be no more than an honorary title whose rights and privileges were unreal. Those privileges definitively ended with the death of the widow of Diego Colón (the eldest son of Columbus) so that the following viceroys had nothing to do with the titles of the family of Columbus.

The central institution for the administration of the colonies were the '*audiencias*'[3] (literally, royal audiences, with judicial and quasi-legislative powers). The oldest was the *audiencia* of Santo Domingo, founded in Hispaniola in 1511. It was organized for the double purpose of serving as a judicial institution for the control of all of the Indies and as a counterweight to the newly appointed governor, Diego Colón. With the founding of the Royal and Supreme Council of the Indies in 1524 (*Consejo Real y Supremo de las Indias*), the *audiencia* of Santo Domingo was subordinated to the new

2 Ibid.

3 Ibid., page 65.

body of control. Later, the *audiencia* of Mexico was founded and, in the course of the 16th century, it was followed by the those of Lima, Guatemala, Compostela (in Nueva Galicia, a province of Mexico), and, finally, Santa Fe in New Granada (modern-day Colombia) in 1550. The creation of the *audiencias* was followed by the establishment of the colonial *provincias* (provinces). While the *audiencias* were administrative units directly linked to the *Consejo de Indias* and the king, the *provincias* seemed, at first, to be an initiative of the conquistadors who, at their own cost, undertook the exploitation of new regions.[4] From the two biggest and most famous enterprises of conquest, those of Cortés and Pizarro, there arose the viceroyalties of New Spain (*Nueva España*) and Peru, whose enormous territory forced the Crown to create many more *audiencias*. In the new regions that had been discovered both within and beyond the vice-royalties more *provincias* began to be created, each with its own governor or, later, captain general, who depended on the closest *audiencias*.

On the religious front, the imperial organization was no less complex. In 1508, the Pope granted the Spanish Crown the right to the Royal Patronage (*Patronato Real*) of all of the Indies; that is, the right to administer and inspect the whole of the colonial Church and appoint ecclesiastical officials. The patronage bull of 1508 stipulated that in the whole of Indies no church, convent, nor other religious center could be erected without a license from the king of Castile. It was a matter of forming 'spiritual armies'. The basic objective was that the missionary orders—the Franciscans, Dominicans, Augustinians, Mercedarians, and Jesuits, being among the most important—would play a key role in the conversion, acculturation, rule over, and, in some cases, the protection of the indigenous peoples. By the end of the government of Philip II of Spain, some 5000 priests, more than 100 each year (most from the Franciscan and Dominican orders[5]), traveled to the colonies.

By the second half of the 16th century the American continent had a political, religious, and territorial organization that consisted of the founding of *audiencias*, bishoprics, archbishoprics, and viceroyalties. All of these organizations had to be supervised by the most important institutions of colonial administration of the Spanish Crown: the *Consejo de Indias* and the *Casa de Contratación*, which were respectively responsible for colonial legislation and the sound functioning of the overseas commerce between the Spain and the Western Indies. These institutions turned into the main priority of the administrations of Charles I and Philip II since they were

4 Ibid., page 143.

5 Martínez, José Luis, 'Pasajeros de Indias. Viajes [...]', op. cit., page 189.

the means by which communications and trade with the New World and its political control had to be ensured.

While a description of these juridical frameworks is an important chapter in imperial history, it is not sufficient for a complete understanding of the mechanisms of control. This political organization faced a major problem: distance. The implantation of any mechanism of control from the metropolis ran into the difficulty of running things from a distance in territories that were little known. Rather than a juridical one, the real problem of long-distance control was one of communication.[6] How was it possible to exert control and proclaim rule over enormous stretches of land, with their respective goods, persons, and souls, on the other side of a great sea? As we mentioned at the start of this book, its objective is to help find an answer to that question by reviewing knowledge and practices in the fields of navigation, cartography, and natural history. Before going on to descriptions of the technical treatises and devices, it might be convenient to speak of some aspects of the history of those institutions that were created to establish a stable communication between the metropolis and the New World: The *Casa de Contratación* and the *Real Consejo de Indias*. It was in the bosom of these institutions that there was implemented a set of actions that, as will become evident below, had a visible impact on the history of the sciences and the political history of modern Europe.

## Seville and the *Casa de Contratación*

The overseas interests of the Crown required a collection center where it could, in a unified way, oversee commercial activities and organize the registering and storage of merchandise: it also required the issuance of norms for the treatment of gold and other goods from the Indies, as well as common rules for the collection of information about the geography and natural resources of the New World.

Madrid was not near the sea, but it was efficiently situated in the center of the Iberian Peninsula and was a privileged point for linking imperial

6 Some detailed analysis about the circulation of information, the configuration of the Empire as a collective and bottom–top endeavor, and the knowledge-gathering of royal institutions in the sixteenth-century Spanish Empire can be found in the recent paper by Adrian Masters, "A Thousand Invisible Architects: Vassals, the Petition and Response System, and the Creation of Spanish Imperial Caste Legislation", *Hispanic American Historical Review*, 98(3), 2018; and in the vast book by Arndt Brendecke, *Imperio e Información: funciones del saber en el dominio colonial español*, Iberoamericana, Madrid, 2012

policy both in and beyond the Mediterranean. It was the seat of the court and, later, of the *Consejo de Indias*: the political center of Spain, we might say. For obvious reasons, however, the center of Spain's maritime operations had to be in the Atlantic and Seville offered the best conditions for that aim.

The discovery did not create the commercial vocation that already existed in the active port city of Seville, but it did drive it forward: between 1498 and 1514 no fewer than 437 Genovese firms operated there, along with 328 Andalusian firms, 72 from Burgos, and 69 Basque, 57 English, 20 Florentine, and ten Venetian ones.[7] In 1503, Seville had nearly 40,000 inhabitants.[8]

From the time Columbus returned, the traffic of ships gradually increased. To get an idea of this traffic: in 1520 alone 71 ships sailed to the New World, with 79 vessels sailing there in 1540. Between 1504 and 1640 a total of 438 convoys departed for the New World and, between 1504 and 1555, the departure of a total of 2865 ships was recorded. The number of ships that returned was notoriously lower, which, in some cases, was due to frequent shipwrecks and attacks by pirates and, in others, to the fact that some ships remained in American territory and continued to operate from there. In the same period, 1946 ships returned to Spain. In the course of the 16th century the number of ships that weighed anchor for the New World continued to grow and by the end of the century nearly 7000 had left.[9]

Seville turned into one of the most important and busiest cities in Spain and Europe during the 16th century. Its river port was situated at a distance of 80 kilometers from the sea, along the Guadalquivir River, which debouches into the Atlantic at the port of Sanlúcar de Barrameda. The Seville–Guadalquivir– Sanlúcar complex was a strategic center of the Indies Run: a river port with a heavy traffic, a collection point, and a natural base for organizing and supplying expeditions to the New World. It was also the place where a large number of bankers, merchants, shipbuilders, cosmographers, sailors, and artisans lived.[10]

7 Domínguez Ortiz, Antonio, 'Sevilla a comienzos del siglo XVI', in: Acosta, Antonio, González, Adolfo and Vila, Enriqueta, (coordinators.), *La Casa de la Contratación y la Navegación entre España y las Indias*, Sevilla, Universidad de Sevilla / CSIC/ Escuela de Estudios Hispano-Americanos / Fundación el Monte, 2003, page 12.

8 Ibid., page 9.

9 Martínez, José Luis, 'Pasajeros de Indias. Viajes [...]', op. cit., page 161.

10 Recently, some authors have carefully examined the importance of artisanal practices and epistemologies behind the Atlantic navigational endeavor in Spain. They show the artisanal manufacturing of the Iberian instruments, as well as the collaboration (trading zones) between cosmographers and artisans in order to calibrate these instruments properly (Sánchez, A. and Leitão, H., "Artisanal Culture in Early Modern Iberian and Atlantic worlds"). Besides, Henrique Leitão recognizes the agency of those instruments and even the emotional engagement (gravitas)

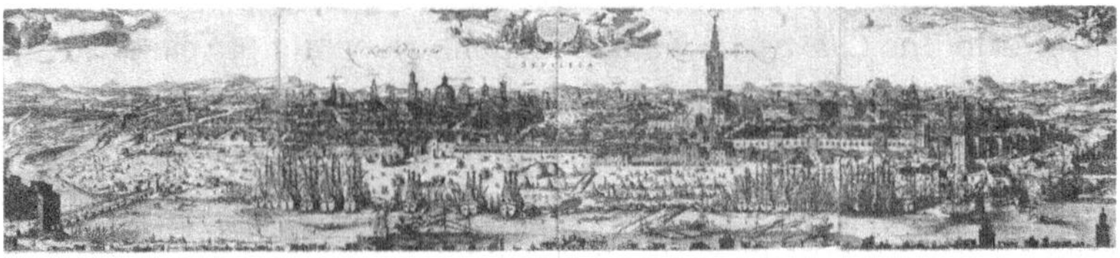

Illustration 2.1. Panoramic view of Seville in the 17th century. At the bottom, the inscription reads: 'Who has not seen Seville has not seen a wonder'.  Naval Museum, Madrid.

Seville was a cosmopolitan metropolis, full of movement and variety: a port open to all news and influences from abroad. On its streets you would find bankers and merchants from Genoa, Venice, and Florence; sailors from Sicily and Greece; pilots from Biscay and Portugal; mulatto gypsies; Black and Indian slaves; and soldiers and adventurers from every corner of Spain.[11]

Fernand Braudel rightly refers to Seville as 'the center of the first world economy'.[12] Not only did people from a wide variety of places and experts in every kind of calling gather there; it was also a place where valuable and strange objects circulated. To its intensive commercial activity and circulation of precious metals should be added that of exotic objects, which were also sold at high prices; hitherto unknown plants with supposed medicinal properties; edible plants; animals that were exotic for Europeans—such as macaws, parrots, toucans, armadillos, and sloths— and indigenous artefacts from the Indies, such as gold work, feather work, weapons, utensils, and textiles.

In Seville, a gathering center was created for everything related to the traffic of ships, people, and merchandises to and from the New World. According to Ernst Schäfer, an anonymous document of 1502, which reads 'it seems that what must be provided to put the business and trade of the Indies in order is the following [...]', might be a first sketch for the foundation of the *Casa de la Contratación de las Indias*.[13] That memorandum proposes

of their use because inadequate use implicated vital risks (Leitão, H., "Instruments and artisanal practices in long distance oceanic voyages"). Other authors have additionally emphasized the Portuguese migration of artisans, practitioners, and cosmographers to Spanish cities and their impact on the Spanish navigational tradition, exploring the cases of Pedro Nunes, Jorge Reinel, Rui and Francisco Faleiro, and Diego Ribeiro (Almeida, B., "Transmitting nautical and cosmographical knowledge in the 16th and 17th centuries: The case of Pedro Nunes"; Collins, E., "Interactions of Portuguese artisanal culture in the maritime enterprise of 16th-century Seville").

11 Domínguez Ortiz, Antonio, 'Sevilla a comienzos del siglo xvi', op. cit., page 85.

12 Cited in: Acosta, Antonio, González, Adolfo and Vila, Enriqueta, (coordinators.), 'La Casa de la Contratación y [...]', op. cit., page 5.

13 Schäfer, Ernesto, 'El Consejo Real y Supremo de las Indias', op. cit., page 31.

that an office be created in Seville for the storage and arrangement of every kind of merchandise that leaves for the Indies or returns to Spain. This administration would be headed by a factor or administrator, a treasurer and two accountants. These Crown officials would not only have to be familiar with the merchandises and knowledgeable about trade; they would also have to inspect the rigging of the fleets and train the crews.

The Crown officials in Seville had to maintain close relations and be in permanent contact with officials who lived in the Indies and had to keep a record of the exchange of merchandises. The same memorandum refers to the way the gold mines in the Indies should be exploited, the correct treatment of the indigenous people, and the way in which the latter should be educated in order to live an orderly life.[14] According to Schäfer, the Catholic Monarchs responded to these considerations by ordering in 'Alcalá de Henares, on January 20, 1503, the founding in Seville of the *Casa de la Contratación de las Indias* and establishing the first ordinances for this authority'.[15] The *Casa de Contratación* was originally housed in the shipyards of Seville: however, that place was prone to flooding and harmful for the merchandises, so it soon became necessary to move it to the Alcázar de Sevilla, a royal palace. There, it was lodged in the Hall of the Admirals until it was definitively relocated in Cadiz in 1717.

More than a place for storage or a customs house, the *Casa de la Contratación* turned into a complex institution, to which a long list of commercial, technical, and judicial functions was assigned. Its general mission consisted of carrying out specific tasks, such as ensuring that the ships were in a navigable state, granting the licenses needed to undertake a voyage, compiling the records or registers of everything that was embarked for the Indies, charging mercantile taxes, organizing and inspecting the fleets, training the pilots, and controlling the use and design of the navigational charts. In addition, the officials of the *Casa de Contratación* acted as judges in all civil and criminal proceedings related to the violation of their norms and the repatriation and adjudication of the properties of the Spaniards who died on the other side of the Atlantic without heirs.[16] As we shall see, all of these jobs had a strong technical component and required a sophisticated degree of social organization.

14 Ibid., pages 31–32.

15 Ibid., page 32.

16 González, Carlos 'La Casa de la Contratación y la historia cultural', in: Acosta, Antonio, González, Adolfo and Vila, Enriqueta, (coordinators.), *La Casa de la Contratación y la navegación entre España y las Indias*, Sevilla, Universidad de Sevilla / CSIC / Escuela de Estudios Hispano-Americanos/Fundación el Monte, 2003, page 546.

The *Casa de Contratación* began to operate under the leadership of the royal factor, Francisco Pinedo; the treasurer, Sancho de Matienzo; and the scribe, Jimeno de Briviesca. From the time it was founded in 1503, it served as an entity that guaranteed an effective communication between the Catholic Monarchs and their officials in the Indies. It was the clearing house for the correspondence sent to Spain by the administrator and governor of Hispaniola, friar Nicolás de Ovando y Cáceres and all of the colonial officials sent to the Indies and the Canary Islands. Likewise, the *Casa de Contratación* turned into the place where the gold was deposited before being minted in the *Casa de la Moneda* (Royal Mint). The different functions assigned to the *Casa de Contratación* likewise required detailed rules for its responsibilities and working hours, records of the arrivals and departures of ships, the purchase of the materials needed to build and maintain these ships, and the issuance of new forms of registering and licensing the passengers. These were the main concerns of the different ordinances that were issued from the foundation of the *Casa* to Charles the V's assumption of the Spanish Crown in 1516.

In the course of the 16th century, the Spanish Crown issued a series of norms to regulate both the *Casa de Contratación* and the *Consejo de Indias* which became ever more complex and detailed. The purpose, which in a certain way was unattainable, was to impose order and a total control on Spain's growing imperial enterprise in the New World. A detailed account of the complex norms that the Crown established to regulate the work of those two institutions is beyond the scope of this book, but a mention of some of those ordinances will help us to understand the enormous task of imperial control, a central theme of this book.

One of the first provisions of the *Casa de Contratación* was to control and administer the records of the value of all of those goods that were sent to the Indies or sent from the Indies to Spain. The 1503 ordinances of the Casa stipulated that any person who wished to send any product to the Indies was obliged to submit a written declaration, sworn and signed, of the things that would be embarked and their value. The aim of this procedure of control was to guarantee the centralized collection of the corresponding taxes. The reports were collated by the accountant in order to make a record of everything that was transported on each ship and galleon that sailed to the Western Indies, whether they were men, animals or things.[17] It is worth recalling that the job of recording such things was, in a certain way, a kind of written account of the New World whose purposes were not very different from those of the cosmographers, naturalists, or chroniclers.

17 Ibid., page 548.

In 1509, Ochoa de Isasaga replaced Francisco Pinedo as the factor of the *Casa de Contratación* and in 1510 new regulations were issued in order to better organize the time and space of the Casa's work and to optimize its mercantile activities. The discipline of the personnel became a necessary condition for the sound functioning of the project. Fixing the working hours of the employees of the office (from ten and eleven a.m. to five and six p.m.) was one of the measures taken to ensure that the officials of the Casa were present during the hours when there was the greatest traffic of ships in and out of the port of Seville. These efforts to consolidate a group of reliable officials who would attend to their duties and be alert to the arrival, departure, and supervision of the fleets were complemented by a series of specific rules for the activities that had to be undertaken for the sound functioning of the institution and the safety of the ships that set sail for the Indies. Among the most fundamental orders introduced by the 1510 reform were: to keep account books for the departure and arrival of royal goods, stored objects, and the purchase of materials; to detail the exact number of merchandises and passengers; to report on shipments of gold; and to require officials to sign the register of the precise amount of metals that reached Spain. This form of regulating the endeavor sought to ensure that trade, navigation, and naval transport would be a technical enterprise under the absolute control of the Crown.

In the same set of rules issued in 1510, Ochoa de Isasaga, in addition to stating his intention to prepare other colonizing expeditions to increase the supply and knowledge of new products, such as Brazil wood, where the Portuguese had certain advantages, asked for the Casa to be directly responsible for the administration of the properties of those who died in the Indies. The aim was to exert a greater control over any object or piece of information that would be useful for the mercantile interests of Spain. In a similar manner, for the purpose of ensuring that its regulations were duly enforced, Isasaga — for the first time in the short history of the *Casa de Contratación* (barely seven years)—asked for the appointment of a qualified judge to administer and ensure the payment of tariffs, oversee the imposition of sanctions, and establish general licenses that would check smuggling .

As can be seen in the reform of September 26, 1511, the Crown published a letter, addressed to all of the authorities of the Kingdom, which granted the *Casa de Contratación* jurisdiction over everything to do with commerce and navigation and also everything that lay in the ambit of civil and criminal law. This norm was created in order to counteract the lack of a register of incoming and outgoing merchandises in the account books of the imperial bureaucracy. The *Casa de Contratación* was given a certain degree of administrative, political, and judicial autonomy, so that it could work independently of any other city or jurisdiction. Once again, the aim was to centralize the management of maritime operations and safeguard their functioning, trying, at all costs, to prevent illegal trade and related obstacles that might oppose the Spanish Crown's ambition to conquer new territories and consolidate an extensive empire. Thus, from very early in the 16th century the *Casa de Contratación*, more than being a storehouse of merchandise and provisions, assumed the role of supervising, licensing ,and regularizing maritime traffic to prevent the clandestine trade of goods between the new and the old continents.[18]

The obligations and tasks of the Spanish Crown in the New World gradually grew. The increase in its colonial possessions was the result of the discovery of new territories, such as Cuba, Jamaica, and the *Tierra Firme*, which was then known as Castilla de Oro: Golden Castile. Those discoveries became even more important with the exploration of the southern sea, which increased the number of travelers to the Indies and, therefore, of the related ships, pilots, instruments, and new licenses. In addition, its functions multiplied when, in 1511, a general permit to travel to the Indies was granted to all the inhabitants of the Kingdom, with the sole condition of registering themselves at the *Casa de Contratación*. The zeal to explore new territorial possessions in America meant that all of the abovementioned responsibilities were assigned to the officials of the *Casa*, who bore the full weight of the whole colonial administration.[19]

## The universal monarchy

In 1519, Mercurino Gattinara, the chancellor and counsellor of Charles V, told the monarch: 'God has placed you on the path toward a universal monarchy'.[20] Those words were not the simple formalism of a royal servant who wanted to flatter his King: the idea of a Christian global empire now seemed like a plausible mission.

On February 9, 1518, nearly fifteen years after the founding of the *Casa de Contratación*, Charles V was crowned king by the Court of Castile, meeting in

18 Schäfer, Ernesto, 'El Consejo Real y Supremo de las Indias', op. cit., pages 36–37.

19 Ibid., pages 40–41.

20 Cited by Lynch, John (ed.), *Historia de España. Edad Moderna. El auge del imperio 1474–1598*, vol. 4., Barcelona, Crítica, 2005, page 340.

Valladolid. Heir of the Habsburgs, Charles V also inherited other kingdoms, which turned him into the monarch of territories that were unmatched in Europe. From the paternal branch, he received the Netherlands: Artois, Flanders, Brabant, Luxembourg, and the Franche-Comté. On the maternal side, he obtained Aragon and Castile, with its dependencies in Italy—Sardinia, Sicily and Naples—along with his enormous American possessions. From his grandfather Maximilian he acquired the Kingdom of the Habsburgs, that is, the Archduchy of Austria.[21] The largest territories of the empire, however, were very far from Europe.

The conquest of the New World was an enterprise that was constantly growing, and it was not possible to concentrate its organization in Seville, which is the reason why the '*Consejo de Indias*' was soon created to control the information that had to reach the King of Spain. It was initially made up of a group of members of the Council of Castile who were charged with the affairs of the Indies, and it was not until 1519 that the name '*Consejo de Indias*' appeared in a Royal Decree (*Cédula Real*) as yet another body of royal administration. The work of this 'Council' grew without check with the growing conquests on the American continent—among which the most outstanding were those of Hernán Cortés—and with the maritime conquests following the return, in July 1522, of Elcano and the other survivors of Magellan's round the world voyage. In 1524, due to these increased possibilities for undertaking new economic ventures, non-Spanish subjects of the Crown were authorized to engage in commercial activities with the Indies.

In the same year, a supreme and autonomous authority for the colonies was formalized: The Royal and Supreme Council of the Indies (*Consejo Real y Supremo de las Indias*), which, from the start, had its own president and councilors and whose work was independent of the Council of Castile. This new Council was initially made up of five councilors, two secretaries, a financial official, an accountant, and its president, the missionary friar García de Loaiza. In addition to the juridical work of settling civil lawsuits in the Indies and appointing governors and royal officials in the colonies, one of the main responsibilities of this body was to supervise the technical work of the *Casa de Contratación*. In addition to its administrative labors, the *Consejo de Indias* became the scenario for the settlement of profound political, legal, and moral debates. Its regulatory duties implied the formulation of policies for the treatment that should be given to the natives of America, the possession of new territories and the use of natural resources.

21 Schäfer, Ernesto, 'El Consejo Real y Supremo de las Indias', op. cit., pages 40–41.

Once the *Consejo de Indias* began operating, the issuance of norms for colonial administration continued to be a concern for the imperial bureaucracy. An important development occurred in 1543, when Charles V issued a set of regulations with forty chapters, which, in addition to detailing the functions of the *Casa de Contratación*, stipulated that it should be subjected to the oversight of the *Consejo de Indias*. This decision resulted from scandals concerning the bribery of many officials of the *Casa de Contratación*, who failed to keep accurate records of the entrance and exit of merchandises and thus allowed the trade with the Indies to get beyond its control. Thus, the administrative autonomy of the *Casa de Contratación* came to an end, and it was subordinated to the strict authority of the Supreme Council of the Indies.

In 1552, with the same aim of regulating the imperial venture, the Crown issued another set of norms, which this time comprised more than 200 Chapters. Among many other subjects it specified, in detail, the functions and duties of officials, the requirements for administering the properties of those who died in the Indies, and listed the kind of people who were prohibited from traveling to the Indies—mainly, Moors, New Christians (converted Jews), and the descendants of those who had been punished by the Inquisition.[22] For the purposes of this book, the norms on the technical tasks of the *Casa de Contratación* are especially interesting. The Chapters, which go from Article 127 to Article 143, refer to the navigational charts (*cartas de marear*) and then to the responsibilities of the *piloto mayor* and the cosmographers. The job of the latter officials was enormously important since the constant updating and correction of nautical charts depended on them along with the legal endorsement of all maritime maps and navigational instruments. The section from Chapter 144 onwards laid down the rules for sailing to the Indies, which included very detailed norms about the required crew, in accordance with the size of the ships. Ships of 100 tons—the minimum legal size—might carry up to 32 men, while those of 250 or more tons might carry up to 64 men, including officers.

Of major concern were the inspections and reviews of the ships before they sailed. The first inspection was done by qualified officials, known as '*visitadores de navios*' (literally, 'visitors of the ships'), who checked on the stability and tonnage of the vessels and fixed the waterline. The second visit was done by the accountant of the *Casa de la Contratación* once the ships were loaded,[23] and the third by the 'visitors' once again. The final inspection took

22 Ibid., page 105.

23 Ibid.

place just before the ship sailed off from Sanlúcar de Barrameda[24] and required the presence of one of the judicial officials responsible for the departure of whole fleets. For the ships that returned to Seville from the Indies, there was only one inspection, the aim of which was to check on the clandestine import of valuable objects, mainly silver and gold. Despite this enormous normative effort, smuggling never stopped being a frequent practice.

As we can see, in the course of the whole history of the colonial administration regulation was inseparably linked to bureaucratic, commercial, juridical, nautical, cartographic, and cosmographic concerns. The Crown needed to formalize a theoretical knowledge of cosmography to ensure its ability to control and communicate with the colonies at a distance. In that regard, a major step toward the standardization of that knowledge took place in 1552, with the creation of the post of the royal cosmographer, which was first held by Jerónimo Chávez. The office of cosmographer and the chair of cosmography gave the *Casa de Contratación* a more marked technical character, in that one of its main functions was to teach and spread naval and cosmographic knowledge.

In fact, the long list of ordinances that was published in 1552 bore the signature of Prince-Governor Philip, who at that time carried out the bureaucratic functions of his father. With the signing of those ordinances on August 11th of that year, the future Philip II made his first decisions about the work of the *Casa de Contratación* and the regulation of navigation in the Indies before he formally assumed the Spanish Crown. The first edition of the ordinances was printed on the presses of Martín de Montes de Oca, of Seville, in March 1553 in a folio volume of 50 pages. The number of copies that were printed must have been large since the book had to be consulted by all the officials of the Indies administration and put aboard each and every one of the ships.

When Philip II assumed the Crown of Castile and Leon in 1556 due to the poor health of his father, the American colonies covered a territory that was bigger than his already vast dominions in Europe. In addition to the governance of the lands discovered by Hernán Cortés in Mexico and the administration of the *audiencias* of Santo Domingo, Mexico, Panama, Guatemala, Lima, Guadalajara, and Santa Fe de Bogota, Philip II had to maintain an intensive correspondence with the royal officials lodged in the new *audiencias* of La Plata de las Charcas (1559), Quito (1563), Chile (1563, 1606), and Manila, the Philippines (1586). Thus, the Crown of Philip II had to keep in contact with around twelve *audiencias*, as well as the governors of

24 Ibid.

the 35 provinces and the clerics of the 41 bishoprics that existed during his reign (1556–1598).[25] In fact, as the years passed, unanswered documents piled up in the offices of his secretaries and, in such a vast empire, the problem of communication and control at a distance became an unprecedented technical and bureaucratic challenge.[26]

The responsibilities of the *Casa de Contratación* continued to grow, but it centered its attention on naval affairs while the other concerns of colonial governance were spread among the different councils in Spain. The *Casa de Contratación* became part of the new institutions at the service of the Spanish Empire. Colonial legislation fell into the hands of the *Consejo de Indias*; financial operations were the responsibility of the *Junta y el Consejo de Hacienda* (Board and Council of Finance); a *Junta de Guerra de las Indias* (Military Board of the Indies) was in charge of safeguarding navigation; and the officials of the institution in Seville—the treasurer, accountant, factor, cosmographer, chronicler, and others—began to be supervised by the *visitadores* and controlled by the *Consulado de Sevilla* (Consulate of Seville).

In 1571, Philip II appointed Juan de Ovando to be the president of the *Consejo de Indias* and, in 1573, the 'Ordinances for the Drafting of a Book of Descriptions of the Indies' were issued. The purpose of these norms, organized by Ovando, was to systematize all the available information that was required to govern distant lands. With the aim of systematizing and standardizing an accurate and useful knowledge of the colonies, the office of cosmographer–chronicler of the *Consejo de Indias* was created, whose job was none other than to compile and arrange trustworthy facts that would be useful for the administration of the Empire.

The final five articles of the ordinances that were approved in 1571, that is, Articles 117 to 122, gave a detailed definition of the responsibilities of the cosmographer–chronicler:

> to make and order the cosmographic tables of the Indies, laying down in them, in accordance with their longitude and latitude and number of leagues as prescribed by the art of geography, the provinces, seas, islands,

25 The statistical data have been taken from: Schäfer, Ernesto, 'La labor del Consejo de Indias en la administración colonial', in: 'El Consejo Real y Supremo de las Indias [...]', op. cit., pages 385–526.

26 Briggs, Asa and Burke, Peter, *De Gutenberg a internet. Una historia social de los medios de comunicación*, Marco Aurelio Galmarini (trans.) Madrid, Taurus, 2002. These authors echo the following description, by Braudel, of the administrative and bureaucratic work during the reign of Philip II: 'a colossal enterprise of transport by sea and by land [...] the daily dispatching of hundreds of orders and reports', page 26.

> rivers and mountains and other places which have to be portrayed and pictured in accordance with the general and particular descriptions which are made of those places [...] proceed with what you are responsible for entering into the general book of descriptions which there must be in the council.[27]

Article 118 specified the procedures for gathering this type of geographical information. It explained, for example, how to determine longitude by means of the observation of eclipses. In addition to the data on geographical positions, the cosmographer–chronicler had to compile information on the inhabitants of the New World, and their 'customs, rituals and antiquities and memorable events'. He also had to compile data on 'herbs, plants, animals, birds and fish', and the final article stated that he had the obligation to assemble information about the itineraries and routes of voyages to or within the Indies.[28] These ordinances defined the specific functions of the cosmographer–chronicler, but another document, 'Title of the descriptions', laid down instructions about the procedures for compiling and organizing the information needed to improve the work of the colonial administration.[29]

The cosmographers and pilots assigned missions in the Indies were ordered to describe and make a written record of their voyages and, on their return to Seville, hand over an authenticated copy of their notebook of entries to the cosmographer and pilot major of the *Casa de Contratación*, with another copy to be sent to the cosmographer of the Council.[30] The official instructions for compiling the information were not only addressed to Crown officials; it was thought that information provided by any traveler or even by the native population would be useful.

The group of norms issued by Philip II in 1573 included a section titled 'Ordinances for the discovery, new settlement and pacification of the Indies', which asked travelers:

> To make a daily commentary and record of everything they see and find and of what happens to them in the lands which they discover, and they should set it all down in a book and, after it is laid down, read it out in public every day before those who were at the discovery in order to find

27 Cited in: Portuondo, María M., *Secret Science: Spanish Cartography and the New World*, Chicago, University of Chicago Press, 2009, page 122.

28 Ibid., page 123.

29 Ibid., page 12. See Table 3.2. 'Articles concerning the post of cosmographer-chronicler of the Council of Indies, 1571', page 127.

30 Ibid., pages 131–132.

> out more about what is happening and be able to confirm the truth of all of it, and it should be signed by some of the principal persons: that book shall be guarded with great care, so that, when they return, they will take it and present it to the *audiencia* which granted them the license to go there.[31]

And, in 1575, the pilot major and pilots were ordered to:

> make a description and keep a diary on each voyage of everything which happens on it: recording the days when they left and entered the ports, the routes and courses which they sailed along every day, the winds of the Sea and the Earth which carried them; the calms, storms and hurricanes which they survived; the currents, sightings of land, islands, reefs, shallows, rocks and obstacles and other dangers and problems which presented themselves; the markers, entrances, exits, depth, ground, size, length, width, water and firewood and all the other features of the ports which they touch at and enter, of which there would otherwise be no description and take a special written account of all that and deliver it to the Master Navigator and Cosmographer of the Casa de Sevilla.[32]

The contents of this grand record of the territories, their inhabitants, and their history were divided into five books: *Cosmography*, *Hydrography*, *Geography* (which included chorography and topography), *Natural History*, and *Moral History*. The books on natural history and moral history were meant to be inventories of natural resources and events.

The project had a clear methodology and purpose. It had to be based on facts confirmed by the direct experience of trustworthy subjects—that is, on 'known and investigated things'[33]—and it also had to be organized by people with the requisite training. With the publication of the Ordinances, this job fell into the hands of the official who now held the post of cosmographer–chronicler of the *Consejo de Indias*, Juan López de Velasco. The result of this responsibility for compiling and arranging an extensive history of the Indies was the *Geography and Universal Description of the Indies* (*Geografía y descripción universal de las Indias*[34]). López de Velasco

31 González, Carlos. 'La Casa de la Contratación y [...]', op. cit., pages 563–564.

32 Ibid., page 564.

33 Portuondo, Maria M., 'Secret Science. Spanish cartography [...]', op. cit., page 137.

34 For a description of that book, see the study by María Portuondo, ibid., Chapter V, 'The Cosmographer at Work', pages 172–209.

was responsible for a second book, known as *El Sumario*. Both *Geography* and *Sumario* were accompanied by an extensive collection of maps.[35] Due to the policy of keeping this body of information secret, these enormous works were only circulated in manuscript copies to a limited readership but, during the subsequent review of their contents, they were subjected to careful assessments and set off scientific debates in the bosom of the imperial government.

Possibly with the hope of overcoming the flaws in a nearly impossible enterprise, Velasco began a series of projects aimed at improving the updating, veracity, and precision of information from the Indies. His idea was not to insist on a single man being responsible for such an ambitious project but to make a more effective use of the colonial bureaucracy, which would directly compile the information. The new effort led to two of the most ambitious programs of scientific research in the 16th century: the questionnaires of the *Relaciones Geográficas de Indias* (Geographical reports of the Indies) and the systematic observation of lunar eclipses to produce an ever more precise measurement of longitudes.[36]

Instead of Ovando's project, Velasco implemented a program for compiling geographical, ethnographical, and natural information that was even more ambitious than the previous one. Instead of the big books of descriptions, which combined data from many sources into a single narrative, it encouraged the compilation of many short and direct collections of data. These requests for information were shaped into specific sets of questions and instructions for undertaking the astronomical observations needed to determine geographical coordinates.[37] The answers to these questionnaires gave rise to the information found in the *Relaciones Geográficas de Indias*. The questionnaires were designed to collect information in an orderly and comparable way. The aim of the program was to assemble data on places of interest, settlements, and cities with their names and geographical location; important geographical features, such as rivers, lakes, mountains, and volcanoes; ports and coastal cities, along with their hydrography; and natural history and useful resources, such as medicinal plants and animals, and minerals of possible use of value. It thus seems that Velasco broke the project down into an orderly collection of data and observations by many travelers, working with common frames of reference and ensuring that all the explorers answered the same questions following the same methods.

35 Ibid., page 197.
36 Ibid., page 209.
37 Ibid., pages 210–211.

In the final stage of Philip II's reign, the Court's cosmographical studies were taken over by Juan de Herrera (1530–1597), the architect of the royal palace of San Lorenzo de El Escorial.[38] In 1582, Philip II created the Mathematical Academy of the Court, under the leadership of Herrera, another institution that had a strong interest in matters of cosmology and navigation.

Meanwhile, the *Casa de Contratación* continued its work. In the 1580's, the cosmographical activities of the institution revolved around Rodrigo Zamorano, who had a university education and was interested in developing fields of knowledge other than navigation, such as astrology and natural history. Zamorano grew American plants in his own garden in Seville and exchanged botanical specimens with the Dutch naturalist Charles de L'Ecluse. Zamorano published the *Compendium of the Art of Navigation* (1581), following the format previously established by Pedro de Medina and Martín Cortés, and faithfully adhering to the statutes which had created the chair of cosmography at the *Casa de Contratación* in 1552.

With Juan López de Velasco's retirement from the *Consejo de Indias* in 1588, the work of compiling information continued to operate within the same parameters Ovando had established, but the activities that the cosmographer–chronicler carried out were separated. On one side, the cosmographers focused on the study of mathematical, astronomical, cartographical, nautical, and hydrological knowledge, while on the other the chroniclers of the Indies were responsible for the descriptive aspects of geography, ethnography, and natural history.[39]

Perhaps the most important feature of this change is that the knowledge of cosmography was no longer kept secret. The transition from Philip II to Philip III seemed to have begun a new stage in the relation between the Crown and its men of science. Philip III looked on the work of these cosmographers as he did that of the great writers and artists of Spain's *Siglo de Oro* (Golden Century) and saw them as an opportunity to show the world the power of his Empire. For that reason, these scientific descriptions of the Spanish dominions began to be a badge of prestige and authority.[40] Thanks to Philip III's interest in making this body of geographical knowledge known to the public, the Crown's cosmographical publications became a way for him to boast of the vast extent of his Empire.[41] Since the responsibilities of the cosmographer were now limited to astronomical, cartographical, and

38 Ibid., page 79.
39 Ibid., page 257.
40 Ibid., page 260.
41 Ibid., page 298.

navigational practices, Herrera thought that there was an opportunity to focus cosmography on its mathematical aspects.

The data's lack of precision required the assembling of more exact astronomical tables, which, in turn, would require an observatory, better instruments, and more highly trained pilots. Andrés García de Céspedes proposed the creation of an observatory in El Escorial in order to produce more precise tables and thus more accurate observations. Hence, the separation of mathematical from descriptive cosmography became ever more pronounced. Of course, this did not mean that the Council had lost its interest in more descriptive forms of knowledge, that is, natural history and moral history.

We have discussed the regulation and functions of these colonial administrative entities in some detail because a clear definition of the rules of the game is essential both for governance and science: fields that are related and require complex processes of social organization, such as those of imperial Spain of the 16th century we have described. Trade, legal affairs, navigation, history (both moral and natural), and cartography formed part of the same endeavor of control that turned into the gigantic technical enterprise that most historians of modern science underestimated. What is seen here is a process of setting forth the rules of the game for the production of knowledge: the establishment of norms for the organization of information, the definition of common frames of reference, the standardization of data and authoritative measurements, and the classification of experiences. They are all precepts that have both juridical and technical implications as well as epistemological ones. The imperial state was a technical and scientific organization, insofar as science and technology were matters of governance and control.

# 3. The *piloto mayor*[1]: cosmography and the art of navigation

**Abstract**
The third chapter discusses the most important manuals of navigation and cosmography that were produced in Spain in the 16th century, along with their importance for the history of Western science.
Oceanic exploration beyond the confines of the Mediterranean required new navigational techniques, that were highly dependent on astronomy. Portugal and Spain consolidated the new science of celestial navigation (*navegación de altura*), which was compiled in and spread by a set of texts and manuals that were read both in and beyond the Iberian Peninsula throughout the 16th century.

**Key words:** Manuals of Navigation, Cosmography, Navigation, 16th century, Astronomy

As we have seen, the complex effort to build an efficient colonial administration, which was a technical and normative problem in itself, was inseparable from an intense scientific activity for the same purpose of attaining control at a distance. The exploration and conquest of the New World and the subsequent shaping of a new world order were the result of a colossal enterprise of religious and commercial expansion, one that only became possible to the extent that what might have been the greatest political, technical, and scientific endeavor in history was developed and put into practice. The fundamental problem of imperial control was establishing an efficient long-distance communication between the metropolis and the newly discovered lands and, in the 16th century, the only means of transatlantic communication were the sailing ships. As we have already pointed out, crossing the Atlantic Ocean implied unprecedented technical challenges. The crossing of the great ocean

1 It has been usually translated as 'pilot major' in English-speaking historiography.

Nieto Olarte, M., *Exploration, Religion, and Empire in the Sixteenth-century Ibero-Atlantic World: A New Perspective on the History of Modern Science*. Taylor & Francis Group, 2022
DOI 10.5117/9789463725316_CH03

was a voyage to a new world of which there were no trustworthy reports, nor familiar references, nor maps, nor guides. Once past the confines of the Mediterranean world—without previously established paths, routes, maps, or itineraries, and in the middle of an enormous and nearly infinite sea without visible coasts or islands or any other sign that might guide them—the new navigators of the Atlantic had to solve major technical problems. The ships had to be strong to resist a hostile and wild sea; rapid, to cover long distances with limited provisions; agile, to navigate with or without favorable winds; and of a suitable size, to approach unknown coasts and bays. But even more complex was a need for navigational techniques that would allow them to follow safe routes in a sea where 'you can only see sky and water'.[2] The answer precisely lay in the sky, in what became known as *navegación de altura*, based on the stars and the knowledge of the Ancient and Arab Worlds, which enabled sailors to find their location on the terrestrial globe by measuring the height of the stars above the celestial globe. The celestial globe had to be read like a map of the Earth and should enable the observer to measure the time, latitude, and longitude in which he found himself.

The concrete expression of the control of space lay in cartographic representation and the production of charts, which required a common model, a single and faithful map of both the Old World and the New: the *Padrón Real*, the Royal Register or Master Map. The direct responsibility for compiling the required information fell to the pilots of the ships, who were controlled and trained by the *mayor* or Master Navigator in Seville.

## The post of *piloto mayor*: seamanship and cartography

Among the officials of the *Casa de Contratación*, the post of the *piloto mayor* needs to be described in more depth since his responsibilities had a marked technical and scientific character. Those who held that post in the course of the 16th century[3] had a notable importance in the history of

2 Medina, Pedro de, *Arte de navegar en que se contienen todas las reglas, declaraciones, secretos y avisos que a la buena navegación son necesarios y se debe saber*, Valladolid, Francisco Fernández de Córdova, 1545, 'Prohemio', Folio III.

3 Amerigo Vespucci, March 22, 1508; Juan Díaz de Solis, March 25, 1512; Francisco de Coto, July 27, 1515; Sebastian Cabot, February 5, 1518; Ñuño García de Toreno and Giovanni Vespucci, June 26, 1526; Hernando Colón, August 2, 1527; Sebastian Cabot, March 1532; Hernando Blas and Diego Gutiérrez, March 6, 1548; Hernando Blas, September 1549; DiegoSánchez Colchero, June 19, 1550; Alonso de Chaves, July 11, 1552; Rodrigo Zamorano, April 13, 1586; Pedro Ambrosio de Ondériz, September 16, 1595; Andrés García de Céspedes, May 15, 1596; Rodrigo Zamorano,

the Spanish conquest of the New World, seamanship, and cartography. The responsibilities of the *piloto mayor* were clearly defined: on the one hand, to examine the pilots who were going to the Indies and approve their charts, instruments, and courses; on the other, to produce the *Padrón Real*, a model map for navigational charts that would ensure safer routes for journeys to the New World. In the same *Cédula Real* of 1508 that appointed Amerigo Vespucci as the first *piloto mayor*, the reasons why the post was created, and its functions were clearly explained:

> insofar as it has come to our notice and is based on experience, we have seen that, since certain pilots are not as expert as they need to be, nor so strict about what they should know in order to rule and govern the ships which sail on the voyages on the Ocean Sea to our islands and the mainland, we have in the region of the Indies, or due to their faulty knowledge of how to rule and govern and not having the basis for knowing how to use the quadrant or astrolabe for the height [of the stars in the sky] or their calculations, they have fallen into errors, and the people who sail under their command have suffered many dangers, which has been a disservice to Our Lord and our treasury, and the merchants who have business there have suffered much harm and loss [...] and to remedy the abovementioned, and because it is necessary that for such voyages, as well as other voyages, which, with the help of Our Lord, we hope to make in order to discover other lands, it is necessary that there be people who are more expert and better grounded in such navigations, and so that those who sail under them may travel more safely is our mercy and will, and we order that all of the present and future pilots of our realms and estates, who would like to go as pilots in the said navigation of those islands and mainland which we have in the region of the Indies or other parts of the Ocean Sea, be instructed and learn what one needs to know about the quadrant and the astrolabe so that, combining practice and theory, they may take advantage of them on the voyages they make to those parts, and that those who do not know may not go as pilots on those ships, nor earn wages for pilotage, nor may merchants hire them as pilots, nor master mariners receive them on ships without their first having been examined by yourself, Amerigo Vespucci, our *Piloto Mayor*.[4]

April 14, 1598. 'Timeline of those who held the post of Master Pilot at the Casa de Contratación, 1508–1620'. 'Spanish Nautical Cartography in the Renaissance'. In, Woodward, David, *History of Cartography. Cartography in the European Renaissance*, vol. 3, Chicago / London, University of Chicago Press, 2007, APPENDIX 40.2, page 1141.

4 Pulido Rubio, José, *El piloto mayor de la Casa de la Contratación de Sevilla, pilotos mayores del siglo XVI.* [Biographical information], Seville, Tip. Zarzuela, 1923, page 461.

From 1519 onwards, the *Casa de Contratación* employed specialists in the making of instruments and maps and in 1523, it appointed its first official with the title of cosmographer, who shared with the pilot major the responsibility for keeping the *Padrón Real* (Master Map) up to date. The pilot major was charged with the great responsibility of guaranteeing the technical and material conditions that would allow the ships to sail safely. For that reason, those who held the post had to demonstrate their skills in several fields: on the one hand, they had to be experienced sailors and, on the other, have a command of the astronomical techniques of celestial navigation. The *cédula real* of 1527 ordered that:

> anyone who would like to be a pilot should prove, before witnesses, that he had navigated for six years in the Indies; that he had been on the Spanish Main and New Spain and Hispaniola and Cuba; and that he had his navigation chart and knew how prick off his distance between each change of course on it and record the courses and lands and ports and the most dangerous shallows and the places he must shelter in; and the places where could find a supply of water, firewood and the other things needed for such voyages; that he had an astrolabe to measure the height of the sun and a quadrant for the north and knew the use of both things, both for taking the height and adding or subtracting the declination of the sun and the rise and fall of the star together with the knowledge of the hours there are at any time of the day and night.[5]

This need to teach the art of navigation and cosmography to the pilots became even more evident in 1552 with the creation of a 'Lecture for the Art of Navigation and Art of Cosmography', held by Jerónimo Chaves, the son of the then pilot major, Alonso de Chaves. With the same aim of teaching the pilots, Alonso de Chaves wrote his *Quatri Partitu en Cosmographia Practica*, otherwise known as the Mariner's Mirror.[6] It is worth noting that, in addition to this text, Pedro de Medina published his *Regimiento de Navegación* (Rules of Navigation) in 1552, an enlarged edition of which was published in 1563. The contents of those books met the requirements

5 Martín-Merás, Luisa 'Las enseñanzas náuticas en la Casa de la Contratación', in: Acosta, Antonio, González, Adolfo and Vila, Enriqueta, (coordinators.), *La Casa de la Contratación y la navegación entre España y las Indias*, Seville, Universidad de Sevilla / CSIC/ Escuela de Estudios Hispano-Americanos / Fundación el Monte, 2003, page 676.

6 Chaves, Alonso de, *Quatri Partitu en Cosmographia pratica y por otro nombre llamado Espejo de Navegantes*, [1518–1538], Castañeda, Paulino; Cuesta Domingo, Mariano, and Hernández Aparicio, Pilar, (eds.), Madrid, Instituto de Historia y Cultura Naval, 1983.

for the instruction of pilots laid down in the establishment of the Lecture on navigation: the tracking and correct interpretation of the heavenly bodies—the Sun, Moon and stars—and the sound use of the astrolabe, navigational chart, and other instruments used to orient oneself in a sea without roads.[7] The Lecture was meant to provide a theoretical training in notions of cosmography and, as part of that, included the following lessons:

> The rules are to do with the height of the sun and how you determine it, and the height of the Pole and how you determine it, and everything else which appears in the rules. The use of the chart and how to prick off the distance between each change of course on it, and always know the place where the ship is. The use of the instruments and how they are made, to determine if they have made mistakes and what the compass, astrolabe, quadrant, cross-staff are, and what the compass needles mark so that you know, in any place, whether you are north-easting or north-westing, which is one of the most important things to know, due to the equations and leeways which must be calculated when navigating. The use of the general diurnal and nocturnal clock and that you know by memory and in writing, at any day throughout the year, the movements of the moon, in order to know when and at what time there will be tides for entering the rivers and bars [shoals] and other things to do with practice and use.[8]

It is evident that the project of training trustworthy pilots for the endeavor of imperial expansion and control had to combine two sources of knowledge that were sometimes in conflict: the experience of the sailor and the theoretical knowledge of Renaissance cosmography. In that regard, the aim was clear: to turn the experiences of all the pilots into a 'universal' knowledge, that is, a fixed language that would allow for the data to be summarized within a common frame of reference.[9] The Crown's concern for training pilots in the new art of navigation was seen in a collection of manuals that

7 See: Medina, Pedro de, *Regimiento de Navegación. Contiene las cosas que los pilotos han de saber para bien navegar y los remedios y avisos que han de tener para los peligros que navegando les pueden suceder*, Seville, 1563, 'Proemio'.

8 Joseph de Veitia y Linaje, *Norte de Contratación en las Indias Occidentales*, cited by Martín-Merás, Luisa 'Las enseñanzas náuticas [...]' op. cit., page 677.

9 Barrera, Antonio, *Experiencing Nature: The Spanish American Empire and the Early Scientific Revolution*, Austin, University of Texas, 2006, page 45; and Sandman, Alison, 'Mirroring the World: Sea Charts, Navigation, and Territorial Claims in Sixteenth-century Spain'. in: Smith, Pamela and Findlen, Paula (eds.), *Merchants and Marvels: Commerce, Science, and Art in Early Modern Europe*, New York, Routledge, 2002, pages 83–108.

were written and published in Spain, mostly in Seville, in the course of the 16th century. These are the main source of this book and will be dealt with in greater detail in this chapter and in the following chapters.

## The navigation manuals

The notorious absence of Spain from the most widely read studies of the birth of modern science is even more surprising when you look at the scientific and technical achievements of the voyages of discovery and conquest of the New World in the 16th century, especially cosmography, navigation, and natural history. As will be shown here, this body of knowledge was not only abundant and notable; it was also definitive for the history of modern Europe.

The manuals of navigation are a unique genre, which were a response to the needs of the explorers of the Atlantic. More than a dozen Spanish authors and treatises dealt in detail with the knowledge of astronomy and cosmography, shipbuilding, the manufacture and use of instruments, the handling of ships, the different callings of the crew, supplies, artillery, dangers at sea and the causes of shipwrecks, meteorology, geography, currents, winds, and all the knowledge that was needed to meet the new challenge of transatlantic navigation.

The contents of the manuals were closely related to those of the Portuguese *Roteiros* (the equivalent of the 'Rutter' in English: a handbook of sailing directions). Before the voyages of Columbus and those who followed him to the New World, the Portuguese had already made calculations of latitude by using astrolabes and measurements of the height of the sun made on their voyages to the south of Africa. The *Roteiros* already included declination tables and instructions on the handling of the astrolabe to measure the height of the heavenly bodies.[10] Henry the Navigator recruited astronomers and mathematicians to further his interests in the sea, and celestial navigation, such as the use of astronomical instruments, was a mainly Portuguese achievement of the 15th century.[11] The voyages to the

10 Portuondo, María M., 'Secret Science. Spanish cartography [...]', op. cit., page 50.

11 It must be said that the Portuguese were truly influential in the Spanish navigational endeavor concerning the artisanal fabrication of instruments, cosmographical theoretical developments, and Portuguese pilots' expertise in some sea routes. Several Portuguese pilots led Spanish expeditions, such as Magallanes. There were similar problems in Spain and Portugal, such as the territorial demarcation in maps, the mathematization of science, and a significant closeness between the academical discussions between the Universities of Salamanca and Coimbra. Some interesting approaches to these topics can be found in Collins, E., "Interactions of Portuguese

southern coast of Africa and its eventual circumnavigation entailed a high degree of precision in calculating latitude. The *Regimiento do Astrolabio e do Quadrante* (Rules on the use of the Astrolabe and Quadrant), published in Lisbon in 1509, may be the oldest manual on celestial navigation[12] (though there may have been a previous edition in 1495).

Following the classical models of Ptolemaic cosmography and with an arsenal of new findings by travelers, a scientific tradition with a strong theoretical component emerged in Spain to meet the concrete needs of the Empire. The political importance and military value of the knowledge contained in the manuals explains why not all of them were printed. However, several *were* printed and handed out to Spanish pilots and navigators, translated into other languages, and went through several editions.

Seville was thus the center for the creation of a set of writings that combined classical cosmography and the new challenges of transatlantic navigation. Among those that were widely circulated, we find: Martín Fernández de Enciso, *Suma de Geographia* (Seville, 1519, 1546); Alonso de Chaves, *Quatri Partitu en Cosmographia Pratica*, otherwise known as the *Espejo de Navegantes* (Mariner's Mirror, manuscript, 1518–1538); Francisco Falero, *Tratado del Esphera y del Arte de Marear* (Seville, 1535); Pedro de Medina, *Arte de Navegar* (Valladolid, 1545); Johannes de Sacrobusto, *Tratado de la Sphera* (Spanish translation by Jerónimo Chaves, Seville, 1545); Jerónimo Chaves, *Repertorio de los Tiempos* (Seville, 1548); Martín Cortés, *Breve Compendio de la Sphera y de la Arte de Navegar* (Seville, 1551); Pedro de Medina, *Regimiento de Navegación*, (Seville, 1563); Juan de Escalante de Mendoza, *Itinerario de Navegación de los mares y tierras Occidentales* (Seville, 1575); Rodrigo Zamorano, *Compendio de la Arte de Navegar* (Seville, 1581); Diego García de Palacio, *Instrucción Náutica* (Mexico City, 1587); Baltasar Vellerino de Villalobos, *Luz de Navegantes* (Manuscript, 1592); and Andrés García de Céspedes, *Regimiento de Navegación* (Madrid 1606).

## Manuals for the Empire

The manuals fulfilled one of the main responsibilities of the *Casa de Contratación*: to ensure the safety of transatlantic voyages, an enterprise that

artisanal culture in the maritime enterprise of 16th-century Seville" and Almeida, B., "Transmitting nautical and cosmographical knowledge in the 16th and 17th centuries: The case of Pedro Nunes".

12 Taylor, Eva G. R., *The Haven-finding Art: A History of Navigation from Odysseus to Captain Cook*, New York, Abelard-Schuman Limited, 1957, page 162.

also entailed the instruction of pilots and making charts and navigational instruments.

The first of these treatises was the *Suma de Geographia* (Summary of Geography), by Martín Fernández de Enciso, who published it in Seville in 1519, in Alcalá in 1530, and in Seville again in 1546.[13] As the author explained, the purpose of the *Suma* was to provide pilots in the service of the Crown with the geographical and astronomical information needed to continue with the conquest of the New World. To achieve this, he compiled the rules, precepts, and observations that appeared over and over again in treatises with similar aims and contents in the course of more than a century.[14] As in most of the treatises that followed it, the opening lines of the manual had a dedication to the King and to God: 'the most high and very powerful King, the prince and lord don Carlos, by the grace of God King of Spain, Naples, the Two Sicilies, Lord of the Western Indies, Archduke of Austria, Duke of Burgundy, Count of Flanders [...]'.[15] The title pages, dedications, prologues and prefaces are interesting documents since they clearly reveal the aims of these manuals and the context in which they were written. Thus, all of the manuals explicitly referred to their relation to an imperial context and were addressed to and dedicated to the King or high royal officials.

*La Suma de Geographia* followed the geographical model of the earliest Roman geographer, Pomponius Mela, but it departed from the classical model in certain fundamental aspects, such as the traditional division of the world into three parts: Europe, Asia, and Africa. Instead, Enciso divided it into two parts: East and West. The eastern part corresponded to the known world and the western part to the New World. The geography of the eastern part followed the classical authors, without major changes, but it was necessary to include hitherto unknown information in the description of the new lands.[16]

In addition to an explicit intention to serve God and the Empire, the treatises usually included some remarks on the importance of navigation, both in the history of mankind and in that historic moment in the Spanish

13 Fernández de Enciso, Martín, *Suma de Geographia que trata de todas las partidas y provincias del mundo: en especial de las indias y trata largamente del arte de marear juntamente con la sphera en romance con el Regimiento del sol y del norte*, Seville, Casa de Andrés Burgos, 1546.

14 Haring, Clarence H., *Trade and Navigation between Spain and the Indies in the Time of the Hapsburgs*, Cambridge, Harvard University Press, 1918. page 386.

15 Fernández de Enciso, Martín, 'Suma de Geographia que trata [...]', op. cit., Folio II.

16 Portuondo, María M., 'Secret Science. Spanish cartography [...]', op. cit., page 51.

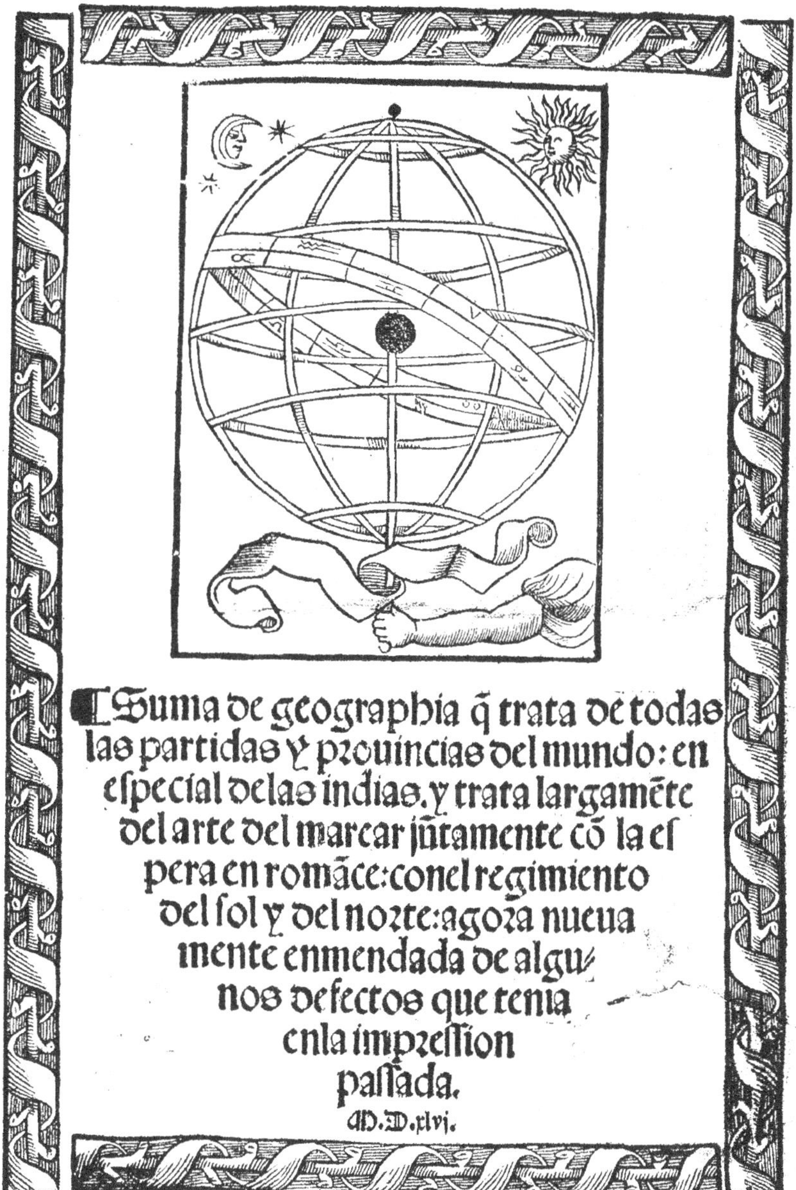

Suma de geographia q̃ trata de todas
las partidas y prouincias del mundo: en
especial delas indias. y trata largamẽte
del arte del marear jũtamente cõ la es
pera en romãce: conel regimiento
del sol y del norte: agora nueua
mente enmendada de algu-
nos defectos que tenia
enla impression
passada.
M.D.xlvj.

Illustration 3.1. *Suma de Geographia*, by Martín Fernández de Enciso, 1530 © Obras clásicas de náutica y navegación / José Ignacio González-Aller Hierro (comp.) Madrid: Fundación Histórica Tavera, D.L. 1998. Copy taken from the Fundación Mapfre. The drawing shows the celestial sphere held up by a man's hand, which suggests the powerful idea that mankind has the control and dominion of the universe.

Regimiẽto de nauegaciõ
Contiene las coſas que los pilotos hã
de ſaber para bien nauegar: y los remedios y auiſos que han de
tener para los peligros que nauegando les pueden ſuceder.
¶ Dirigido a la Real Mageſtad del Rey don
Philipe nueſtro Señor.
Por el Maeſtre Pedro de medina vezino de Seuilla.

Illustration 3.2. *Regimiento de Navegación*, by Pedro de Medina, 1563. © Obras clásicas de náutica y navegación / José Ignacio González-Aller Hierro (comp.) Madrid: Fundación Histórica Tavera, D.L. 1998. Copy taken from the Fundación Mapfre . The subtitle explains the purpose of this book: 'It has the things which pilots have to know to navigate well; and the remedies and warnings which they must have for the dangers which might occur when sailing'. The title page also has the dedication: 'To the very high and very powerful Lord don Philip, King of Spain and Lord of the New World'.

Empire. As an example, it is worth citing the initial paragraphs of Pedro de Medina's book at length:

> Among the things of high quality which human ingenuity invented for the sustenance of men, and a very important one, was to build different kinds of vessels and discover the art which governs them and enables one to navigate in the sea. Some moved by sails and with the force of the winds and others moved by oars with the force of men. Due to that navigation, we have many important goods and advantages. One is that is it has spread and spreads the doctrine of Jesus Christ and the preaching of the Holy Gospel through the universe, mostly in the Indies of Your Majesty, which we call the new world where more than 8000 leagues of the sea coast have been navigated. Where so many kingdoms and provinces have been discovered [...] and a great number of nations and different peoples are found and an almost infinite number of them have been converted and are converted to our very holy Faith. By navigation, lands are supplied, and people are comforted. Navigation enables a surplus in one province to be sent to another where it is lacking. And, navigating, what is born in abundance in some parts is taken to where there is a need for it. Thanks to navigation, we have these and other things which are very advantageous for the life of men.[17]

The *Instrucción Naútica* (Nautical Teaching), by Diego García de Palacio, began with a discussion of how the dangers of the sea are overcome by human arts and how, despite the fact that the sea is not a natural medium for man, navigation has brought about the prosperity of the Empire. Montañés, one of the characters in this manual, written in the form of a dialogue, insisted on the close links between navigation and imperial expansion: 'it is so ancient and widely used among we Spaniards that, for that reason, it has won many kingdoms, done feats, achieved so many victories and has brought such good events'.[18] He added that 'without it the advantages some gain from others would not be possible, nor could we link what some lands produce with that which others produce, their arts and their manners of living'.[19] Once again, the relationship between God and the Empire is explicit: 'Your majesty, the King, don Felipe, our lord, with your fleets and

17 Ibid. I have been faithful to the narrative but have made some changes in spelling to make it easier to read.

18 García de Palacio, Diego, *Instrucción Náutica, para el buen uso y Regimiento de las Naos, su traza y gobierno conforme a la altura de México* [1587, Pedro Ocharte, Mexico], Madrid, Editorial Naval, 1993, Book I, page 2.

19 Ibid., page 4.

surrounded by your army which gird you every year against the world for the grandeur of God and His name, which make navigation possible'.[20]

For the Spanish authors, the historical importance of navigation had a religious explanation and they often referred to passages in the Bible where trade and religion are combined. Thanks to the art of navigation, Saint Paul was able to reach Rome; King Solomon to ship treasures on the sea to build his Temple; Noah to save the creatures of the universe in a mighty vessel and, without navigation, the propagation of the faith would not have been possible. The same García de Palacio further remarked that:

> And if we consider the universal advantage which has derived from the navigation which our natives have undertaken to the islands and mainland of the Ocean sea from the South to Peru and islands of the setting Sun and many other kingdoms and provinces: the treasure of pearls, gold and silver and precious stones never before heard of or seen, the merchandises and rich things which they bring and have brought from them and (what is even more splendid) the infinite number of idolatrous men who have converted and daily convert to our holy Catholic Faith.[21]

The way the sixteenth-century Spanish authors insisted on that shows that we cannot ignore that their obligatory references to the King were always associated with his religious plan or mission, and imperial expansion was always understood as a crusade to spread the Catholic faith. Thus, spreading the doctrine of Jesus, saving souls, conquering lands, and exploiting material riches seemed to have been part of a single endeavor embodied by the King.

From a historical standpoint, more than the frequent praise of the role of navigation throughout the history of humanity, the explicit political and religious meaning these authors gave to Spanish seamanship in the 16th century is the interesting part. They were not simple manuals for managing ships in the sea. We might say, instead, that they were 'manuals for the governance of the world' since they described the basic tools for the successful universal expansion of the Christian faith and the Empire.

Charles V and Philip II, the monarchs for whom these treatises on cosmography and seamanship were written, had titles which proclaimed them to be sovereigns of a large part of the world, but always by the grace of God. That religious meaning given to imperial expansion has often been reduced to a rhetorical flourish or simple pretext by later historians who,

20 Ibid.

21 Ibid., page 5.

in their zeal to find secular explanations, have subordinated the religious aspect to interests of a political and commercial nature in an age of profound spirituality. The project for a Christian empire, the conquest of the Atlantic, and the evangelization of the Indians of the New World were battles that were not different from the conquest of Jerusalem and the war against Islam and heresy. In any analysis of the history of the 16$^{th}$ century, the religious aspect cannot be limited to the field of rhetoric and, as will be evident throughout the sources we cite, the Iberian expansion of the 16$^{th}$ century and the history of the Habsburg monarchy cannot be understood without acknowledging the presence and role of the spiritual, faith, and God.

It is not only obvious that the religious content of the manuals was intended to serve the King and God, but that they also expressed a profound religiosity in their conception of nature and man. Their teleological and theological approach was evident in the description of a natural world created for man, where the will of God and divine providence are the ultimate cause of the natural order.

In the preface to Pedro de Medina's *Regimiento*, the idea that the natural order is providential and religious is clear, and therefore science, in the manner of the hermetic tradition of the Italian Renaissance, was presented as the knowledge of a divine order:

> The delicacy and subtlety which the navigation of the sea entails is so great and so exalted that it is agreed that it is governed by the celestial bodies. That is: by the Sun/Moon and stars: that they are the guide which shows the path where the men who navigate blaze trails, there, where nature has denied them, because there are no roads on the sea [...] God set a watch in the sky which never fails: it has no defect/it is the clock of the North, by which at any time and place man knows this and understands what time it is at night. From the abovementioned the great subtlety which sailing the sea has is shown.[22]

Alonso de Chaves, when he spoke of the compass, declared that 'the virtue and strength of this instrument is divine and it is not human nor made by men; when it is freed, it is governed by its natural inclination alone, which is the celestial influence'.[23] As we shall see below, the writings of the sixteenth-century chroniclers and naturalists showed similar notions about the divine order of a nature at the service of man.

22 Medina, Pedro de, 'Regimiento de Navegación. Contiene las [...]', op. cit., 'Proemio', Folio III.

23 Chaves, Alonso de, 'Quatri Partitu en Cosmographia pratica [...]', op. cit., Book I, Second Treatise , Chapter I, 'Que trata del instrumento llamado aguja de marear, y de su utilidad y provecho', page 104.

It is in this political and religious context that the concrete purpose of the treatises should be understood: to guarantee the safety of the Spanish transatlantic missions. This was an aim which required a growing number of pilots and crewmen to be adequately taught their trade, so that the texts we refer to here are fundamentally materials to instruct navigators. The texts were in fact dedicated to the king, but they were thought up in accordance with the need to train pilots and sailors: that is, with a clear pedagogical content that is obvious and explicit in their very titles: *Arte de Navegar* (The Art of Navigating),[24] *Regimiento de Navegación que Contiene las Cosas que los Pilotos han de Saber Para Bien Navegar* (The Rules of Navigation, Which has the Things Which Pilots Must Know to Navigate Well),[25] *Compendio de la Arte de Navegar* (Compendium of the Art of Navigation),[26] *Luz de navegantes* (The Light of Navigators),[27] *Espejo de Navegantes* (Mariner's Mirror),[28] among others.

Some authors, following classical models of reasoning, used the format of the dialogue, as in the case of the *Itinerario de Navegación*, by Juan de Escalante.[29] The author imagined a wide-ranging conversation between a young man who is eager to learn the art of navigation and an experienced pilot. The idea was to teach his readers about the routes of the voyages between Spain and the New World, the knowledge of the pilots and the duties of the crew, as well as the material means needed for the crossing, the building of ships, and the use of instruments and navigational charts. The book has a practical tone that seemed to be based on the author's long experience at sea. From a young age and under the sponsorship of his uncle—also a ship captain in Seville—Escalante grew up very close to the world of the sea, so that by the age of 18 he already commanded his own ship and had won a reputation for his skillful fights against the attacks of pirates. Diego García de Palacio chose the form of the dialogue to strengthen the pedagogical purpose of his *Instrucción Náutica*, but this time his interlocutors were a man from Biscay (in the Basque country) and a *montañés* (a man from the port city of Santander, in the region of Cantabria).

This shows that the manuals were conceived of and written as textbooks and directly met the Crown's need to teach its pilots. This is also true of Medina's

24 Medina, Pedro de, 'Arte de navegar en que [...]', op. cit.

25 Medina, Pedro de, 'Regimiento de Navegación. Contiene las [...]', op. cit.

26 Zamorano, Rodrigo, *Compendio de la Arte de Navegar,* Seville, Alonso de la Barrera, 1581.

27 Vellerino de Villalobos, Baltasar, *Luz de navegantes donde se hallaran derrotas y señas de las partes marítimas de las indias, islas y tierra firme del mar océano,* [1592], Madrid, Madrid Naval Museum / Universidad de Salamanca, 1984.

28 Chaves, Alonso de, 'Quatri Partitu en Cosmographia pratica...', op. cit.

29 Escalante de Mendoza, Juan de, *Itinerario de Navegación de los mares y tierras occidentales,* Sevilla, 1575 (Colección Clásicos Tavera. Fundación Mapfre).

*Regimiento, El Espejo de Navegantes*, by Chaves, and Zamorano's *Compendio de la Arte de Navegar*. Due to the pedagogical form in which they were composed, they served as textbooks in the Spanish schools of navigation.[30]

The contents of these manuals covered a wide range of subjects: the principles of Aristotelian/Ptolemean astronomy; the application of astronomical principles to navigation and cartography; the use of astronomical tables to determine one's geographical position; the management of instruments useful for navigation, such as the astrolabe, compass, hourglass, and plumb line; instructions on the making and use of navigational charts; and ideas about the structure, functioning, and building of ships. In addition, the manuals contained precise information about the crew and its duties; the principles of meteorology; information about currents, winds, and handling the ship in difficult conditions; the supplies needed for the crossing; and geographical details about itineraries, ports, and war tactics, to mention only the most important subjects. You might almost say that the manuals amounted to a new scientific genre, which combined the cosmographical and theoretical traditions characteristic of Renaissance humanism with the practical problems of transatlantic navigation.

We will deal with the contents of the manuals in detail in the following chapter, but before ending this section we need to examine some general aspects of their publication and circulation, as well as their relationship with classical authorities and the clash between theory and practice.

## Publications, dissemination, and secrecy

One of the ideas that has mostly strongly marked the recent historiography of science has been the need to understand the problem of knowledge as

30 There is a current academical debate about the real use of these manuals. As Leitão remarks, the manuals were generally written in vernacular languages (Spanish, Portuguese, etc.), closer to the spoken language, while classical treatises were written in Latin. It could be related to a pedagogical purpose of teaching pilots and sailors in their own terms. ("Instruments and artisanal practices in long-distance oceanic voyages", page 198) However, Pérez-Mallaína elaborated a detailed exposition of the high rates of illiteracy and "natural intelligence" (in opposition to theoretical training) of sixteenth-century seamen and pilots, which could lead to an interpretation of few real use of those manuals as texts books (*Spain's Men of the Sea: Daily Life on the Indies Fleets in the Sixteenth Century*, pages 58–62). Considering these arguments, one can say that some figures must be considered: "expert mediators" (as Sandman and Ash identified in "Trading Expertise: Sebastian Cabot between Spain and England" *Renaissance Quarterly*, 57(3), 2004, pages 813–846) who taught seamen orally beyond a direct reading of manuals, and transferred expertise bidirectionally between seamen and cosmographers.

a problem of communication. In the end, any episode in the history of science that does not pay careful attention to the media of communication, the public, and the circulation and use of scientific ideas will be a history of ideas with obvious limitations. In that regard, one of the key aspects that has to be taken into account to understand the scientific practices of sixteenth-century Spain is the tension between the necessary spreading of knowledge and the Spanish authorities' efforts to keep its nautical and cartographical knowledge secret.[31] For military and political reasons, the Crown chose to limit the circulation of geographical information and maps and descriptions of the New World. In the hands of its rivals, information about the location of the new territories, their resources, and access routes might endanger its intention to monopolize the exploitation of those lands.[32]

Despite the official zealousness about this knowledge, some of the manuals we have mentioned were widely circulated, even beyond Spain.[33] Those that were mostly widely read in Spain and Europe include the manual by Enciso, which was published in Seville in 1519, Alcalá in 1530, and in Seville again in 1546; Pedro de Medina's *Arte de Navegar* (1545); and the *Breve Compendio de la Sphera y de la Arte de Navegar* by Martín Cortés (1551). Among these, the best known was Pedro de Medina's *Arte de Navegar*, which was translated into Italian, French, Flemish, and English. In addition, several editions that were used as textbooks in France[34] appeared later in the 16th century. With the same pedagogical aims, Medina published another version of his book, titled *Regimiento de Navegación*, in 1552 and 1563, and wrote another treatise, *Suma de Cosmographia*, in 1561. Following the course of those manuals and their translation is vital to understand the Iberian influences into other navigational traditions, such as the English.[35]

In the same period, Martín Cortés published his *Breve Compendio de la Sphera y de la Arte de Navegar* in Seville, which was also translated into English and published in London in 1561. Medina's book was widely circulated

31 See: Portuondo, María M., 'Secret Science. Spanish cartography [...]', op. cit., page 103.

32 Ibid., page 7.

33 On the spread of the manuals, see: López Piñero, José M., 'El arte de navegar en la [...]', op. cit., Chapter II.

34 Haring, Clarence H., 'Trade and Navigation between Spain [...]', op. cit., page 386.

35 One interesting example of the transmission and reinterpretation of the Spanish navigational tradition in other countries is the case of Sebastian Cabot, who was a Venetian cosmographer–pilot that reached the post of Pilot Major of the Spanish Kingdoms, and ended his life in England as a leader of a pilot-teaching effort. This case is widely explained in Sandman, A. and Ash, H., "Trading Expertise: Sebastian Cabot between Spain and England".

and went through 20 editions, most in French, whereas the English seemed to prefer that of Cortés, of which six translations[36] are known.

Another manual of undeniable importance was the *Regimiento de Navegación*, written by Andrés García de Céspedes, the Royal Cosmographer, in 1606. García de Céspedes, who had a theoretical education in mathematics and astronomy, reviewed and updated previous treatises and made a compilation of the knowledge, accumulated until then, of nautical instruments and celestial navigation and it became the paradigm of the science of navigation in the 17th century.[37]

The first of these treatises printed in the New World was Diego García de Palacio's[38] *Instrucción Náutica, Para el Buen Uso y Regimiento de las Naos, su Traza y Gobierno, Conforme a la Altura de México*, published by Pedro de Ocharte in Mexico City in 1587. This was one of the most exhaustive treatises on shipbuilding and it is interesting that it was published in Mexico at a time when there were no publications of that kind in Europe. García de Palacio was an eminent jurist who might have studied law in Salamanca. In America, he studied and tested the characteristics of the trees there and the properties of American timbers that might be used in shipbuilding, and was responsible for the manufacture of sails and cordage woven in cotton. He also supervised the building of two1000-ton galleons.[39]

Not all of the manuals came to be printed, although most were circulated in manuscript form and had a visible impact. Juan Escalante de Mendoza's *El Itinerario de Navegación*, written in 1575, was submitted to the *Consejo de Indias* and approved by the cosmographers and pilots of the *Casa de Contratación*. However, Escalante de Mendoza did not obtain a license to print it and the same *Consejo de Indias* prohibited its publication because it was afraid that, in the hands of foreigners, the information compiled therein would threaten Spain's dominion over the new routes, lands, and riches of the New World. The main reason why that book could only be published four hundred years later was that, in contrast with other manuals, it included a detailed geographical description of routes, distances, and sailing conditions: information that could not fall into the hands of Spain's enemies.

36 Portuondo, María M., 'Secret Science. Spanish cartography [...]', op. cit., pages 52–53.

37 García de Céspedes, Andrés, *Regimiento de Navegación*, Madrid, Juan de la Cuesta, 1606. Also see; Haring, Clarence H., 'Trade and Navigation between Spain [...]', op. cit., page 390.

38 In 1580 he went to live in Mexico. In January 1581, he obtained a doctoral degree at the Universidad Capital Virreinal—where he later became the rector—and held many official posts. He was also the author of another book on military affairs: *Diálogos militares de la formación e información de personas, instrumentos y cosas necesarias para el buen uso de la guerra*, Mexico, 1583.

39 García de Palacio, Diego, 'Instrucción Náutica, para el buen [...]', op. cit., page 22.

The book written by Alonso de Chaves, *Quatri Partitu en Cosmographia Practica* (also known as the Mariner's Mirror), was never published for the same reasons: because the contents of Chapter IV 'deal with all that which concerns navigators in all parts of the Indies'. However, in its manuscript version, this manual was disseminated in order to teach pilots in Seville. Another treatise which remained in manuscript was *Luz de Navegantes Donde se Hallaran Derrotas y Señas de las Partes Marítimas de las Indias, Islas y Tierra Firme del mar Océano* , by Baltasar Vellerino de Villabos, which, as its name (Light of Navigators, where you Find Routes and Signs of the Maritime Parts of the Indies, Islands and the Mainland of the Ocean Sea) indicates, gave a detailed description the routes ships took to and from the Indies in the 16th century and was accompanied by 115 drawings of islands, bays, and ports of interest. As in the previous cases, it is understandable that its publication would be a risk for the Crown.

## Humanism and the classics

A discussion of whether sixteenth-century Iberian science did or did not have a modern character is beyond the scope of this book, but a study of these manuals forces us to inquire into the role Spain had in Western science. The manuals turned into a 'scientific' genre of literature, where the traditions of classical learning were put at the service of a concrete aim and the combination of these two traditions resulted in the emergence of renewed forms of cosmography and natural history.

The Crown of Castile would never have turned into the center of such an extensive empire if it had not appropriated the philosophical and technical traditions of the Arabs, or incorporated the techno-scientific traditions of the Ancient Greek and Roman Empires and made them its own. Without the knowledge of the Arabs and Jews and the experience of the Portuguese, Catalans, Galicians, and Basques, Castile would never have been the center of commercial empires with global pretensions.

It is worth recalling that a large part of the Iberian Peninsula had been Arab territory and the centers for the translation of classical texts from Arabic to Latin were found in medieval Spain. Toledo, for example was the epicenter for the collection and translation of classical texts.[40]

40 In the 12th and 13th centuries, cities like Toledo were important centers for the translation of Greek works. The most important of these translators from Arabic to Latin was Gerard de Cremona (1114–1187), who traveled from Italy to Spain to search for Ptolemy's *Almagest*. Once there, he decided

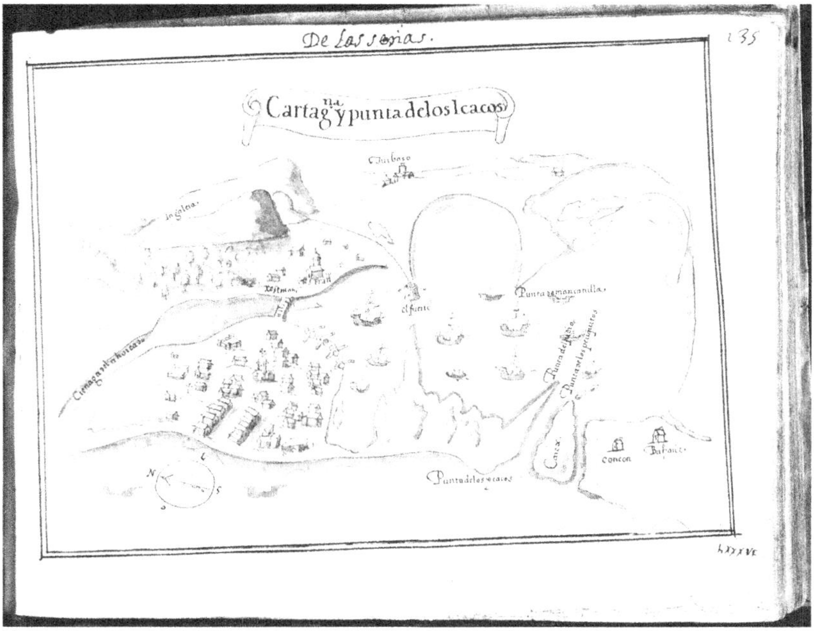

Illustration 3.3. Cartagena and Punta de los Icacos. In: *Luz de navegantes,* by Baltasar Vellerino de Villalobos. Book II, Folio LXXXVI. © All rights reserved. Universidad de Salamanca. This map of Cartagena shows the detail given to the depiction of certain places that were important for the Spanish Empire.

As in the rest of Europe, the humanism of the Spanish Renaissance was closely linked to Italian humanism. An emblematic figure of Spanish humanism was Antonio de Nebrija, who, educated in Italy, was the leader of a scholarly tradition in Salamanca with a strong philological component and a precise knowledge of the Latin language of the classical authors. In the first half of the 16th century, the universities of Alcalá de Henares and Salamanca were the center of an intensive philological activity revolving around the Polyglot Bible and the study of the texts of classical antiquity. Hernán Núñez de Guzmán (1473–1553) and Juan Ginés de Sepúlveda—official historian to Charles V and tutor of Philip II—did translations and studies of the works of authors like Pliny the Younger, Aristotle, and Pomponius Mela—the latter, who wrote around 43 A.D.,[41] was the author of a three-volume geographical compendium, titled *De Chorographia.* The works which Spanish humanism focused on, that is, the Holy Scriptures

to remain in order to learn Arabic. He later translated books on algebra; the writings of Galen; texts by Aristotle about topics such as physics, logic, and meteorology; and Euclid's *Elements,* among others. See: Lindberg, David, *Los inicios de la ciencia occidental. La tradición científica en el contexto filosófico, religioso e institucional (desde 600 a. c. hasta 1450),* Barcelona, Paidós, 2002, Chapter VII.

41 Coroleu, Alejandro, 'Humanismo en España', in: Kraye, Jill (ed.), *Introducción al humanismo renacentista,* Cambridge, Cambridge University Press, 1998, pages 295–330.

and the writings of the great thinkers of antiquity, were important components of the books of Catholic writers of the Iberian Peninsula in the 16th century.

One can recognize the influence of different classical traditions in Spanish humanism, especially the natural philosophy of Aristotle, a general framework for understanding the natural world, as well as Euclid's geometry and Ptolemy's *Geography,* which provided tools for coming up with a mathematical representation of the universe. In addition, in the fields of geography and natural history, Pomponius Mela's, *De situ Orbis*, Pliny the Elder's *Natural History*, and Strabo's *Geography* served as models for incorporating the human, animal, and vegetal kingdoms into the cosmos.[42]

It might be interesting to remember that Nicolaus Copernicus' *On the Revolutions of the Celestial Spheres* appeared in 1543 and, while a few European universities put it on their curricula (Salamanca being among them), it should not surprise us that during the 16th century the cosmological systems were still faithful to the basic precepts of Aristotle and Ptolemy. That should not be taken as a sign of the backwardness of Spanish science, nor as a measure of the degree of modernity of Iberian writers. As is well known, ensuring the general acceptance of Copernicus' ideas required defenders as powerful as Galileo Galilei and Johannes Kepler, and this process was not only undertaken in the face of a strong ideological and religious opposition—it also entailed major technical difficulties. There is a need to insist on the idea that the 'Copernican Revolution' was not a sudden paradigm shift but that it took a whole century to occur. As part of Renaissance humanism, the Iberian authors operated under a cosmological system that was based on the work of the classical authorities of Ancient Greece, mainly the *Physics* of Aristotle (384–322 B.C.) and the work of Claudius Ptolemy (c. 100–c. 170 A.D.).

Perhaps the navigators were not very aware of the fact that the dissemination of Ptolemy's *Geography* in the 15th century was as decisive for them as it was for the cosmographers and cartographers. Nevertheless, due to its leading role in the cosmography of the Renaissance, it is worth recalling the main features of the dominant cosmological system, those of Aristotle and Ptolemy.[43]

According to the classical idea of the cosmos in the Middle Ages, the Earth is a motionless sphere at the center of the universe surrounded by the moon, sun, planets, and other heavenly bodies, which revolve in circles around it at a constant speed. The universe consists of five elements. The terrestrial

42 Portuondo, María M., 'Secret Science. Spanish cartography [...]', op. cit., page 21.

43 For a complete introduction to the work of Ptolemy, see: Lloyd, G. E. R., *Greek Science After Aristotle*, New York, W. W. Norton & Company, 1975; and Lindberg, David, 'Los inicios de la ciencia occidental', op. cit.

sphere—or, rather, the sub-lunar sphere—is made up of four elements: earth, water, air, and fire. These four elements naturally move in a straight line: earth and water toward the center of the universe, and air and fire in the opposite direction. The other heavenly bodies, including the stars, planets, and the moon, are made up of *aether*, that is, a fifth element with a greater degree of perfection, which naturally moves in a circle around the Earth.

In astronomical terms, Ptolemy's model is a complex system that requires eccentric orbits, epicycles, and equants that enable one to predict the irregular movements of the sky. That great heavenly sphere is defined by geometrical coordinates, a north–south axis, an equator, the tropics of Cancer and Capricorn, and parallels and meridians that divide the sphere into 360 degrees. Claudius Ptolemy's most important work is his *Almagest*, the Arabic name by which his studies of astronomy became known and the most complete treatise on astronomy that survives from antiquity. In practical terms, Ptolomy's Almagest has to do with the use of instruments of observation and measurement, such as the gnomon, astrolabe, the mural quadrant, and the sun clock. Ptolemy was also the author of a treatise on astrology, the *Tetrabiblos*. The separation between astronomy and astrology occurred long after Ptolemy nor was it clear during the Renaissance since both fields were concerned with undertaking a mathematical description of celestial phenomena and related to the capacity to predict future events.

This cosmological idea is closely related to geography and cartography since the same heavenly coordinates served for the geometrical systematization of the representation of the Earth, and because this was the major paradigm that dominated astronomy and geography during the Middle Ages.

Notions of geographical positioning that use meridians and parallels were already known before Ptolemy, but it was the work of this author, who compiled the geographical and astronomical knowledge of antiquity, that came to be the basis of a cartographic tradition of crucial importance in the 15th century. The Latin translation of Ptolemy's *Geography* by Jacopo d'Angelo (1406) created more interest in the subjects of astronomy and geography, to the point where its first printed edition (1475) became a model for the cosmography of the Renaissance.[44]

In contrast with Strabo (64 B.C.–24 A.D.), who was interested in giving a detailed description of the places of the inhabited Earth, Ptolemy was devoted to a mathematical description Earth as a whole . In Ptolemy's book, cartographical practice is inseparable from geography, since the latter is thought to be nearly the same as the art of making maps of the Earth when following the principles of classical geometry.

44 Portuondo, María M., 'Secret Science. Spanish cartography […]', op. cit., page 22.

Illustration 3.4. *Arte de Navegar*, by Pedro de Medina. Book I, Folio I, 1545. © Obras clásicas de náutica y navegación / José Ignacio González-Aller Hierro (comp.) Madrid: Fundación Histórica Tavera, D.L. 1998. Copy taken from the Fundación Mapfre. This is one of the manuals that devoted a whole chapter to explaining the order and composition of the world. The illustration places the Earth in the center, in line with the classical notion of the universe.

Illustration 3.5. The sphere and the four elements. In the *Itinerario de Navegación*, by Juan Escalante de Mendoza. Book II  Ms. 2519. Digital Copy: Biblioteca Virtual de Andalucía, p.165. The pilot explains to Tristán why navigators call the Earth a 'sphere', saying that 'it is divided into spheres because the earth and the water are joined and united, and because the surface of both is a single thing'.

The most notable characteristic of the Ptolemaic model is the use of degrees of latitude and longitude to place the geographical features of the Earth on a system of simple geometrical coordinates. The mathematical representation of the globe divided into parallel lines and meridians is a formidable language for representing the whole of the world in an orderly and intelligible way. The design of a plane makes it possible to capture the whole of the Earth in a single image.

The sphere is divided into 360 degrees, then into 120 degrees, and once more into 60 and once more into two sets of 60 degrees again. In the Latin translation of the book, these units are called *partes minutae primae* and *partes minutae secundae*, from which the minutes and seconds of the arc still used today derived. While they were not invented by Ptolemy, thanks to the wide circulation of his work in the 15th century, the notions of latitude and longitude as basic geometric coordinates became the obligatory paradigms of modern geography and cartography.

This mathematical representation of the world is complemented by other conventions that are very familiar and almost natural in the Western world; however, little has been said about their important political implications. From then on, the north was placed at the top of the map and therefore the east lay to the right of the observer. The 'correct' and 'natural' position of the globe presupposes that the inhabitants of its northern hemisphere—Europe, at that time—were placed on the upper part of the globe. In geographical terms, the globe had three parts: Asia, Europe, and Libya. Europe was the most important part of the globe and was usually shown to be separated from Libya by the Pillars of Hercules and from Asia by the river Tanais.[45] The Ptolemaic representation of the world made the Euro-Asian continent look very long, which, as was mentioned, helped Columbus to justify the possibility of a transatlantic voyage since it meant that the coasts of Europe and the Far East were relatively close.

The Ptolemaic definitions of chorography and geography turned out to be crucial in this respect. The purpose of the first was to provide a detailed description of specific regions of the world, while the second aimed at a knowledge and representation of the globe as a single and continuous unity, only taking the aspects of borders and general outlines into account. Geography's pretension to know the whole of the world was an achievement of the Renaissance, when the human study and control of nature became a real possibility.[46] We cannot fail to point out the relation between these holistic views and their significance for the projects of universal empires, like that of Spain in the 16th century.

These classical notions of geography and cosmology found their fullest expression in the *Tractatus de Sphaera* (On the Sphere of the World) by Johannes de Sacrobosco and the *De Triangulis* (On Triangles) by Regiomontanus,[47] books that had a direct influence on the sixteenth-century Spanish treatises and manuals on cosmography. In his *Tractatus de Sphaera*, Sacrobosco not only explained the cartographical function of the principle of projecting

45 Tanais is the Greek name for the river Don (in Russian: *Дон*), a long river located in European Russia that flows into the Sea of Azov.

46 Portuondo, María M., 'Secret Science. Spanish cartography [...]', op. cit., page 23.

47 López Piñero, María José, 'El arte de navegar en la [...]', op. cit., page 31.

Illustration 3.6. Ptolemaic map. Based on Ptolemy's Geography, 1507. No original maps of Claudius Ptolemy are conserved. However, in the 15th century, his cartographic techniques were used in maps of the Earth such as this one. In addition to the geographical conventions employed to project a sphere onto a plane and the use of coordinates for latitude and longitude, the faces which symbolize the winds on Earth are typical of such 15th century Ptolemaic maps.

the coordinates of the heavenly sphere onto the Earth, he also defined the zones of the Earth and those that are habitable.[48] From his standpoint, only those that lie between the Tropic of Cancer and the Arctic Circle, and between the Tropic of Capricorn and the Antarctic Circle, are habitable. The other zones are either too cold due to their closeness to the poles or too hot because they lie in the torrid zone, where the rays of the Sun reach the surface of the Earth directly.

As an example, it is worth looking at Fernández de Enciso's description of this cosmos in his *Suma de Geographia*:

> This round sphere is divided into nine spheres which are called heavens [...] the ninth sphere is the surface of all and continually moves from east to west, moving all the other spheres along with it and it makes a complete revolution every 24 hours. In the eighth are the stars, which are divided by the zodiac and below them are found the other seven spheres where the seven planets are: the moon, mercury, Venus, the Sun, Mars, Jupiter and Saturn.[49]

48 Ibid.

49 Fernández de Enciso, Martín, 'Suma de Geographia que trata [...]', op. cit., Folio III.

Los cuatro elementos: Tierra, Mar, Aire, Fuego.

Illustration 3.7. The four elements: earth, water, air, and fire. In: *Itinerario de Navegación*, by Juan Escalante de Mendoza, page 91.  Naval Museum Archive, Madrid.

In a similar manner, Diego García devoted the first chapter of his *Instrucción Náutica*, titled 'On the Sphere', to an explanation of the physics and cosmology of Aristotle and defined the main geographical coordinates: the equinoctial line, the zodiac, the horizon, the Tropics of Cancer and Capricorn, the Poles, cartographic zones, parallels, and meridians.

All of the authors who wrote about seamanship in the 16th century belonged to this scientific tradition and several of the treatises we mention here include detailed explanations of the classical notions of cosmology

Esfera celestial.

Illustration 3.8. The celestial sphere, with the Earth in the center, around which the Sun, planets, and stars revolve. In: *Itinerario de Navegación*, by Juan Escalante de Mendoza, page 93. © All rights reserved. Naval Museum Archive, Madrid.

and astronomy, which, in turn, were a necessary introduction to geography and cartography.[50]

The illustrations in Escalante de Mendoza's *Itinerario de Navegación* summed up the basic notions of the geocentric cosmos in which the new navigation was framed.

In addition to the obvious influence of Aristotle and Ptolemy, the humanism, art, and science of the Italian Renaissance were marked

50 Ibid., Folio XXIII.

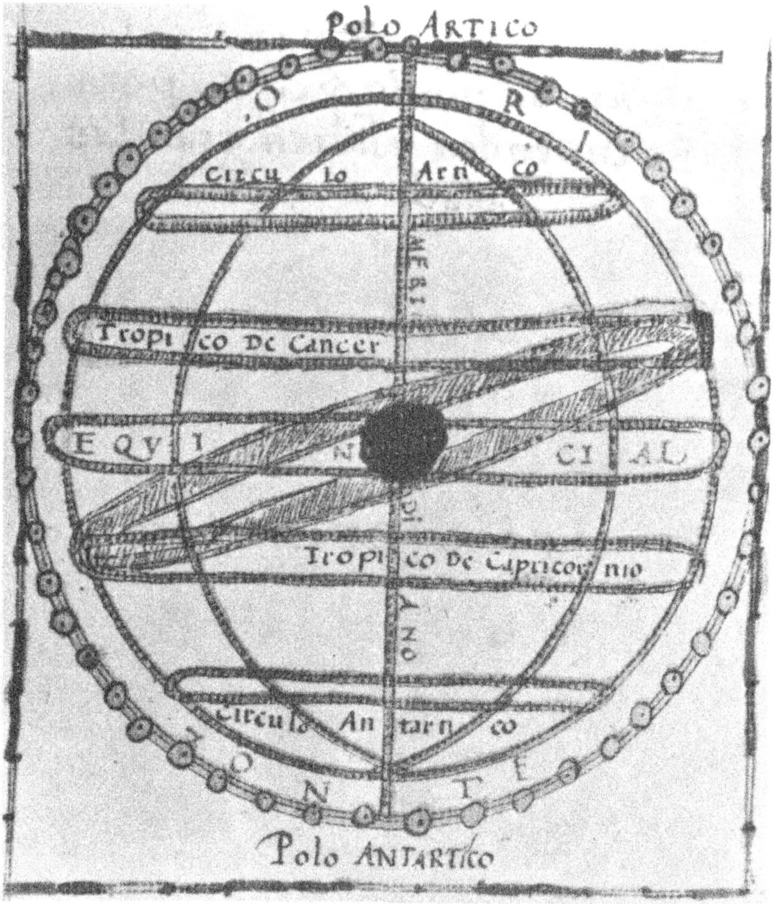


Polos del Mundo.

Illustration 3.9. Poles of the world and the basic coordinates of the heavenly sphere. In: *Itinerario de Navegación*, by Juan Escalante de Mendoza, page 94.  Naval Museum Archive, Madrid.

by the resurgence of the philosophy of Plato. Neoplatonism was characterized by its obsession with depicting nature in accordance with an aesthetics faithful to the mathematical principles of order, harmony, proportion and unity. In accordance with Neoplatonism, geometry, and mathematics are a divine language, one suitable for understanding and representing the work of the Creator. Geometrical principles and the mathematic treatment of space equally dominated the aesthetic and

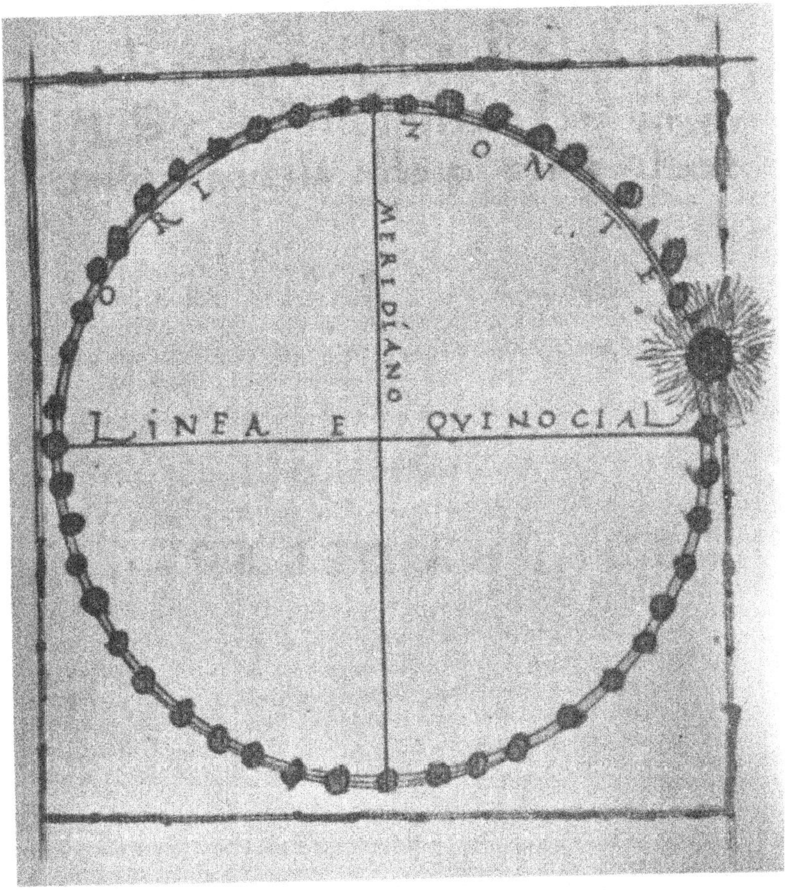


Altura del Sol.

Illustration 3.10. Height of the Sun above the equinoctial line. In: *Itinerario de Navegación*, by Juan Escalante de Mendoza, page 95.  Naval Museum Archive, Madrid.

epistemological principles of cosmography, cartography, astronomy, painting, architecture, and music.

The cosmography of the Renaissance depended on geometry: in that regard, Alfonso de Chaves wrote an interesting essay in which he reminded his readers that geometry is none other than the measurement of the earth:

> this science is one of liberal arts which are most important and necessary in the world: It has its instruments and rules which govern it [...] Through

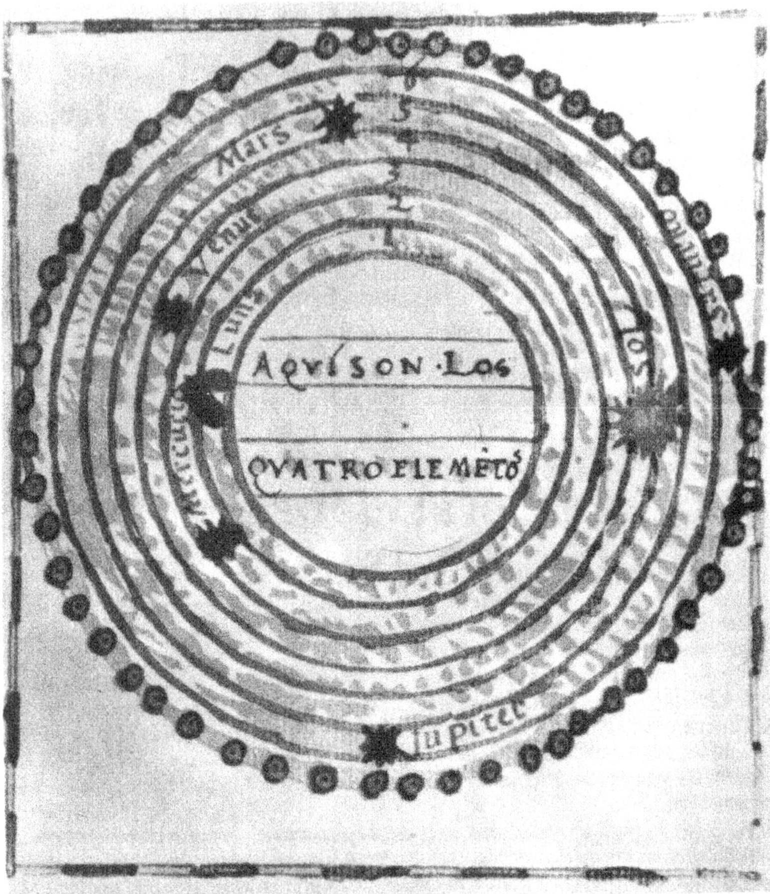


Los siete Planetas

Illustration 3.11. The seven planets. In: *Itinerario de Navegación,* by Juan Escalante de Mendoza, page 97. © All rights reserved. Naval Museum Archive, Madrid.

> this science we can know the distances from place to place [and] the longitude, latitude and altitude and depth of any thing, because everything is covered by straight and oblique lines and measurement and weight. It is with this science that all buildings, devices and instruments in the world are understood and manufactured, and none can be made nor understood without it.[51]

51 Chaves, Alonso de, 'Quatri Partitu en Cosmographia Practica [...]', op. cit., Book I, Second Treatise, chapter VIII, 'which discusses the geometric instrument or altimetric scale and its advantages and use', page 132.

This language was common to all the major protagonists of the Italian Renaissance, like Leon Battista Alberti, Filippo Brunelleschi, and Leonardo Da Vinci, as well as the traditional heroes of what became known as the 'scientific revolution', such as Kepler, Galileo, or Newton.

In the Iberian cosmography and seamanship of the 16$^{th}$ century, mathematics, geometry, and, in part, the Italian humanist education found in Salamanca and Alcalá played a fundamental role. In the second half of the century, personages such as Juan de Herrera, the architect of El Escorial and founder of the Academy of Mathematics in 1583, taught the importance of mathematics as the basis of a new science of cosmography. However, it is worth noting that, in the context of sixteenth-century Spain, and in contrast with the philosophy of Plato as well as those mystical variants of Neoplatonism of a particularly aesthetic character, mathematics was a tool at the service of the needs of the Empire. The academic centers of Spain, mainly Salamanca and, a little later, Alcalá, were the places where classical texts circulated and the knowledge and problems that came from the New World were compiled. In short, Spain was the place and the 16$^{th}$ century the time in which new sources of information and knowledge converged with the needs of a great empire in expansion.

## Experience and authority

Aristotle's natural philosophy, the philosophical nature of Platonic humanism, and the constant references to the classics did not seem to have much practical utility for pilots in the New World; however, they would be a key aspect of the theoretical foundation of the new science of sixteenth-century navigation and cartography. The first purpose of the nautical texts was to fight against the dangerous ignorance of aspects of the sea on the part of the pilots. Nevertheless, these manuals also amounted to a continuous effort to establish a body of theories and standardize the principles of a new form of navigation. This tension between theory and practice is a key to understanding the manuals and their historical importance. Manuals such as Escalante's seem to have based their contents on the author's own experience of the sea while others, such as Zamorano's, trusted more in education and theoretical erudition. However, the tension between one's personal experience and the general theory was evident, as was the eagerness to reconcile them. The two strategies of argumentation, the empirical and the erudite, proclaimed an unquestionable and universal authority.

María Portuondo writes that 'by 1550 the Spanish cosmographers had developed an autonomous scientific tradition and a new genre: the manual of navigation'.[52] From the epistemological point of view, these manuals are enormously interesting documents because they sought to link the theoretical principles of mathematical science with concrete experiences and the needs of a great imperial enterprise on the Atlantic. Classical cosmography was thus a tool for the navigators of the New World. The manuals nourished themselves on the ancient traditions of Renaissance humanism and, in particular, the mathematical rationalism characteristic of the Ptolemaic conception of space, whose principles confronted new information and experiences and some very concrete practical demands.

References to the classics were part of the attempt to lend credibility to the new science, but they were not sufficient. Despite their obligatory citing of and dependence on classical authors and the humanist tradition of Spanish academics and cosmographers, the Spanish navigational treatises of the 16th century showed clear and striking differences from the authorities of antiquity. Up to the 16th century, the world of the Aristotelian–Ptolemaic and Christian tradition was made up of three large parts: Europe, Asia, and Africa. Therefore, the existence of a fourth part of the world led to a radical revolution in geography. Another geographical notion found in writings such as those of Claudius Ptolemy or Pliny the Younger, and thus disseminated by and rooted in the cultured traditions of the Middle Ages and early Renaissance, was the assumption that life on Earth was only possible in the strips between the Tropic of Cancer and the North Pole, and between the Tropic of Capricorn and the South Pole. The only places on the planet that were suitable for human beings to live in were the temperate climates since the torrid zones, with their high temperatures, were not a place apt for life and even less so for civilization. This was one of the great truths of classical cosmography; a truth that only began to be questioned on the basis of the evidence provided by the explorers of the Atlantic. As Fernández de Enciso put it:

> experience shows that what the astrologers say about the great heat of the torrid zone seems to be the opposite, because the Indies, Spanish Main and Western islands which the very Catholic monarchs don Fernando and Doña Isabel, of blessed memory, discovered, are very populated and they are below the torrid zone.[53]

52 Portuondo, María M., 'Secret Science. Spanish cartography [...]', op. cit., page 17.

53 Fernández de Enciso, Martín, 'Suma de Geographia que trata [...]', op. cit., Folio IIII.

The evidences of a continent that was new and inhabited and where nature was exuberant were obvious reasons for questioning the classical authors' ideas about geography and natural history, and even more problematical, the authority of the Holy Scriptures.

As we have mentioned, the Iberian effort to establish a stable link with the New World implied an enormous technical challenge so that, from the standpoint of the history of science, it is particularly interesting to examine the tensions and links between a geographical, astronomical, and mathematical tradition of a theoretical nature, and the experience and empirical knowledge of the men of the sea.[54] The *Casa de Contratación* made strenuous efforts to consolidate a reliable body of knowledge for the conquest of the New World, which involved the difficult job of ensuring that the sailors who were experienced at sea but alien to the world of learning would speak the same language as the cosmographers of the royal academy, men with a university education but little or no experience aboard ships.

The tension between theory and experience is evident in the sixteenth-century Spanish manuals. In the dialogue written by Escalante—one of the authors who did have an ample experience of the sea—the problem is dealt with in a detailed and explicit manner. Tristan, the character who has the role of the apprentice, remarks:

> I have seen, sir, that in everything you have said so far you have not referred to the authority of any person who dealt with it before you [...] which is the custom among wise and learned men, who, when they say anything important in their lectures, usually cite the author who said or wrote it, so it will be more trustworthy and more so, for these subjects which are so important, which require so much certainty and seem to have so little.

The pilot and teacher in this dialogue answers:

> It seems that you are right, sir, but that rule has its place in other sciences, which does not apply to this art of navigation in which a long experience [combined with] a prudent judgment and a cautious understanding is the most certain and truest teacher and usually the greatest authority when it comes to trusting in the rules which were derived from them. And since what I have said about these subjects almost entirely consists of the

54 Debates about this tension are found in the works of Alison Sandman, María Portuondo, and Antonio Barrera.

> evidence and clear experience of he who knows, and who has acquired so much experience in the course of nearly thirty years, it is worthy of the greatest trust it is possible to have.[55]

Later on, Escalante is even more categorical, pointing out that the classical authors deceived themselves because they never had a direct experience of seamanship. Referring to Virgil, Hesiod, Pliny, and Flavius Vegetius, he remarks: '[although] they were so learned in their schools, I do not know if they were correct about everything they wrote about this subject due to the theoretical nature of their sciences and [opinions about] matters they were not familiar with, since they did not attain and understand [what they wrote about] from their own personal and long experience at sea'.[56] In addition to this assessment of the relation between knowledge and experience, Escalante de Mendoza was concerned about the fact that sailing to the New World had lost its healthy customs in the training of a sailor, who, it was supposed, had to start as a ship's boy before ascending to the rank of *grumete* and then work as a seaman before qualifying for the rank of shipmaster or pilot.[57] In a similar way, other authors, such as Pedro de Medina, insisted on the importance of empirical knowledge and explained that their knowledge was the result of their own experience at sea.

This tension between theory and practice, which Alison Sandman has studied in depth in the case of cartography, is apparent in all the fields of knowledge related to the exploration of the New World. It is found in the navigation manuals, the study of the climate, the use and making of maps, the handling of instruments, naval engineering, and natural history.

As we will see in the following section, the tension between experience and theory is evident and very interesting in subjects related to climate, winds, and tides. Climatic matters are closely related to cosmographical theories, but the manuals emphasized the need for a practical study of meteorology applied to transatlantic navigation.

55 Escalante de Mendoza, Juan de, 'Itinerario de Navegación de los [...]', op. cit., Book III, dialogue: 'the same dialogue continues, and they speak of things which are very important for those who beat to windward', chapter, 'On the Currents of the Sea', page 187.

56 Ibid., Book III, dialogue, 'dialogue which deals with the winds and their changes and shifts and the signs which must be seen to sense and foresee that there will be storms at sea', chapter, 'Authors Who Wrote about the Signs and Proofs Which Indicate the Kind of Weather Which Is in Store', page 220.

57 Ibid., Book III, dialogue, 'The fleet having entered the port of Havana and all the ships having been supplied there during five days, the dialogue begins about the way in which the master pilots and sailors and ship owners should keep accounts of the cargo, especially those who sail in the western seas and lands', chapter, 'The skill and incompetence of sailors', page 192.

## Man against the sea: shipwrecks and meteorology

The behavior of the tides, winds, currents and climate in general is part of the essential knowledge of the good seaman and it would seem that such knowledge is only acquired at sea. However, for the Iberian cosmographers, meteorology, such as cartography, was a field closely related to astronomy and the mathematical sciences. Thus, the Spanish manuals dealt with the relations between astronomy, the climate, and the tides. Escalante de Mendoza, for example, regarded the study and understanding of the tides as a problem that is closely linked to the lunar cycles: 'The tide, *señor* Tristan, is the daily launching of the waters of the sea on its coasts and shores, which we understand to be in accordance with the movement the moon makes when it rises in the west, forced by the violence of the prime mover, and that movement makes the waters rise in some parts of the earth and decline in others'.[58] The purpose of making this a theoretical problem is clear: 'all of the things which it is most important for a sailor to know and understand to correctly practice his calling of a good pilot have their logic and reason, sir'.[59] A knowledge of the causes and behavior of the currents of water had to be sought in classical cosmography, according to which the natural movement of the seas from east to west was explained as the work of the 'prime mover' in order to keep the lower heavens in movement and, as a result of the movement of the moon, a heavenly body with an especially strong influence on the waters.[60]

Escalante de Mendoza also showed a close interest in the nature of hurricanes and the places, regions, and times where they usually occurred. He likewise spoke of how to predict their arrival and the precautions and tasks one had to take before a hurricane arrived and what to do in the midst of it.[61] His manual devoted a large section to the treatment of 'the shifts and reversals of the winds and the signs you have to notice to

58 Ibid., Book III, dialogue, 'the following table serves to determine the concurrent observations for each year and using the same concurrent, find out the day when there is a conjunction of the Moon', chapter, 'To determine the conjunction of the Moon with certainty', page 233.

59 Ibid., page 234.

60 Ibid., Book III, dialogue, 'The same dialogue continues and speaks of things which are very important for those who beat to windward', chapter, 'On the ignorant pilots who put on too many sails when beating to windward', page 182.

61 Ibid., Book II, dialogue, 'The interlocutors being in the port of Ocoa, the dialogue begins which deals with hurricanes and the time, altitude and place where they are usually common, and the natural cause of them, and how one should be warned of them and of how they act, and it discusses other things about them which are worth knowing and remembering', chapters, 'In

suspect and prognosticate that there will be storms at sea'.[62] In the third chapter of book three of his manual, 'On Rustic [basic] Astrology', Diego García, for his part, explained how to 'forecast the change of weather, that is: calm, rain, winds, storms, cold, etcetera'.[63] Once again, one can see that the experience of the sailors was a frequent input for the manuals that instructed new pilots:

> that no pilot should fail to know of these signs which the inquisitive, with their studies, works and vigils, have taken from many authors, such as Alberto Magnus, Aristotle, Ptolemy, Pliny, Virgil, and many others, along with some experiences I have had.[64]

Diego García likewise wrote of the different signs from nature that the navigator had to recognize and sought to establish rules that would enable him to understand the climate. For example, he mentioned the signs produced by the Sun, the Moon, the stars, fire, water, and even the human body. Further, as indicated by its title, Chapter Seven of book three of his manual dealt with the 'rules of tides' and Chapter Eight with the '[r]ule by which you can work out high tide and low tide on any day at any time'.[65]

One notes an effort, among the Spanish cosmographers, to support their knowledge with a 'science' of rules based on astronomical methods. As is evident from the treatises we discuss here, nature itself—the currents, tides, winds, and storms—form part of this story. In the end, the first requirement of the art of navigation is none other than the mastery and use of natural forces. The most interesting question is how man's control over these natural forces is proclaimed and the answer lies in the power of the human arts, that is the practical and theoretical knowledge that enables man to understand, predict, and dominate nature.

The ultimate aim of the manuals was to guarantee the success of the imperial enterprises, which meant preventing the losses caused by shipwrecks.

what places and regions do hurricanes usually occur' and 'What one should do when dealing with hurricanes', pages 138–139.

62 Ibid., Book III, 'Dialogue which deals with the winds and their changes and shifts and the signs which must be seen to sense and foresee that there will be storms at sea', pages 217–222.

63 García de Palacio, Diego, 'Instrucción Náutica, para el buen [...]', op. cit., Book III, 'On basic astrology', Chapter I, pages 66–67.

64 Ibid.

65 In a similar way, the fifth book of Medina's *Regimiento* discusses '[h]ow to calculate, from the moon, the arrival of the waxing and waning moons at sea'. Fernández Enciso devotes a chapter to 'meteorology predictions from observations of the sun, moon and other signs'.

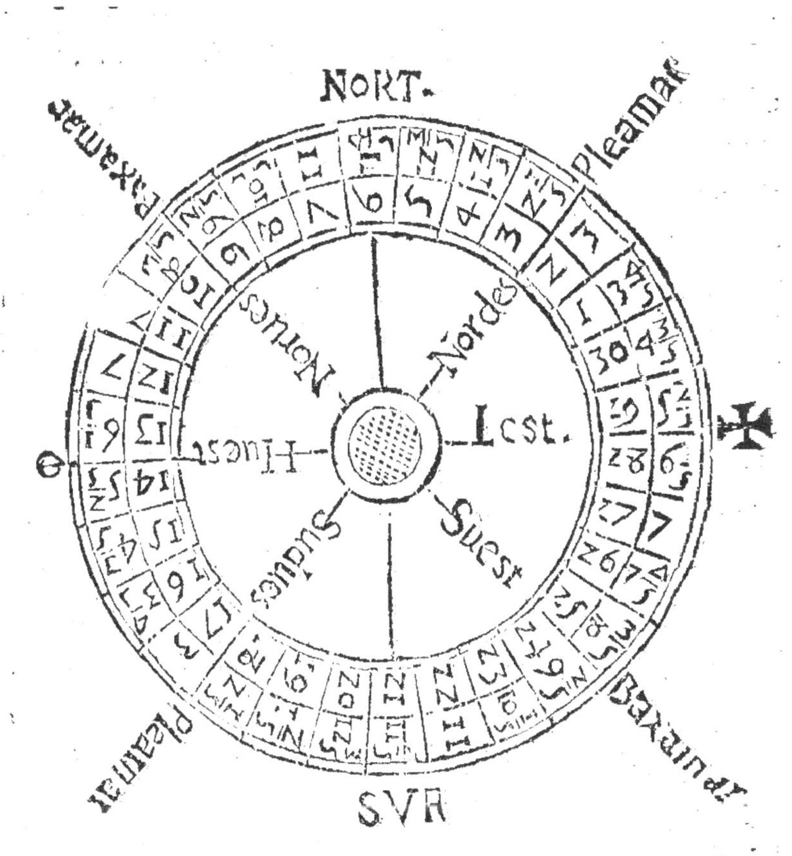


Illustration 3.12. Rules of the tides. In: *Instrucción Náutica*, by Diego García de Palacio. Book II, chapter VII, p. 59, 1587. © Obras clásicas de náutica y navegación / José Ignacio González-Aller Hierro (comp.) Madrid: Fundación Histórica Tavera, D.L. 1998. Copy taken from the Fundación Mapfre. This diagram summarizes the behavior of the tides on the Earth.

All of the technical, commercial and religious endeavors of the Indies Run were blocked by these frequent disasters. Shipwrecks were a fact of life at sea: a possibility that was real and quite probable.[66] They were also an interesting subject insofar as they were the events that most clearly revealed man's struggle against the forces of nature.

As an embodiment of the failure of 'the machines of empire' and the stoppage of the work they did, shipwrecks can help us to attain a better understanding of the crucial aspects of the navigation that led to the

66 On shipwrecks on the Indies Run and particularly in the Colombian part of the Caribbean, see: Romero, Luis René and Pérez, Juan Felipe, *Naufragios y puertos marítimos en el Caribe colombiano del siglo XVI al XVII*, Mexico, Siglo XXI, 2005.

¶ Auiso. xiiij. De quando la nao con fuerça de tiempo va garrando con peligro de perderse: q̄ se deue de hazer para la remediar.

Cosa es q̄ muchas vezes acōtesce q̄ndo las naos tienē demasiado tiēpo y garrādo y cō grā peligro d̄ perderse sino se prouee el remedio cō el saber del piloto lo q̄l puede ser ē las dos maneras siguiētes/ vna es quādo la nao esta en puerto/ y otra q̄ndo va nauegādo. Pues quādo lo tal acōteciere ora estādo la nao en puerto sobre sus amarras/ o yendo nauegando y reparando sobre sus anclas: en qualquiera destos dos casos q̄ se ofrescā tener se han los remedios siguientes.

¶ A lo primero digo que quādo el tal tiempo sobreuiniere y la nao estuuiere surta en baya / o en puerto ancho donde puede reboluerse/ quando [illegible]do con dos Anclas / o con mas / o con menos. Deue [illegible]

Illustration 3.13. Navigational instructions XIIII: 'On when the ship is grasped by the force of the weather and is in danger of being lost: what one should do to correct it'. In: *Regimiento de Navegación*, by Pedro de Medina. Second part, Folio LXX, 1563 © Obras clásicas de náutica y navegación / José Ignacio González-Aller Hierro (comp.) Madrid: Fundación Histórica Tavera, D.L. 1998. Copy taken from the Fundación Mapfre The drawing shows the tension between the forces of Nature, in this case the wind, and technology, or the way in which the art of navigation can turn this natural force into a powerful ally.

European conquest and the role of the ships therein. The testimonies and narratives of these frequent and dramatic events, which occurred in circumstances in which their protagonists confronted death and whereby the enormous efforts that went into building expensive ships loaded with costly merchandises were lost, tell us much about the final meaning of the imperial enterprise. Preventing shipwrecks was the main aim of the nautical treatises and the techno-scientific work done by the *Casa de Contratación*. The training of pilots, the making of maps, the manufacture of instruments, and the inspection of vessels were all meant to avoid tragic and costly losses at sea.

Medina presented a long list of the questions put to the pilot in different circumstances pertaining to navigation: how to sail in unfavorable winds, how to fix the course, how to get back on course, how to plumb the depth at sea, how to read the currents, how to spot the meteorological signs, what to do in a storm, what to do when the *nao* ships water, and when the rudder breaks and the ship can no longer be controlled, among others.[67]

The main causes of shipwrecks seem to have been of a natural kind, like storms and tempests, but the very purpose of those treatises was to make it clear that, even in the worst circumstances, a skilled pilot had to have the knowledge needed to save the vessel, its crew, and its cargo. Even though the will of God and force of nature seemed to be beyond the rule of mankind, shipwrecks were the result of human error in these treatises and, therefore, they were subject to man's control. In his book, Chaves writes: 'It should be noted that the pilot and shipmaster should always look for the natural causes of the dangers or general misfortunes that often occur in sailing due to the winds or tempests or mighty storms'.[68] The manuals provided instructions for surviving the hostility of nature at sea, and, by the same token, spoke of the causes of and remedies for shipwrecks, a subject that was also suitable for exalting the bravery and heroism of the sailors who served the King.[69]

The religious character of the sixteenth-century mariners cannot be placed in doubt and references to divine causes are especially evident in their accounts of shipwrecks and frequent brushes with death. Here too, however, there is a tension that enables us to understand the meaning of the books on seamanship of this period.

On the island of Bermuda, Tristan, the apprentice of the navigator, asks the pilot about the stories that blame bad weather on a 'legion of demons and

67 Medina, Pedro de, 'Regimiento de Navegación. Contiene las [...]', op. cit., Book II, 'navigational warnings', Folios LVII–LXXII.

68 Chaves, Alonso de, 'Quatri Partitu en Cosmographia Practica [...]', op. cit., Book III, Third Treatise, Chapter IV, 'which deals with the dangers and fortunes which often happen at sea and the solution for each one', page 234.

69 Ibid., pages 233–237.

Segunda parte.

¶ Auiso .xj. De quãdo se va nauegando y la nao haze mucha agua: que remedio aura para la tomar.

COstumbre es q̃ se tiene quando la nao va naue-
gãdo a cada guardia requerir la bõba para sa-
ber si va la nao estanca lo q̃l siẽpre se deue assi ha-
zer. ¶ Mas si acõtesciere q̃ se descubre algũ agua
la qual dado a la bomba no se acabare de echar fuera: mã
de el piloto tomar el agua a otra bomba y escotar cõ dos:
y si el agua fuere mucha vaya vn hõbre abaxo y mire el a-
gua si llega ala Carlinga: o si esta encima: y si estuuiere
encima den cõ ambas bombas hasta q̃ la agoten: y agota
da miren por el ampolleta si a tres ampolletas corridas dã
vna a la bõba: y si assi fuere el agua es sufrible q̃ se puede
nauegar cõ ella dẽde Canaria a las indias y bolner cõ ella
sino crece de alli. ¶ Mire se tãbiẽ q̃ si a dos ampolletas se d[illegible]

Illustration 3.14. Navigational instructions XI: 'On the measures [remedy] which should be taken when the vessel is sailing and ships a lot of water'. In: *Regimiento de Navegación*, by Pedro de Medina, 1563 © Obras clásicas de náutica y navegación / José Ignacio González-Aller Hierro (comp.) Madrid: Fundación Histórica Tavera, D.L. 1998. Copy taken from the Fundación Mapfre. The author includes, for example, a detailed explanation of how to 'remedy' a situation where the ship takes in a lot of water, in accordance with the place where it occurs and its extent.

the belief in the protective powers of St. Elmo, the patron saint of sailors'. The pilot responds:

> this illusion of the common sailors who think and believe that it is St. Elmo who goes around and appears in the ships and the sea really shows the plain ignorance of those who say and believe it, because what the sailors call St. Elmo is nothing but an incorporated coupling of the elements of air, water and fire, which, with the mighty and furious storms, freeze and conceive in the midst of the region of the air'.[70]

Speaking of the lights that appear on the crosspieces above the masts and which many attribute to the presence of St. Elmo when he comes to ease the storms, the pilot declares: 'there is no miracle of any kind but all are natural things'.[71] Referring to St. Elmo again, Tristan asks:

> Is there anything to confirm what you say, sir, written in some book by some reliable author or in the account of some person, because what you uphold goes against the opinion of all those who most sail and have sailed up to the present time. It is only fair that you tell me how and in what way you know it or have come to understand it.[72]

The pilot then answers: 'I have not read, *señor* Tristan, but I have seen and experienced, so that I can entirely vouch for it, although I confess that I sailed for more than fifteen years with the same ignorance as everyone'.[73]

As many other sixteenth-century narratives show, a belief in the divine will and the influence of demons was commonplace among sailors[74] but what stands out in the treatises on navigation is the effort to find causal explanations that are based on experience and do not rely on the notion of divine intervention.

70 Escalante de Mendoza, Juan de, 'Itinerario de Navegación de los [...]', op. cit., Book III, 'Having sailed during the night with furious winds and storms, Tristán then speaks to the pilot in the morning and the dialogue between them begins, which deals with the lights which often occur on stormy nights at sea, which sailors call Saint Elmo's Fire, and some important things about how to sail well, as follows', page 213.

71 Ibid., page 215.

72 Ibid., page 216.

73 Ibid.

74 See Chapter 4 of this book ('Life on Board'), specifically page xx ('Men of the Sea and Men of God').

## Routes and chorographic descriptions: The New World within the new global order

The nautical manuals did not restrict themselves to cosmography and astronomy. They also included information about chorography, natural history, and in some cases, ethnography, which, to a greater or lesser extent, are found in most of the Spanish books on seamanship. Before discussing the most important aspects of the manuals, which concern deep-sea navigation, it is worth pausing for a moment to speak of a subject of enormous military importance, which was often the reason why these treatises were kept secret: their chorographical descriptions, itineraries, and routes. The treatises usually included detailed descriptions of islands, coasts, bays, currents, rivers, and climates, and even references to the natives of the New World. Usually, at the end, after the subjects of cosmology and seamanship were dealt with, there were geographical descriptions that explained the whole route a vessel should take to the Caribbean Sea, including its islands and coastlines.

Fernández de Enciso includes a '[d]escription of the Provinces and Parts of the World', which are divided into the following descriptions: Europe, Asia, Biblical history, Africa, the Second India, and, finally, the Western Indies. The description of the world found there is of vital interest since it offers a clear view of the Christian global order centered in Western Europe. The world is divided into two big parts: 'the line of the diameter passes through El Hierro, the westernmost island of the Canaries and that which divides the whole universe into two parts: one toward the East and the other towards the West'.[75] The starting point and reference point for everything that is described is Europe and Spain. According to Enciso, 'Europe is the westernmost and begins at the Strait of Gibraltar where the town of Tarifa is',[76] while being the westernmost part of Europe, Spain, in turn, is the starting point of his geography. The position of Spain in the universe seems to be justified in the most profound way. In the opinion of Enciso, the history of Spain is related to Biblical history: 'I would say that Tubal, the grandson of Noah, was the first to populate Spain' and, after that, his line continues a long genealogical tree to reach the then monarchs of Spain.[77] This geography rests on the privileged order of the Christian world, a notion that regards the whole process of

75 Fernández de Enciso, Martín, 'Suma de Geographia que trata [...]', op. cit., Folio XXIIII.

76 Ibid.

77 Ibid., Folios. XXIIII–XXV.

imperial expansion as the justification for these studies of cosmography and seamanship as natural.

The *Suma de Geographia* of Fernández de Enciso closes with a chapter on the 'Indias Occidentales', which is really the novel feature of the book.[78] That section includes a detailed description of large part of the Caribbean Sea; the islands of Hispaniola, Santo Domingo, and Jamaica, the coasts of the Spanish Main; the navigable ports and coasts of the regions of Santa Marta; and many references to big rivers such as the Marañón.

The purpose of the manuals continues to be clear: to serve as guides for travelers. The detailed descriptions of the itineraries precisely serve that purpose and an exemplary case in this regard is the book *Luz del Navegante,* by Baltasar Vellerino de Villalobos, which traces a route, accompanied by more than a hundred drawings, which act as a guide for navigators on their American voyages.

The second book of Escalante's text gives an equally detailed geographical description of the crossing to the Western Indies, covering the whole voyage from the vessel's departure from the Barra de Sanlúcar to the Canary Islands to the Indies, with precise information about the distances, latitudes, and winds of the whole Atlantic crossing.

There are also descriptions of the coasts on the mainland of America, with their ports, respective latitudes, and the precise distances and times of the voyage. To get a flavor of the information in all its detail, it is worth citing another passage of the dialogue between Tristán and the pilot. Tristán asks:

> How many leagues, señor pilot, are there between the island of Dominica to the port of Nombre de Dios, and what other ports and towns are there in the province and the coast of the Spanish Main, and what neighboring towns and farms are in each and at what latitude and setting do you find them?

And the pilot answers:

> From the island of Dominica to the port of Nombre de Dios on the straightest route, sir, there are 300 and 98 leagues, calculated as follows: there are 100 and 75 leagues between island of Dominica to the Punta de Coquivacoa, as I already said. And 35 from the same point to the Cabo de la Vela [and] to the Cabo de la Aguja. And a league from the Cabo de la

78 Ibid.

> Vela to the port of Santa Marta and nine leagues from the port of Santa Marta to the Punta de la Canoa and two from the Punta de la Canoa to the port of Cartagena.[79]

All of these tips are integrated into a story that continues to specify the coordinates, distances, and sailing times between the main ports and prominent places on the seas and coasts of the Caribbean.[80]

The fourth book of the treatise of Alonso de Chaves is another example of a geographical guide, 'which deals with all that touches the back and forth voyages and navigation in all parts of the Indies, islands, and Mainland of the Ocean Sea'.[81] This final book of that treatise starts with an alphabetical list comprising more than eighty entries, which records the names of all known places . The chapters are arranged as follows: 1) On Sailing from Spain to the Indies, 2) On the Island of the Cannibals, 3) On San Juan, 4) Hispaniola, 5) Jamaica, 6) Cuba, 7) The Lucayans, and 8) Coast of the Spanish Main. In Chapters 8 to 25, Alonso de Chaves describes, in different degrees of detail, the eastern coast of America from the Strait of Magellan to North America.

As we have noted, the confidential nature of this kind of information meant that neither Escalante de Mendoza or Chaves was allowed to publish his treatise, since the Crown was afraid of making such vital economic or military matters public, let alone circulating the information.

It is important to point out that the treatises were representations of the globe, which, when they spoke of a 'New World', did not exclusively refer to the territory of America but rather a new world of which a fourth continent was now a part. These treatises include detailed 'textual maps', that is, written versions of the Master Map (*Padrón Real*, also translated as 'Royal Register') that incorporated the New World into a Christian map in the same way that the new continent was incorporated on the maps of the New World .

79 Escalante de Mendoza, Juan de, 'Itinerario de Navegación de los [...]', op. cit., Book II, dialogue, 'The rainstorm having passed and the order given to raise the topsails, the dialogue between the interlocutors continues, which deals with sailing along the Spanish Main and a description of it and its ports, with the altitude and site of each one', page 131.

80 Ibid.

81 Chaves, Alonso de, 'Quatri Partitu en Cosmographia pratica [...]', op. cit., Book IV, page 251.

# 4. Machines of the empire

**Abstract**
As the central concern of the book, Chapter 4 reveals how a single ship in the transatlantic journey is a microcosm in which we can recognize the complex sum of elements necessary to connect the Iberian Peninsula with its colonies. The knowledge and prowess required to sail a ship to its destination imply an articulated combination of skills and functions. First, we have the shipbuilders and the manufacture of powerful ships and navigation instruments, and second, once at sea, the ships had to be operated by a crew with multiple and specific crafts. This chapter therefore describes the ships and life on-board on a transatlantic journey.

**Key words:** Ships, Instruments, Pilot, Navigation, Life on-board

'Planet Earth' is a strange name since two thirds of its surface is covered by water, here, from a geographical point of view, 'Planet Ocean' would be more appropriate. However, the physical composition of the Earth has implications that are much more interesting for history than its name. Human expansion and the conquest of the world, an encounter with distant cultures, the creation of great commercial systems, and the consolidation of a global empire, only became possible because of one of the most surprising human achievements: dominion over the sea.

The final book of Gonzalo Fernández de Oviedo's *Historia General y Natural de las Indias* is devoted to the misfortunes of sailing, and he introduces the subject with an interesting reflection on the importance of navigation in the history of mankind. Quoting Pliny, who wrote about sails made of linen or canvas in book nineteen of his *Natural History*, he remarks: 'what greater miracle can there be than to have a plant which thus makes Egypt a neighbor of Italy'.[1] For Oviedo, on the other hand, it was not until the

1 Fernández de Oviedo, Gonzalo, *Historia general y natural de las Indias*, [1535], 5 vols., Madrid, Atlas, 1959, vol. 5, Book I, Chapter I, 'On the Father and Son Who Floated in the Sea on a Plank until the Father Died, and How the Son Escaped', page 306.

Nieto Olarte, M., *Exploration, Religion, and Empire in the Sixteenth-century Ibero-Atlantic World: A New Perspective on the History of Modern Science*. Taylor & Francis Group, 2022
DOI 10.5117/9789463725316_CH04

16$^{th}$ century that the real power of using sails was really appreciated. While Pliny thought that linking Italy with Egypt was one of the great technical achievements of humanity, Fernández de Oviedo believed that it could not be compared with the new maritime routs of the 16$^{th}$ century:

> And that same linen and those same sails enabled captain Sebastián Elcano and the *nao* Victoria to depart from Spain, because that ship sailed off from the Guadalquivir River and rode around the globe and roundness of the world, and follow the sun along the whole of the parallel (on which the ship I mention descended the world), heading West and turning to the East, and returned to the same Seville [...] there is no comparison between the voyage of that ship, which sailed so far, and anything which Pliny says about [...] [the subject].[2]

The history of the conquest of the seas goes back much farther than the 16$^{th}$ century and its crucial antecedents took place far from Christian Europe. The development of navigational techniques was a matter of gradual modifications,[3] so the idea that there was a revolution or great paradigm shift that began in the 15$^{th}$ century is not defensible. On the other hand, one could certainly say that the establishment of a maritime empire on a global scale *was* unprecedented in world history and that it took place in the Iberian Peninsula at the end of the 15$^{th}$ century and continued throughout the whole of the 16$^{th}$ century. As Carlo M. Cipolla has rightly stated: 'the secret of this sudden and rapid European predominance lay in the skill the Atlantic nations acquired in the use of sailing ships'.[4]

Traveling on land is a very different experience from voyaging at sea. On the ocean, man is in a medium that is not his own and, therefore, humans need an enormous amount of technology in order to safely travel across water. It is difficult to keep to the right course at sea as it is to keep a precise account of the distance one has traveled. In the end, answering the question 'where am I?' will entail a different problem for the land traveler than for the sailor. The former can always count on some points of reference or visual orientation and can often ask other travelers or the local inhabitants. The latter, by contrast, does not have any of those resources. Perhaps the

2 Ibid.

3 A very thorough and detailed history of navigational techniques, although with an obvious Anglo-centric slant, is the very often quoted book of: Taylor, Eva G. R., 'The Haven-Finding Art: A [...]', op. cit.

4 Cipolla, Carlo M., 'Las máquinas del tiempo y [...]', op. cit., pages 179–180.

journeys that most resemble sea crossings are those across the great deserts. However, the transatlantic voyages bring up many more problems, not only of orientation, but also the survival of passengers confined for weeks or months in small ships with a limited supply of food and drinking water. In that regard, the seas are the biggest deserts in the world.

As we have pointed out, the motives for undertaking voyages on unexplored seas were powerful, but there were not a few fears and difficulties. To sail on unknown seas implies dealing not only with the challenges of navigation and distance but also with human fears and beliefs about the unknown. Classical geography assumed that certain places on Earth, such as the Poles and the tropics, are zones prohibited to mankind. Before Columbus, the literature on travels—for example, the Greek myths and the accounts of Marco Polo and Sir John Mandeville—were full of descriptions of worlds full of enormous riches and creatures and places that are sometimes strange and sometimes terrifying. In the Christian world, there were frequent references to the promised land, the holy land, or the earthly paradise, but those that speak of inhospitable and hostile places, as full of monsters and dangers for men, were also common.[5]

The technological systems that made navigation possible—the ships and their components, compasses, astrolabes or cross-staffs, and hourglasses and sounding lines, among other devices rigorously handled by the crew—were elements that, taken as a whole, began to make an enterprise of long-distance communication safe.

Sailing ships and the new techniques of oceanic navigation played a central role in profound changes in the world political and economic order. They brought about the European conquest of enormous territories all over the planet; the mastery and exploitation of millions of human beings of different races and culture; the enslavement and trade of millions of Africans; the mining and global commerce of gold and silver; the establishment of large plantations of agricultural products for European markets, such as sugar and tobacco; and a long-distance commerce that facilitated an unprecedented accumulation of wealth, capital, and political and military power.[6]

In *The Many-Headed Hydra: Sailors, Slaves, Commoners, and the Hidden History of the Revolutionary Atlantic*,[7] Peter Linebaugh and Marcus Rediker

5 Mandeville, Sir John, *The Travels of Sir John Mandeville*, C. W. R. D. Moseley, (trans.), Harmondsworth, Penguin Books, 1983; and Polo, Marco, *Libro de las maravillas*, Barcelona, Ediciones B / Grupo Zeta, 1997.

6 Rediker, Marcus, *The Slave Ship: A Human History*, New York, Penguin Books, 2008, page 43.

7 Linebaugh, Peter and Rediker, Marcus, *La Hidra de la Revolución. Marineros, esclavos y campesinos en la historia oculta del Atlántico*, Mercedes García Garmilla (trans.), Barcelona, Crítica, 2005.

Illustration 4.1. *Columbus the first discoverer.* Theodor de Bry, 1594. This picture of Columbus, which widely circulated in Europe in the 16th century, shows the struggle of man, with his skills, against the forces of nature, under the protection of God.

call the ships that crossed the Atlantic 'machines of empire', the origin of the title of this book. However, the 'machines of empire' to which we refer are not limited to a large and complex artefact or a sum of apparatuses, but to complex technological systems[8] whose functioning is only understandable by considering the overlapping relations between humans, devices, and nature.

A powerful GPS or cell phone of today would be no more than pieces of plastic and metal without a network of satellites, huge telecommunications companies, regulations, public policies, and a combination of social practices revolving around the use of such devices. By the same token, without astronomical tables, geographical coordinates, and calculation references, an astrolabe would be no more than a lifeless piece of wood or metal.

Man's dominion of the sea is only possible when the ship, rudder, and sails are governed by a trained crew. The pilot and the captain, along with the sailors, make use of the different components of the ship in order to make it move in the right direction. Likewise, the captain, without his devices

8 Hughes, Thomas, 'The Evolution of Large Technological systems', in: Bijker, Wiebe K., Hughes, Thomas and Pinch, Trevor, *The Social Construction of Technological Systems: New Directions in the Sociology and History of Technology*, Cambridge, MIT Press, 1989, page 51.

and the total of jobs and skills undertaken by a large and well-trained crew, would be completely defenseless at sea.

With this in mind, it is indispensable to redefine the notion of an 'actor', which has been traditionally thought of as a social and human agent alone. The complex networks that we set out to describe in this chapter include devices, natural objects, divine agents, and human beings. To understand such scientific practices and their political significance, we need to acknowledge the role both of animate and inanimate agents such as those we have mentioned. This chapter therefore discusses the components of the 'machines of empire' that are directly related to the sea: the ships, the instruments, the crew, and everything to do with life onboard a sailing ship crossing the Atlantic.

The peoples of the Mediterranean and North Atlantic notably advanced the means for navigation in rivers, coasts, and inland seas. While they built cargo vessels of six hundred to one thousand tons called *urcas* and *carracas* (carracks), the most efficient ships were the ancient galleys, which moved by combining sails driven by the force of the wind and oars that used manpower. Those long and light ships were also useful in wars, but their use began to decline in the 15$^{th}$ century until they almost disappeared in the 16$^{th}$ century, when the surviving vessels were reassigned to operations within the Mediterranean.[9] The great oared galleys had certain advantages since they did not depend on the wind or its direction, and they also had a big cargo capacity. However, for ocean voyages on the Atlantic or Pacific they turned out to be useless because the provisions for such a large crew could not be carried nor conserved on long crossings.

When the Iberians sailed out on the Atlantic Ocean in search of a new route toward the Indies, new ships and new navigational techniques were needed. In their enterprises in the Orient and the Americas, both the Portuguese and the Spanish needed ships that would be strong enough to endure the rough conditions in the high seas and agile enough to cover long distances, with a carrying capacity to house the crew and supplies needed for long voyages. At the same time, the ships would have to be shallow-draught to enter ports or cross bars whose depth was unknown. But the exploration of the Atlantic was essentially a conquest achieved by sails and the most powerful agent in this story is the wind or, rather, the human and technical capacity to transform the natural force of the winds into a propelling force for the hulls of ships. The sails, particularly triangular

9 Martínez, José Luis, *Cruzar el Atlántico*, Mexico, Fondo de Cultura Económica, 2004, page 11; and Haring, Clarence H., 'Trade and Navigation between Spain [...]', op. cit., Chapter XI.

ones—the lateen sails—seem to have solved this problem since they could turn the wind into an efficient motor even if the crew was small. The sails caught the force of the wind, which, in turn, was transferred to the mast and from the mast to the hull of the ship that was then into motion and overcame the resistance of the water. A sixteenth-century ship could sail in the same direction as the wind or with side winds, and the action of the latter allowed it to follow courses in the opposite direction to the wind, moving in zigzags and employing other parts of the ship, such as the rudder and keel. A sixteenth-century ship, with the right crew and instruments, could even overcome and take advantage of headwinds and counter currents, and thus keep to almost any route it wished to keep.

The idea of thinking of ships as complex networks of heterogeneous elements is not an invention of twenty-first century sociologists. The sixteenth-century navigational manuals clearly expressed a similar idea since, in fact, their purpose was to link the technical, human, and natural aspects of sailing.

In his *Instrucción Náutica*, Diego García de Palacio devotes a chapter to shipbuilding, and it is interesting to see that the first part of the book is a discussion of cosmology. That part of his book sets out to describe the different elements that make up the sublunary sphere—water, earth, air, and fire, according to medieval cosmography—and to show how a sound knowledge of the composition of nature is necessary so that mankind, with its hard work, can turn the forces of nature to its advantage:

> thus, in navigation, the winds, the measurements, and account of the sky; everything is natural, the sails, rigging, adjusting them and controlling [things], with a thousand other minutiae: all are a matter of human industriousness. And where the need to get so many things to agree occurs, it is clear that learning and ingenuity are needed.[10]

In a clear reference to the natural philosophy of Aristotle, Vizcaíno and Montañés, the two characters of the dialogue that *Instrucción* uses for didactic reasons, are spokesmen for the following ideas about ships at sea: in the first place Vizcaíno refers to them as complex social organizations, 'and so the ship, with everything it should have in it in the way of people and accessories, may be compared to a harmonious and well-ordered republic'.[11] And, later

10 García de Palacio, Diego, 'Instrucción Náutica, para el buen [...]', op. cit., Book IV, Chapter I, 'On the calculations and other things which pertain to the measurements of the Hull of any Ship', page 88.

11 Ibid., page 89.

on, he tries to explain its functioning as the one of a complex organism, making use of ideas characteristic of the natural philosophy of Aristotle:

> since there is a soul and body in it and powers applied to all of the tasks needed for its conservation. And it has actions and movements needed for its aims and its vegetative [functions] are adjusted to its sensitive [functions] and those to its intellectual ones [...] so that we are familiar with and understand this machine or labyrinths of the *naos*, officers and other supporters which they have.[12]

Speaking in the same Aristotelian language; Montañés, the interlocutor of Vizcaíno, complements those thoughts:

> the materials are like the body; the timbers, like the bones; the rigging and ropes, like the nerves; the sails, like so many handkerchiefs; and the tendons there are in the hatch, like the mouth. It also has a stomach and other places where it purges and cleanses itself, like those which a man has. The people are like its spirit, the main officers are like the rulers.[13]

Finally, the government of a single vessel resembles the government of the empire as a whole: 'thus all those in the *nao* [generally, a carrack] and all of their callings follow the orders the ones give to the others and they have to come to an understanding and willingness, as is seen in the councils and hearings of his Majesty'.[14]

The best way to understand the functioning of a sixteenth-century ship is to use, as a guide, the manuals of seamanship and cosmography of the period. Just as we have pointed out, while the 'machines of the empire' were more than an accumulation of isolated vessels, each ship, in itself, was a unit in which many of the factors that made the sixteenth-century imperial ventures possible were combined, added to each other and linked.

## The ships

In his *Itinerario de Navegación de los Mares y Tierras Occidentales* (Itinerary of Navigation in the Western Seas and Lands, Seville, 1575), Juan de Escalante

12 Ibid.
13 Ibid.
14 Ibid.

Auiſos

que con el ſe hiziere. Dos coſas deuen tener los ynſtru-
mentos dela nauegacion / vna que ſean ciertos / y otra
que ſean polidos: y muy bien hechos: y que el piloto ſe
precie de tenellos tales. Pues el ſer ciertos le es grā pro
uecho / y ſer polidos y muy bien hechos da contento.

Auiſo .ij. Como el piloto ha de conoſcer el na-
uio en que ha de nauegar y ſaber las
mañas que tiene.

En la nauegacion de la mar andan muchos na-
uios: y no todos de vna manera mas de difer
tes hechuras y maneras: y aſſi vnos ſon de
porte y otros de otro [illegible]

Illustration 4.2. Navigational instructions II: 'How the pilot should know the ship in which he has to sail and be familiar with the peculiarities it has'. Second part, Folio LVII. Regimiento de Navegación, by Pedro de Medina, 1563 © Obras clásicas de náutica y navegación / José Ignacio González-Aller Hierro (comp.) Madrid: Fundación Histórica Tavera, D.L. 1998. Copy taken from the Fundación Mapfre.

de Mendoza presented a general panorama of the different kinds of ships and their strengths and weaknesses. He said of the Venetians and their big carracks that they are good for war; that the French mostly have small vessels, which enable them to enter and leave their small and shallow ports; that the Flemish, with their great flat *urcas* (cargo ships), can navigate with a heavy cargo in the shallow waters of the Flanders canal; that the English build very small vessels suitable to their ports and their trade with Europe; that the Portuguese and their powerful *naos* can sail to the remote Western Indies, loaded with passengers and merchandise and equipped for war; and, finally, that the Spaniards [*castellanos*], with their growing naval industry and large and small *naos* of every kind can navigate 'all the seas of the world'.[15] In the opinion of Escalante de Mendoza, the most famous ships were those built in the Bilbao canal in the province of Vizcaya, where, despite having neglected their trade to build ships to sell to foreigners, the shipwrights still counted on the best maestros and the finest materials—seasoned timber, ship nails, tars and hemp—for the fabrication of *naos*.[16]

The Atlantic was crossed by a wide variety of vessels. For example, they used *urcas*, which were ships built to transport cargo, as well as large Portuguese carracks and many lesser ships like fly-boats (*filibotes*), pinks (*pingues*), polacas (*polacras*), and *pinazas,* as well as other ships of modest size known by the generic name of *pataches* (a cross between a brig and a schooner). Nevertheless, the ships which dominated the Indies Run were the caravels, *naos*, and galleons.[17]

The first and most versatile transatlantic ships were the caravels. These were light vessels characterized by a slim line and were equipped with triangular sails, which allowed them to sail at a high speed: they opened up the possibility of routes close to the wind and approaching unknown, deep-water coasts more safely. That was not true of the bigger *naos*, with their rectangular sails and deep keels.

The strength of oared ships such as galleys lay in their capacity to sail when there was no wind or even against the wind. Their weakness was the large crew they required since a galley had to enlist at least 150 oarsmen, in addition to the crew, which was ten times the number of sailors in a caravel and even more than the number of sailors in a three-masted galleon.

15 Escalante de Mendoza, Juan de, 'Itinerario de Navegación de los [...]', op. cit., Book I, dialogue, 'The pilot speaks to the captain of the ship the following things', chapter, 'Characteristics of the ships of each of the Christian nations', page 36.

16 Ibid., page 37.

17 Haring, Clarence H., 'Trade and Navigation between Spain [...]', op. cit., Chapter X.

Illustration 4.3. Displays of a ship that sails against the wind and into the wind. In: *Itinerario de Navegación*, by Juan Escalante de Mendoza. Book 1, page 57, 1575 © Obras clásicas de náutica y navegación / José Ignacio González-Aller Hierro (comp.) Madrid: Fundación Histórica Tavera, D.L. 1998. Copy taken from the Fundación Mapfre. In this section, Escalante explains the time of the year when ships should depart in order to take advantage of 'benevolent' winds and avoid 'furious' ones. The latter may cause shipwrecks.

Any sailboat, whether it be an ancient Phoenician one or a modern twenty-first century yacht, faces the same difficulties and follows the same principles. The great technical challenge of sailing vessels is to take the biggest possible advantage of wind power, which depends on an efficient combination of sails and aerodynamics, as well as the hydrodynamics of the hull. It might be useful to recall the basic principles of navigation by sail. A boat with an adequate design and crew can move in any direction, except, of course, against the wind. Large square sails are the best when sailing with the wind behind the sails, blowing in the same direction in which the boat is headed. Triangular sails, for their part, are more efficient at driving ships in side winds. 'Sailing through' (*navegación 'al través*), that is, the movement of the ship in a perpendicular to that of the wind, is possible due to the shape of the sails and the resistance of the hull, which stops the ship from drifting off course and being dragged away by the wind. While it is not possible for any sailboat to advance with its bow to the wind, it can luff or sail close to the wind; that is, it can maintain its course in a direction close to that of the wind, undertaking variable angles with the ship and the sails.

In that way, if the wind is blowing in a different direction to or even against the desired course, the sailing vessel can still navigate and gradually approach its destination with zigzags. The ships which 'sail through' or close to the wind can reach speeds on the surface of the water that are even faster than those sailing with the wind in their sails. Pedro de Medina speaks of caravels that covered three or four leagues in an hour,[18] which means that, advancing at the average speed of a sixteenth-century ship, they could travel more than seventy leagues in 24 hours. That leads one to think that the ships traveled at a speed of a little more than three knots. According to the naval historian Samuel Eliot Morison, in the best climatic conditions a caravel in the days of Columbus that was well built, with the right rigging and a good crew, could reach speeds of up to twelve knots.[19] A speed of three to twelve knots, equivalent on average to five to twenty kilometers per hour, would have enabled such ships to cover distances much longer, in a shorter time, than sixteenth-century century traveler on land.

The modern reader, who is probably familiar with the idea of enormous transatlantic ships, will be surprised by the modest size of the first

18 Medina, Pedro de, 'Arte de navegar en que [...]', op. cit., Folio XXXIII, cited by Morison, Samuel Eliot, 'El Almirante de la Mar [...]', op. cit., page 51.

19 Ibid., page 52.

transatlantic vessels. Thus, for example, it seems amazing that *La Niña* and *La Pinta* were more or less twenty meters long, six meters wide, and three meters deep. It is known that *La Santa María* was a little larger and could house around 40 sailors, while *La Niña* and *La Pinta* each had a crew of 25. However, these small vessels had at least two advantages over the larger ones that were used for trade in the Mediterranean: their low cost and their greater ease in approaching coasts. That characteristic was ideal for undertaking voyages of exploration in which the depth of the sea near the coasts was unknown.

Due to their size, speed, and versatility, caravels such as *La Niña* and *La Pinta* seemed to be ideal ships for exploration. The vessels used to navigate the Rio Tinto were caravels with two masts and combinations of triangular and square sails of a modest size and a capacity of between fifty and sixty tons. The Portuguese caravels used to explore the southern part of the western coast of Africa were solely equipped with triangular sails. For their part, the Spanish caravels might combine a square sail in the forward mast (the foresail) with triangular sails in the mainmast (the main sail) and aft mast (the mizzen sail or yard.)[20] Later on, the name 'caravel' (*carabela*) was used to refer to almost any kind of ship of a low tonnage.

Deep-draught ships, usually official ones, showed obvious advantages due to their cargo capacity and military might. They were symbols of power and security, but they also entailed certain limitations because of their meager versatility and difficulties in approaching the mainland. Ships of small tonnage could raise their sails at the first gust of wind and, therefore, their advantage was speed. While vessels with a capacity of up to a 1000 tons were built, those with around 200 tons were the most commonly used to explore the Atlantic in the early 16th century.[21] The growing demand for ships with more space for cargo and passengers, along with the obvious military advantages afforded by large vessels, meant that small ships gave way to the *naos* and galleons that dominated the Indies Run from the mid-16th century onwards. The name 'galleon' comes from the ancient galleys, which were shorter than the medieval galleys. In addition to their larger load or carrying capacity, the galleons had three masts and a combination of sails, which enabled them to reach their destination without depending on the direction of the wind or requiring manpower at the oars, and independently of the size of the ship.

20 Haring, Clarence H., 'Trade and Navigation between Spain [...]', op. cit., page 328.

21 On ships, see: Braudel, Fernand, 'El Mediterráneo y el mundo [...]', op. cit., pages 399–406 and 801–813.

In the second half of the 16th century, the tendency was to build larger ships, but it soon became necessary to limit their draughts because of the difficulties of reaching the port of Seville. The port of La Barra de Sanlúcar was not deep enough to anchor ships of more than 200 tons, which meant that the ships had to go through the arduous and expensive job of transshipping the cargo.[22] The growing size of the ships meant that the Crown had to issue some royal decrees (*cédulas reales*; 1522, 1552, 1577) to regulate their size, the number of crewmen, and the number of passengers.[23]

A sixteenth-century *nao* had three masts, round sails on the foremast and the main mast, and a triangular sail on the mizzen mast, along with complementary sails. These were robust vessels, arranged in accordance with the following proportions: one=beam, two=keel and three=deck length.[24] The fore castle and stern castle stood out on the main deck. The first was used to store the rigging, sails, and tools, and was also the place where the mariners slept. The upper part of the stern castle, on the other hand, held the rudder and the binnacle, the housing for instruments such as the nautical needle, compass rose, and compass. The remaining space held the officers' sea chests, passengers' trunks, and bedding. Above the roof (or 'awning') of the stern castle, at the rearmost part of the ship, there was the cabin reserved for the captain or privileged passengers. The roof of that cabin, which was completely surrounded by railings, was the highest part of the deck and was also known as the poop deck or bridge. There were few free spaces on the main deck since it had to hold the hatches for loading and unloading cargo, bilge pumps, artillery pieces, the iron stove used for cooking on deck, the sails, cordage, and ladders. Below the deck there was another mezzanine floor, with the hold or storehouse for the cargo, water and wine barrels, food, and provisions. In some of the bigger ships cabins for officers and passengers were built. Under the hold there was the bilge, the lowest point of a ship's inner hull, which was filled with a ballast of gravel. That was the area where all of the oily wastes that oozed out of the equipment of the ship were deposited and thus it was often a focal point for all that was fetid and rotting. Finally, the anchors were tied to the sides of the prow and the ship's lantern lay at the rearmost part of the vessel.

This brief description of the components of a well-equipped ship shows its complexity and give us an idea of how shortage of space was a big problem on ocean crossings. It is difficult to imagine how the crew, marines and

22 Martínez, José Luis, 'Pasajeros de Indias. Viajes [...]', op. cit., page 73.

23 Ibid., page 16.

24 Ibid., page 72.

Illustration 4.4. *Sailing ship with the Fall of Icarus*, by Peter Brueghel.

passengers fitted into a ship when they had had to share the limited space with the provisions, merchandise, and even live animals, which were often part of the cargo or supplies.[25]

## Shipbuilding

With the aim of improving the conditions of the voyage and the speed, safety, and capacity of the vessels, the conquest of the Indies set into motion an active industry of naval engineering, marked by a race for innovation in technology and navigational techniques. Among the main Spanish shipbuilders, the Bazán brothers, Álvaro and Alonso, and Pedro Menéndez de Avilés, stood out. The '*galeza*' or '*galeaza*', for example, was an attempt by the Bazán brothers to build ships that would combine the advantages of oared and sailing vessels, but without much success. With the aim of meeting the need to increase the size of ships, Pedro Menéndez de Avilés organized the building of new ships known as a '*galeoncetes*', which were equipped with keels that were longer in relation to the beam or width of the ship;

25 Martínez, José Luis, 'Cruzar el Atlántico', op. cit., page 22.

the innovations of these vessels were a first step in the development of the frigate, which would dominate the seas in the second half of the 17$^{th}$ century.

As Escalante de Mendoza pointed out, the biggest and best Spanish ships were built in the Basque provinces: the Crown and the Andalusian merchants often looked there for the ships and the riggings for their ventures to the Indies. Since pine trees with timbers of little resistance predominated in the south, the shipbuilders coveted the oaks and other strong woods found in the north for their vessels.

The main concern of the sixteenth-century nautical manuals were navigational techniques, although many included sections on ships and their components and main characteristics and even some aspects of naval engineering. For example, Alonso de Chaves devoted part of his book to naval engineering and the different kinds of ships, even though he illustrated his arguments by only referring to *naos* of two hundred tons. He referred to the parts of the ships, in alphabetical order in a glossary of more than a hundred terms, with their respective definitions.[26]

The first part of Escalante de Mendoza's *Itinerario de Navegación* provided a good description of the technical complexity of building a ship, including such subjects as 'what species of woods and trees [should be used] and at what times should they be felled [...] what nails [should be used] and what tow to caulk them with [...] what tars to apply, of what tree should the masts and lateen yards be, what canvas or cloth should be used for the sails, what is the best cordage for the rigging'.[27] Escalante de Mendoza gave specific indications about the kinds of wood that should be used for building the different parts of the ship, the right time to cut the trees, and the best way to treat them. To obtain a detailed understanding of the different bodies of knowledge one had to have when selecting the best materials for the ship, it is worth citing the pilot's answer to those questions at length:

> The keel, sir, is the main and first timber of any *nao*, the one on which it is assembled and founded. It must be of very straight oak and if it can be of a single piece, it will be best. And for all the other timber, which is cut from any tree, not only for a *nao* but also any other structure which is made of wood, it is best that it be cut when the leaf and fruit from the

26 Chaves, Alonso de, 'Quatri Partitu en Cosmographia Practica [...]', op. cit., Book III, Third Treatise, Chapter I, 'which deals with the ship and its parts and the terms used in navigation', pages 210–223.

27 Escalante de Mendoza, Juan de, 'Itinerario de Navegación de los [...]', op. cit., Book I, dialogue, page 37. 'The pilot speaks to the captain of the ship in which the following travel', chapter, 'Characteristics of the ships of each of the Christian nations'.

same tree which you wish to cut have just fallen, at the start of the days of the second quarter of the waning moon, because the trees have less moisture and are better seasoned and more apt then. The most suitable plank, which should be cut in the same way, is of *cajigo* [*quejigo*],[28] a certain species of oak which should be planted and later transplanted for the best results, like the chestnut tree you want for the best chestnuts. And the ribs which a ship has inside should be of *roble bravo* or holm oak. The rudder must be made of beams from the oak logs which are very dry and very straight. Those timbers should be cut when seasoned and cured in the sun, as I have said, and spend at least a year [like that] before they are used to build and manufacture the *nao* which you set out to make. And for the upper parts of the *naos*, which we mariners call the dead parts, pine from the villa de Utrera, a place in the city of Seville, is very good wood or another of the same species; and those are the timbers which in these, our regions of Spain, we are experienced with and are approved for the manufacture of any ships, for being the most esteemed and hardest and least likely to rot, although in other parts and regions there may be timber of other genera and species which will be no less suitable and effective. And the best tow to caulk the *naos* with is a bunch of cleaned hemp. And the tar of Biscay is very good, mixing it with a lot of whale fat.[29] The nails should be of iron that is not very brittle, although, if they were not more costly, bronze ones will be better, because they are very durable and do not wear out or rust so readily. And the ships which have to sail in the southern or western ports should not be fastened with wooden pins, as they are usually fastened outside of Spain, in Flanders, France and England and other places, because for the ships which sail on voyages in such places, they usually burn them to preserve them from the mist which usually surrounds them, burning the heads of the same pins, and in the course of time they loosen and fall off and the water enters, which means that the ships are at a great risk of foundering, as I have seen in some they have made. The masts and yards, for which God always breeds trees of all kinds, will be very good if made of the pine which is brought from Flanders; as experience shows, there is no timber more suitable for them, especially the pine which the Flemish call *prusa*. The topsails (sails of the mainmast) must be as light as possible, so that they do not give too much weight or slant to the *nao*, which is what should always be looked at and sought, and thus, the mainmast should be made of the lightest

28 *Quercus cerroides*, a species of oak, with a hard wood, which is abundant is in the Pyrenees.

29 Fat or oil of sardines and whales.

> wood. The best topsail is the one made from the hemp, which is said to come from Calatayud and further, the thread should be tarred before it is tied and twisted. The thinner, tougher and closely-woven, the better the sails will be and the best canvas you find for them is that from Pondavid and after that, the Villa do Conde, in Portugal, the precise evidence for which we have seen in our voyages.[30]

The dialogue continues with the details and explanations of anchors, cables, and artillery, as well as the correct measurements and proportions of a *nao*.[31] Obviously, in discussing the subject of the kinds of wood, their origin and treatment, and the cordage, sails, and nails, the nautical manuals represented an accumulation of complex experiences and bodies of knowledge that were difficult to cover in a single book. The building of ships is, in the end, the sum of artisanal practices where tradition and experience come before theory. However, as in other fields of seamanship, naval engineering has a theoretical component. As we shall see below, the ships had to correspond to very precise measures and proportions.

However, during the whole course of the sixteenth-century imperial enterprise, shipbuilding faced a constant and major problem: finding common, stable, and precise units of measurements. Most of the treatises dealt with the problem and included definitions of lineal measures and measures of capacity, such as 'barley grains, fingers, a twelfth (after a small copper coin called an *uncia*), palms, *dichas*, span, feet, *sexquipies*, degrees or grosses, simple steps, geometrical steps, cubits, the rod or cord, stade, mile, league'.[32] Escalante, for example, defined a ton as the 'equivalent of a barrel with the same 27 and a half *arrobas* [an *arroba* roughly equals 25 pounds] capacity of two of the casks of wine or water which they make in Seville'.[33] These debates and definitions indicated a constant effort to establish equivalencies and underlined the urgent need to establish standard weights and measures,

30 Ibid., Book I, dialogue, 'Then the interlocutors go from the ship on which they traveled to the skiff which came for them and having embarked on the admiral's ship, the dialogue between them begins, which deals with ships and their measurements and sizes, and the other things and characteristics which they should have to sail more safely', pages 37–38.

31 Escalante de Mendoza, Juan de, 'Itinerario de Navegación de los [...]', op. cit., Book I, dialogue, 'After the interlocutors have eaten, they begin the dialogue which discusses the measurements of tons and topmasts and lateen yards and other riggings, maritime instruments and people which any ship, in accordance with its size, must carry in it, in order to sail well', page 42.

32 Chaves, Alonso de, 'Quatri Partitu en Cosmographia Practica [...]', op. cit., Book I, Second Treatise, Chapter IX, 'which deals with the terrestrial measurements used in geometry and cosmography', page 142.

33 Escalante de Mendoza, Juan de, 'Itinerario de Navegación de los [...]', op. cit., page 42.

which is fundamental in naval engineering and, as we shall see, no less important in navigation and cartography.

At the start of the 16th century, most Spanish ships were built in Spain. Very soon, however, the convenience and need to build ships in the colonies became evident. Although shipbuilding in the ports of the Indies was initially prohibited, it became indispensable as the traffic between the different Spanish possessions grew. What is more, it might be said that the American shipyards would come to play an important role in the later navigation of the Pacific Ocean.[34]

The ship which Columbus and his sailors assembled in Hispaniola in 1496, the *Santa Cruz*, was the first European vessel built in America that we know of. To start with, such ships were built with parts salvaged from others, but shipyards were soon established, such as the one in Havana. Ten *bajeles* (a generic name for a ship but usually used to refer to a lighter vessel) came to be built there, some of which were used by Hernán Cortés to explore the Mexican coast. From very early in the 16th century, the shipyards of Havana were busy equipping the expeditions of the conquistador Pedro Menéndez de Avilés and, at the end of the century, even foundries were being established for the manufacture of artillery.

The need to create American shipyards was more urgent in the Pacific since the conquest of Peru would not have been possible without them. The Spanish exploration of the Pacific was sustained by vessels built in the New World. The famous voyage to the Pacific of Miguel López de Legazpi in 1564, for example, was done with ships made in New Spain. The New World might have been a major source of timber but, when it came to the rigging, cordage, sails, metal pieces, and even carpentry tools, the shipyards of America had to rely on imports from Europe. Cargoes of iron fittings, canvas, rigging, and work tools reached the Antilles. Nevertheless, when voyages in the Pacific began, the job became more complex, since everything which was needed to build a ship—the sails, cables, rigging, nails, anchors, resin, grease, tow and oil, among many other inputs—had to be carried on the backs of *indios* to haphazard shipyards and ports in the Southern Sea.[35]

One indication of the importance that the shipping industry reached in the New World was the publication of Diego García de Palacio's *Instrucción Náutica* in Mexico in 1587. In the fourth book of that Manual: 'An Account of and What Pertains to the *La Rosa* (proportions) Of Any *Nao*', he gave a detailed description of the parts of a ship and their complex linkage. That

34 Martínez, José Luis, 'Pasajeros de Indias. Viajes [...]', op. cit., page 86.

35 Martinez, José Luis, 'Pasajeros de Indias. Viajes [...]', op. cit., page 87.

section was a thorough treatise on naval engineering whose aim was to spread the knowledge needed to 'build *naos* with sound calculations and measurements [...] and describe all the necessary parts'.[36] The author also implied that this knowledge had not been published before, since he said that 'I will tell you what has not been written down until now'.[37]

As we explained above, and in line with the book of García de Palacio, a ship of that kind required a complex combination of natural and manmade materials and a specialized manpower, and for their proper linkage 'learning and skill' were needed. To organize all those components, the shipbuilder had to have a clear idea of the units of measurement and the names of the parts, as well as the dimensions and functions of the many pieces the ship comprised:

> to understand [the job] better, it is advisable to know the manner of measuring and the names of some of the most common timbers and masts. Thus, make the calculations for the ships, which is generally done with cubits, [remembering] that two feet or two thirds of a yard [*vara*] make a cubit, and that is how you calculate and instruct the builder of any ship, explaining that the keel must be so many cubits, which is the length, and the mainmast, so many cubits, which is the height, and the beam, so many cubits, which is the width.[38]

After listing the names of the main pieces or timbers of a ship—'stern post, keel, stem, *estamenara* [a certain ship timber], top-timber, *forcazes* (stern crutches), *cinglones* (forward crutches), *baos* (beams), *latas*, sleepers, *cintas*, *madres*, *corvatones*, *aletas*, *llaves*, etc.'[39]—and explaining that the proportions between these pieces may vary, depending on the specific needs of the ship, García de Palacio gave examples of the measurements of each piece for ships of different tonnages. By way of an example, he included some blueprints: 'And because the calculation and measurements are better understood [this way], I will make the following drawing of the body of the ship with the measurements of the beam in three positions for more clarity, which are as follows'.[40]

36 García de Palacio, Diego, 'Instrucción Náutica, para el buen [...]', op. cit., Book IV, Chapter I, 'On the calculations and other things which pertain to the measurements of the Hull of any Ship', page 88.

37 Ibid.

38 Ibid., page 89.

39 Ibid., page 90.

40 Ibid., page 93.

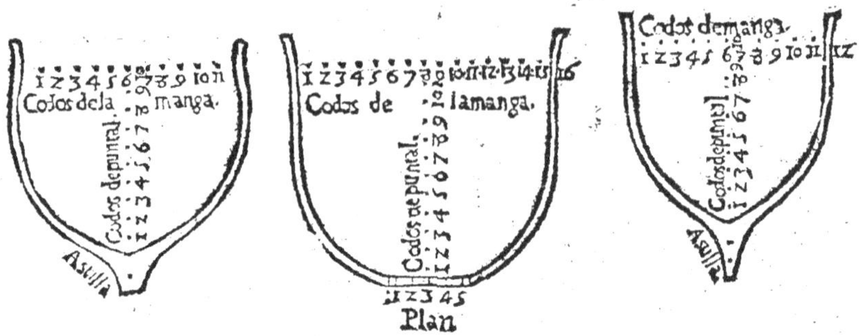


Illustration 4.5. 'On the calculations and other things which pertain to the measurements of the Hull of any Ship'. In: *Instrucción Náutica*, by Diego García de Palacio. Book IV, p. 94, 1587 © Obras clásicas de náutica y navegación / José Ignacio González-Aller Hierro (comp.) Madrid: Fundación Histórica Tavera, D.L. 1998. Copy taken from the Fundación Mapfre.

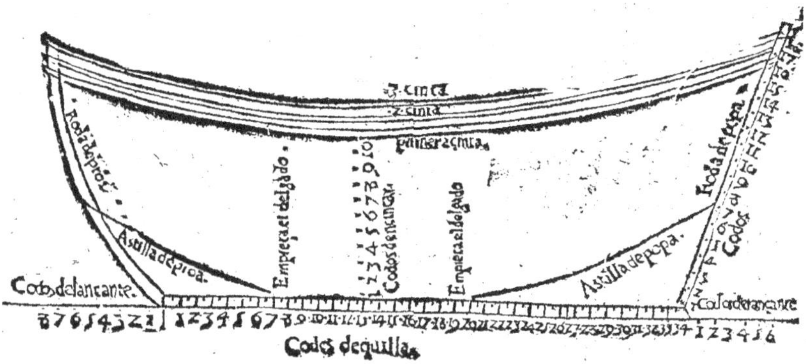


Illustration 4.6. 'On the calculations and other things which pertain to the measurements of the Hull of any Ship'. In: *Instrucción Náutica*, by Diego García de Palacio. Book IV, p. 95, 1587 © Obras clásicas de náutica y navegación / José Ignacio González-Aller Hierro (comp.) Madrid: Fundación Histórica Tavera, D.L. 1998. Copy taken from the Fundación Mapfre.

This book does not intend to give a detailed description of the instructions for the building of ships, but it is worth mentioning the different chapters of García de Palacio's treatise in order to have a better idea of the technical complexity of a sixteenth-century ship. The initial chapters were devoted to explaining the correct arrangement of the key parts of a sailing ship and how each and every one of its pieces must fit into each other in a state of perfect harmony and proportion. The third chapter discussed 'the yards and topsails' and their proportions in relation to ship. The fourth chapter dealt with the 'rigging and tackle of the mainmast'; the fifth, with 'the rigging of the foremast'; the sixth, with the 'rigging of the masthead and fore topsail'; while the seventh chapter illustrated the 'riggings of the mainyard and the foresail'; and the tenth, the 'mainsail'. He also included some helpful diagrams for the manufacture of sails.

There were also sections that dealt with other key parts, such as the anchors and cables. There was even information about details such as 'the pitch, tar, tow and other minor things needed for the ship and its storeroom' and a detailed list of tools and materials used for repairs.[41] Further on, García de Palacio devoted two chapters to the small boats known as the *chalupa* and the *batel*.

The Indies Run and its inevitable armed battles made the naval industry an active and prosperous enterprise. In the 16th century, the Crown did not have a navy of its own, so it had to hire and charter vessels from private shipbuilders and merchants. In principle, this system of hiring ships favored both the Crown and the ships owners but, as time passed, the Crown became so dependent on this system it even had to seek the services of foreign vessels.

Building the ships was cumbersome, and their maintenance was costly. Among the many obstacles and enemies in the way of the conquest of the New World, we cannot omit the teredo worm or naval shipworm,[42] which wrought havoc with the wooden hulls of ships in the Caribbean Sea. Seville was an important place for the maintenance of vessels; in addition, it was also the main center for careening the hulls, that is, repairing or fixing structural damages deleting. That was a very difficult procedure because it involved taking the whole ship out of the water, or, to put it another way, taking the keel out of the water with the help of ballast. To do this, expert and well-paid caulkers and carpenters had to be hired, so that the process was both expensive and time-consuming.

Even though the Crown did not have its own naval industry, it did set out to regulate it. In the 16th century, a complex and detailed maritime legislation was established and laid down rules for everything to do with the capacity, crew, armament, cargo, and provisions of all ships sailing to America. The royal ordinances that governed the characteristics of the ships stipulated, in detail, how each ship had to be equipped: the crew, in accordance with the tonnage; artillery of bronze and iron; the size and number of cannons; the type of munitions and amount of gunpowder that would be needed; and even the number of pieces of armor and small weapons, such as harquebuses, crossbows, and lances.

41 García de Palacio, 'Instrucción Náutica, para el buen [...]', op. cit., pages 307–308.

42 A marine bivalve mollusk in the family *Teredinidae*, called a shipworm because of its general appearance. At the anterior end, it has a small shell with two valves that are specialized at boring through wood. It tunnels into underwater timbers and severely damages the hulls of wooden ships.

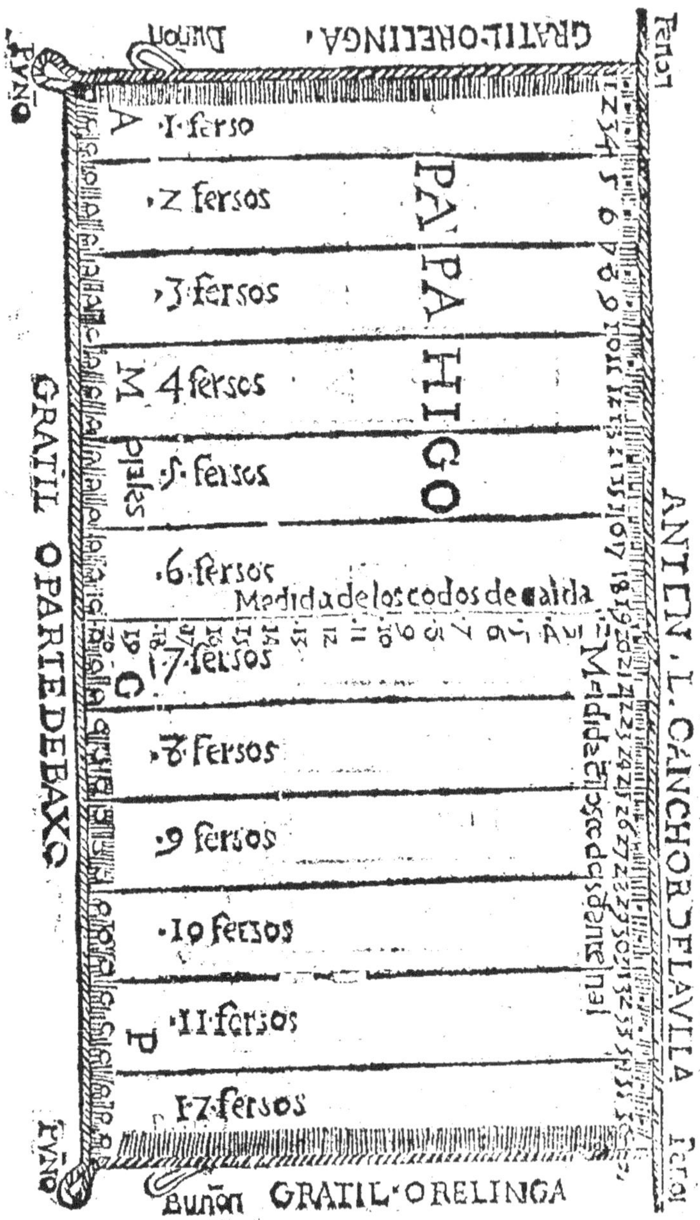


Illustration 4.7. On the mainsail and foresail. In: *Instrucción Náutica*, by Diego García de Palacio. Book IV, chapter X, p. 104, 1587 © Obras clásicas de náutica y navegación / José Ignacio González-Aller Hierro (comp.) Madrid: Fundación Histórica Tavera, D.L. 1998. Copy taken from the Fundación Mapfre.

## War and artillery

Transatlantic sailing presented many hazards: hurricanes, storms, shipwrecks caused by human error, epidemics, onboard diseases, and attacks by pirates or armed enemies. The monopoly of trade with the New World that Spain and Portugal wished to establish and the subsequent exclusion of the rest of Europe and the world from the exploitation of the new continent, added to the testimonies and rumors of the enormous wealth there, were the reasons why the Atlantic became a battlefield for several centuries during which the axes of world power were defined and reshaped.

Carlo M. Cipolla argues that the European dominion of the world resulted from sailing vessels armed with cannons since they were 'the tool which made the European saga possible'.[43] Spain and Portugal's aspiration to monopolize the trade routes to America and the Orient unleashed a long maritime contest with other European nations. Starting in the first decade on the 16th century, pirates arrived on the scene and the middle of the century also saw the start of a clandestine trade encouraged by the colonists of Spain themselves, above all those scattered round the islands of the Caribbean.[44] In addition to the problems of navigation and the constant struggle against distance and nature, the ships of the Indies Run faced dangerous human enemies and thus the ships were also used as weapons of war. The periods of peace were few and short, so that travel to the Indies was always a matter of war. The military equipment of the ships therefore amounted to another complex chapter of engineering and innovation in the conquest of the New World. Cargo ships also had to be equipped with artillery and other weapons but since such equipment was expensive, especially the cannons, the artillery of some ships was really meant for battle, while others only appeared to be so in order to intimidate possible enemies.

This was an indispensable subject in Spanish naval treatises, which included extensive sections or chapters about military matters and fighting off pirates.[45] Some of the manuals had copious information about warfare and weaponry, especially cannons.[46] Alonso de Chaves set down a long list

43 Cipolla, Carlo M., 'Las máquinas del tiempo y [...]', op. cit., page 179.

44 On pirates, see: Lucena Salmoral, Manuel, *Piratas, corsarios, bucaneros y filibusteros*, Madrid, Síntesis, 2005; and Martínez, José Luis, 'Pasajeros de Indias. Viajes [...]', op. cit., page 41.

45 See: Chaves, Alonso de, 'Quatri Partitu en Cosmographia Practica [...]', op. cit., Book III, Third Treatise, Chapter V, 'which deals with that pertaining to the war and battle at sea between one single ship and another', pages 237–248.

46 García de Palacio, Diego, 'Instrucción Náutica, para el buen [...]', op. cit., Book IV, Chapter XXXIII, 'On the war ship', page 120.

of weapons, munitions, and artefacts of war: 'To sail, the *nao* must always be on the alert or well-armed'.[47] Among other things, he mentioned lombards (smoothbore cannons), *pasamuros* (siege cannons), harquebuses, shotguns, gunpowder, crossbows, scythes or sickles, arrowheads, lances, pikes, shields, bucklers, fuses, swords, and daggers.[48]

The teaching of the military aspects of sailing was not limited to artillery but also included the techniques of naval battles. García de Palacio dealt with this subject in two chapters, titled 'On the *Nao* Which Attacks' and 'On the *Nao* Which Defends Itself', which offered detailed instructions about how to keep the artillery and weapons ready for action so that the ship could defend itself at any time.[49]

The bellicose attitude of rival nations and corsairs and pirates had an enormous effect on the Indies Run: the crossing of the Atlantic was not done by single ships, since they would be too vulnerable to attacks by enemies. From the start, the first mechanism of defense on the crossing was the formation of fleets or convoys that had to include armed vessels. Hence, to the complexity of a transatlantic ship, which we have already described, there had to be added the effort to link the ships in convoys readied for battle. Keeping ships with different draughts and sails on long voyages was not an easy job; thus, this involved the improvement of new forms of communication among the ships: flags, cannon shots, and lanterns being the main example. The fleets departed with a leading ship and a captain general, who had to give orders that the others had to follow. In his narrative of his voyage round the world, Antonio Pigafetta spoke of the rules and signals that Magellan laid down so that the ships would not get separated from the convoy or become lost:

> At night, the captain's ship always had to go ahead of the rest, who would follow it with the aid of a small wood-burning torch, called a lamp [*farol*], which always hanged from the stern of his ship. This signal was evident at once. Another fire is obtained with a lantern or with a stub and cord made of rush, called straw [*strengue* or *esparto*], which is submerged in water for many hours and then dried in the sun or in smoke, for the best results. With another signal exactly the

47 Chaves, Alonso de, 'Quatri Partitu en Cosmographia Practica [...]', op. cit., Book III, Third Treatise, Chapter II, 'which deals with the people and provisions which the ship should have and the weapons and munitions it must carry in order to sail well prepared and well-armed', pages 227–229.

48 Ibid.

49 García de Palacio, Diego, 'Instrucción Náutica, para el buen [...]', op. cit., pages 125–129.

> same, they had to respond, so that he [the captain] would know that everyone was following. If, apart from the lamp, two fires were lit, it was, to turn or line up for another course, since it was not convenient to follow the wind or was necessary to slacken the pace. When three fires were lit, then you had to lower the bonnet, which is a part of the sail which is raised below the main sail, when there is fair weather, in order to advance; when it is lowered, it is also easer to take in the mainsail, in a few minutes, in case of a squall. If there were four fires, all the sails had to be lowered, with another flame indicating their stillness afterwards. More fires or the shot of some cannon was the signal of land or shoals. Later on, there were another four fires and that was to raise all the sails again and follow the course always indicated by the torch in the stern. And three fires now meant, raise the bonnet, as there was a breeze. To make sure that all of the caravels followed him and stayed as a group, only the fire which was lit at the start was left, so that all would respond the same.[50]

As we have seen, a single ship was the assemblage of a great many pieces. To sail, however, the combination of a number of artefacts was needed, ones that, without being part of the ship, were nevertheless equally indispensable: the ship's navigational instruments.

## Navigational instruments

Celestial navigation requires a degree of precision in astronomical observations that is impossible to obtain without the aid of devices, which is why nothing like the empires of modern Europe would have existed without those small, and to a certain point simple, instruments. Without astrolabes, mariner's compasses, and hourglasses, Europeans would never have left the confines of the Mediterranean. In his *Espejo de los Navegantes* (the Mariner's Mirror), Alonso de Chaves clearly explained the central role of these instruments in seamanship:

> to me, it seems necessary to deal with and discuss those instruments before any other thing at the start of the book, in view of the fact that everything rests on and is based on them, without the knowledge of which

50 Pigafetta, Antonio, *El primer viaje alrededor del mundo. Relato de la expedición de Magallanes y Elcano*, Leoncio Cabrero Fernández (ed.), Madrid, Dastin, 1985, pages 48–49.

> it would be impossible [...] to be able to understand or use everything which the explanations in this book deal with from here on.[51]

In the last resort, the aims pursued by the 'machines of empire' were the establishment of a continuous communication between Spain and America, the movement of people and goods between the New World and Seville, and, finally, control at a distance. In this enterprise, the instruments had an essential role, one that, to tell the truth, is little discussed by the modern historians of empire. The experiences of travelers—the data, observations, and, if you like, the 'discoveries' of the explorers—were meaningful to the extent that they were portable and could be taken to other places and then returned to the place where the expeditions originated.

Such instruments are fundamental devices in this story insofar as they able to act at a distance as gatherers and bearers of a stable body of information. In order to be gathered, organized, and finally compared, the 'data' must be the result of observations controlled by the firmness and stability that well-calibrated devices guarantee.[52] The reason for that is that the purpose of observations made with common scales and in a disciplined way is precisely to eliminate individual and subjective interpretations, which, in turn, ensures mobility since it makes the accumulation and production of knowledge that come from remote places possible.

Unaided human testimonies are not as trustworthy as the testimonies of humans equipped with calibrated instruments, which eliminate local and circumstantial factors. This depersonalization does not only check the arbitrary nature of the senses, it also creates a homogenous discourse that facilitates the combination of the experiences of different people in different places. This requires both uniform instruments and disciplined observers who follow common rules; that was the reason why the *Casa de Contratación* insisted on laying down instructions about the manufacture of those devices, the manner in which such observations had to be made, the use of manuals, scientific training, and the establishment of common units of measurement.[53] This need for more and better tools led the Crown

51 Chaves, Alonso de, 'Quatri Partitu en Cosmographia Practica [...]', op. cit., Book I, Second Treatise, 'dealing with all to do with the instruments needed for navigation in Spain', pages 103–144.

52 See: Bourguet, Marie-Noëlle, Licoppe, Christian and Sibum, Otto H. (eds.) *Instruments, Travel and Science: Itineraries of Precision from the Seventeenth to the Twentieth Century*, London, Routledge, 2002.

53 Valverde, Nuria, *Actos de precisión. Instrumentos científicos, opinión pública y economía moral en la Ilustración Española*, Madrid, Consejo Superior de Investigaciones Científicas, 2007, pages 42–43.

to organize the fabrication, sale, and control of instruments in Seville, specifically in the *Casa de Contratación*. As part of their preparation, pilots had to acquire good devices, which had to be inspected by the officials of the *Casa de Contratación*. García de Palacio explained that the correct handling of the right instruments was a necessary condition for the training of a good pilot:

> [And for this] one should have good instruments, which are the navigation chart, compasses, an astrolabe which weighs twelve pounds, a wooden quadrant, two clocks of those from Lisbon, two pairs of mariner's compasses, Venetian sandglasses, a copper oil lamp, cotton for wicks and one hundred fathoms of tarred sounding line with six pounds of lead.[54]

As parts of the machines of empire, the different devices and instruments that are described below only perform their functions insofar as they interact with each other, disciplined human beings, and nature. This section will describe them in the following order: the astrolabe, the cross-staff, the compass or compass needle, clocks and different devices to measure time, the plumb line, navigational charts, and astronomical tables or *Regimientos de navegación* (the regime or rules of navigation). After that, we shall briefly refer to other devices and skills that were essential for a mastery of seamanship.

### The astrolabe

The history of the astrolabe goes back to the very beginning of astronomy, long before the boom in transatlantic navigation. As with any complex device, it cannot be attributed to a single inventor: it is an instrument that was used since antiquity and was essential for the formulation of Claudius Ptolemy's *Geography*, on the basis of which geographical positions were defined in degrees of latitude and longitude. The advances in these kinds of devices made by the Arab astronomers were notable, and the first astrolabes used by the navigators of the 16th century were a direct legacy of that tradition. Long journeys through the desert have a certain similarity to voyages along the great ocean since, in the desert, as at sea, there are no visual references or paths and to orient himself the traveler must recur to the stars. They are devices that, in a simple manner, summarize an enormous

54 García de Palacio, Diego, 'Instrucción Náutica, para el buen […]', op. cit., Book IV, Chapter XXII, 'The pilot', page 113.

amount of information that may have different degrees of complexity. The nautical astrolabe is simpler than others and its basic function is to measure the height of the celestial bodies above the horizon, particularly that of the Sun. That method defines the location of the observer in relation to latitude, that is, in relation to the position any body occupies on the terrestrial globe, expressed in the degrees of its distance from the equator, be it to the north or south.

The theoretical explanation of its functions seems simple. The height, in degrees, of a celestial body above the horizon—for example, the Pole Star—enables one to know the degrees of latitude at which the observation is made. However, making these measurements with a certain accuracy at sea is not so simple: in reality, it can only be done in suitable atmospheric conditions and it is probable that several days are needed to obtain reliable data. The constant movement of the ship presents an even greater difficulty: the observations must be made near the mast of the mainsail, the calmest spot on the ship. A good astrolabe must be strong and heavy, but not to the point where a man cannot easily lift it. The astrolabe enables one to calculate latitude by means of observing the position of the Sun, which adds a further difficulty: since the Sun cannot be looked at directly, the use of dark or smoked glasses is necessary.

In his *Breve Compendio de la Sphera y de la Arte de Navegar*, Martín Cortés included instructions on the manufacture of the astrolabe. To appreciate the level of details of these instructions, it is worth looking at a long excerpt of the author's discussion of the making of that device:

> You must take a sheet of copper or brass (which is better for this than any other metal) of the size you wish the astrolabe to be and the usual size it has is a palm in diameter and it be should be as thick as half a finger at least because the heavier it is, the more steady it will be to take the height; that sheet or plate has to be rounded, making a circle in it, [but] leaving room outside of it for a corner in which we shall make a handle, and in this handle we shall open a hole, in which, after tracing the astrolabe, you have to put an armillary sphere with a pointer from which the astrolabe must be hung in order to take the altitude: after rounding and making the handle straight and levelling both sides of the plate so that it is all of the same thickness, one side should not weigh more than the other, which you will examine in the following manner: hang the plate from the armillary sphere or the hole you have made and in the same hole hang a leaden pendant, tied with a horsehair or thin silken thread, and if, while the astrolabe is hanging and free, the thread of the pendant passes

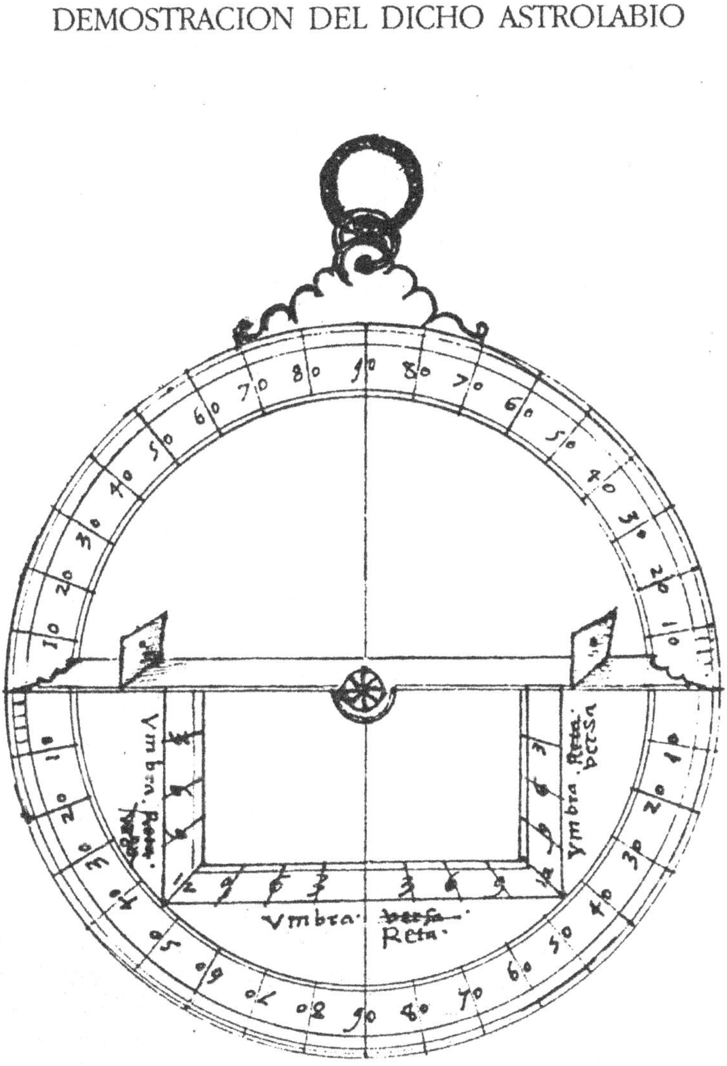


Illustration 4.8. Maritime astrolabe. In *Quatri Partitu en Cosmographia Practica*, by Alonso de Chaves. Book I, Second Treatise, Chapter III, p. 120. © Obras clásicas de náutica y navegación / José Ignacio González-Aller Hierro (comp.) Madrid: Fundación Histórica Tavera, D.L. 1998. Copy taken from the Fundación Mapfre.

through the center of the astrolabe, it will be sound and if the thread moves from the center toward one of the sides, that carving [of the metal] will be thicker or weigh more than the other and has to be thinned until the thread passes right through the center: having done that, you have to make a circle in that center that is a little more within the circumference of the astrolabe: and then you have to trace a diameter from the center of the hole in which the handle is to the center of the astrolabe, crossing the whole circle; and that is called the zenith line: which has to be cut by another diameter on the center, making right angles with it, and this diameter is called the horizon line: these two diameters will divide the circle into four equal parts [...] After that, divide the upper and left ones (when the astrolabe is hanging from the handle before you), first into three equal parts and each part will have thirty degrees and each one of these parts will be divided into another three equal parts and will have ten degrees; and each of these parts into two parts and they will each have five degrees [...] you will then write the numbers of the degrees in the spaces of the lesser circle, beginning at the horizon line: and in that space you will put five and in another ten, etc., until the ninety degrees end in the zenith line. After that, you shall divide each space into five and they will be the ninety degrees. The astrolabe having thus been made, you have to make the alidade, for which you shall take a plate of brass barely as wide as two fingers and as thick as that of the astrolabe and as long as the diameter of the astrolabe and make a line in the middle of it in accordance with the longitude; and in the middle of that line make a circle large enough to touch the sides of this plate. After that, cut from this plate one part of what there is on the right line and another part of what there is on the left line, making the circle perfect. The line which will pass through the center of the circle is called the fiducial line (which is the one which signals the height which is taken in degrees). And, after making two detents or notches in two small metal plates like those of the astrolabe and a little less than the thickness of the alidade and the width of the diameter of the circle of the alidade and some inches high and in the middle of these two tables, according to the height, you will draw a line. After the lines are made equal and all of the angles are right angles in each line, in each that you made make two holes in each tablet, one big enough for a thick pin and these will serve to take the height of the stars; and the other hole so fine, a pricking needle will fit into it and these will serve to take the height of the sun: you have to make it in such a way that the bigger holes are on the outside and the inside [ones] of the size I have indicated. Once these tablets or notches are made, they have to be soldered onto the alidade. At the same time you have to take care that the big hole of one notch is facing the big hole of the other and not skewed: once that is done, you have to drill the astrolabe in the center and the same in the center of the circle of the transom; after that, make

> a pointer (which is a nail of the same brass), which is set precisely in the hole of the transom and astrolabe so that the transom may move around the astrolabe.[55]

The level of detail in these instructions shows the importance of using instruments manufactured under the same parameters, so that the observations made in different places can be compared. Even though the instructions were so thorough, it cannot be assumed that the pilots and sailors were able to produce their own instruments, which usually had to be acquired from specialized manufacturers. More relevant to the specific skills of the pilot were the instructions on using the astrolabe. As follows, we quote the instructions for measuring the height of the sun in Diego García de Palacio's *Instrucción Náutica*:

> Once the point of midday is determined and known (to measure the height you must follow the sun before midday to spot the precise point at which it begins its descent), he who wants to take [the measurement] shall sit down and place himself near the mainmast, which is where the ship has fewer ups and downs and is more still; once the second finger of the right hand is hanging from its ring, you shall make your face and the astrolabe directly face the sun and you will know where it is by the shadow the sun makes. And you shall raise or lower the alidade until the sun equally enters the holes in the sights. That being so, from the astrolabe you shall take the degrees shown by the high point of the pinhole (*penicidio*) and use them for the calculations, in accordance with the following rules.[56]

In the following pages the author gave a detailed explanation of the rules for a correct calculation for the height of the sun: 'when the sun is at its zenith, you will note the declination which the sun has that day, because that amount will be the altitude [height] or separation of the line from the place where you are. And if on the day that you take the sun at its zenith, there is no declination, in that case you would be on the equinoctial line'.[57]

Pedro de Medina, for his part, mentioned five points or things that had to be taken into account to measure the height of the sun: the hour, declination,

55 Cortés y Albacar, Martín, 'Breve Compendio de la Sphera [...]', op. cit. Cited in: Sellés, Manuel, *Instrumentos de navegación. Del Mediterráneo al Pacífico*, Madrid, Lunwerg, 1994, second part, Chapter XVI, 'On different wall and horizontal clocks', Folios XLVIII–LIII.

56 García de Palacio, Diego, 'Instrucción Náutica, para el buen [...]', op. cit., Book I, Chapter III, 'Demonstration of the astrolabe to take the height of the sun and its use', page 25.

57 Ibid.

Illustration 4.9. 'Explanation of the position the sailor must follow when he is sailing on the sea and wishes to know the height [altitude] of the Sun'. In: *Itinerario de Navegación*, by Juan Escalante de Mendoza. Book III, p. 247, 1575 © Obras clásicas de náutica y navegación / José Ignacio González-Aller Hierro (comp.) Madrid: Fundación Histórica Tavera, D.L. 1998. Copy taken from the Fundación Mapfre. The drawing shows how an astrolabe should be used to measure the height of the Sun.

shadow, height, and rule. The first instruction was that the measurement with the astrolabe had to be done at precisely midday. The second was that one should start half an hour before noon and follow the rise of the Sun in order to calculate its declination, that is, to determine that part of the line the Sun moved on and how many degrees and minutes it was distant from that line t, naturally taking into account the year on which the measurement is being taken, which might be the first, second, third, or leap year. Then you had to determine whether the shadows cast by the sun moved toward the north or the south and furthermore , using the astrolabe, find the place on the ship where there was the least movement, where was most sheltered from the wind, and even consider the posture of the observer's body. The final instruction was that the observer should apply each rule to its time and place, that is, strictly follow the tables. Like most sixteenth-century cosmographers, Medina included long sections on the tables and their uses.[58]

The practical difficulties of these measurements justified the very detailed nature of the explanations in the manuals, which usually included illustrations, examples, and exercises.

A simplified version of the astrolabe is the quadrant, which is used for its specific function of measuring the height above the horizon in degrees; indeed the quadrant may be simpler and more practical to use for that purpose. Some manuals, such as that of Pedro de Medina, gave it importance and included instructions on its manufacture and use. However, during the whole of the 16th century, the instruments most commonly used for measuring height were the astrolabe and the cross-staff.[59]

## The cross-staff

The cross-staff is used for a similar purpose as the astrolabe, although its functions are more specific and it is a more suitable instrument for measuring the height of the stars at night, particularly the height of the Pole Star. The cross-staff is a wooden rod with a sliding, smaller crosspiece. The main rod must be graduated so that the sailor can measure the angle of an object's elevation With one eye on the end of the main rod, the observer would point the instrument toward the object he wished to measure and then slide the crosspiece until the lower end of the same rod coincided with the horizon and the upper end with the star. Using the graduated markings on the main rod, the sailor took note of the height in degrees.[60] The observer then applied applied the same principle found in an instrument used by the

58 Medina, Pedro de, 'Regimiento de Navegación. Contiene las [...]', op. cit.; and 'Arte de navegar en que [...]', op. cit.

59 Medina, Pedro de, 'Regimiento de Navegación. Contiene las [...]', op. cit., Book IV, Folios XIV–XLVII.

60 Escalante de Mendoza, Juan de, 'Itinerario de Navegación de los [...]', op. cit., Book I, 'Instructions on the use of the cross-staff', page 107.

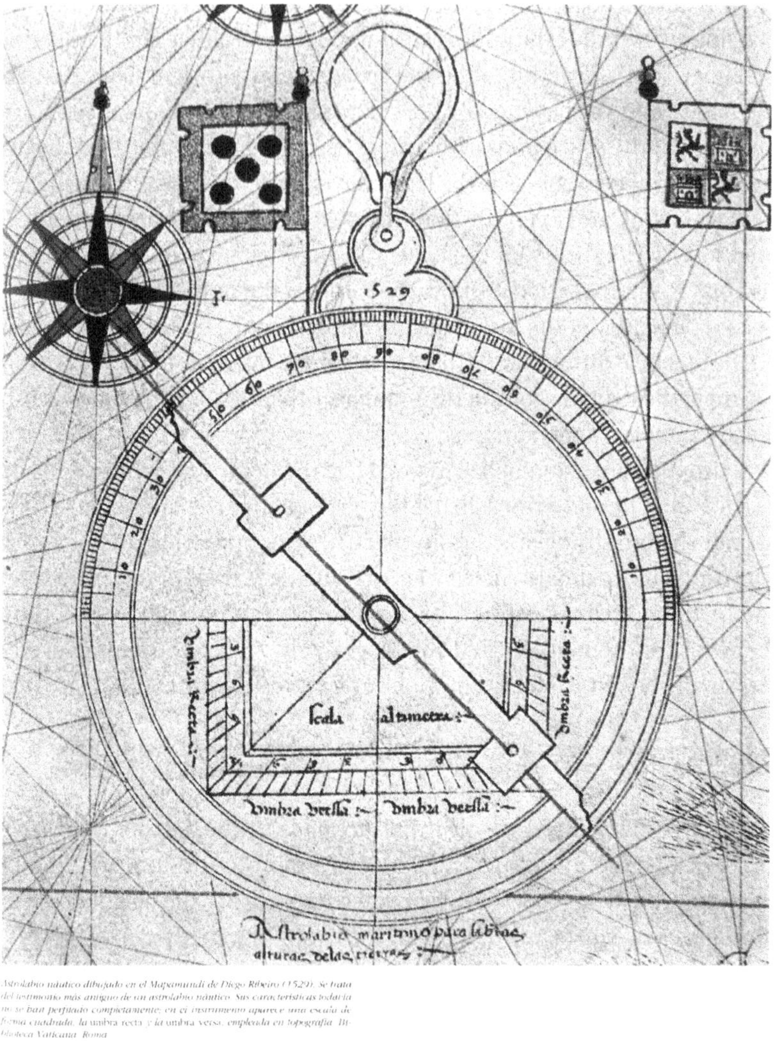


Illustration 4.10. Nautical astrolabe. Detail of the mappa mundi of Diego Ribero, 1529. © All rights reserved. Naval Museum, Madrid.

Arabs, called the '*kamal*', which was also employed to measure the height of a star above the horizon.[61]

Once again, the care with which both the manufacture and use of these devices were described shows the importance of having calibrated instruments and disciplined sailors: a necessary alliance between man and artefact

61 Taylor, Eva G. R., 'The Haven-Finding Art: A [...]', op. cit., page 129.

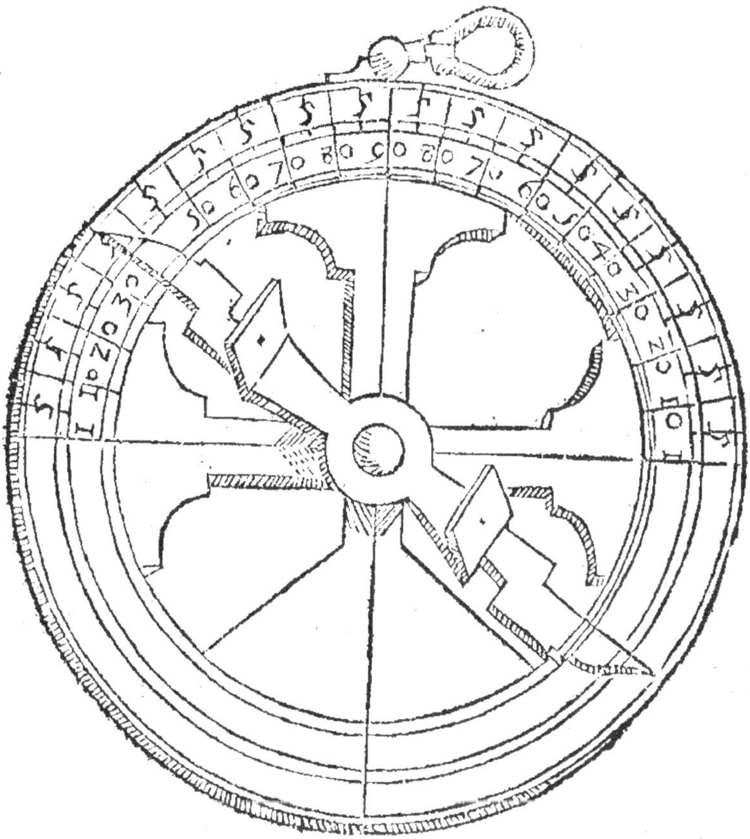

Illustration 4.11. Explanation of the astrolabe for taking the height of the Sun and its use. In: *Instrucción Náutica*, by Diego García de Palacio. Book I, chapter III, p. 26, 1587 © Obras clásicas de náutica y navegación / José Ignacio González-Aller Hierro (comp.) Madrid: Fundación Histórica Tavera, D.L. 1998. Copy taken from the Fundación Mapfre.

to attain stable observations. As in the case of the astrolabe, the cross-staff, along with the astronomical tables and the body of a trained mariner, made up a complex system of observation and geographical localization.

Tristán, the young apprentice in the dialogue of Escalante de Mendoza, spoke of the difficulty of calibrating the instrument: 'I have seen that all the other cross-staffs which sailors use are made with the same design and measurement, and it seems that, so long as they are not wrong in the main [aspect of] its adjustment, they have no need to be experienced, and in case it is required'.[62] But the pilot responds:

62 Escalante de Mendoza, Juan de, 'Itinerario de Navegación de los [...]', op. cit., Book II, dialogue, 'The time having passed when Tristán was busy looking at and seeing the tables for

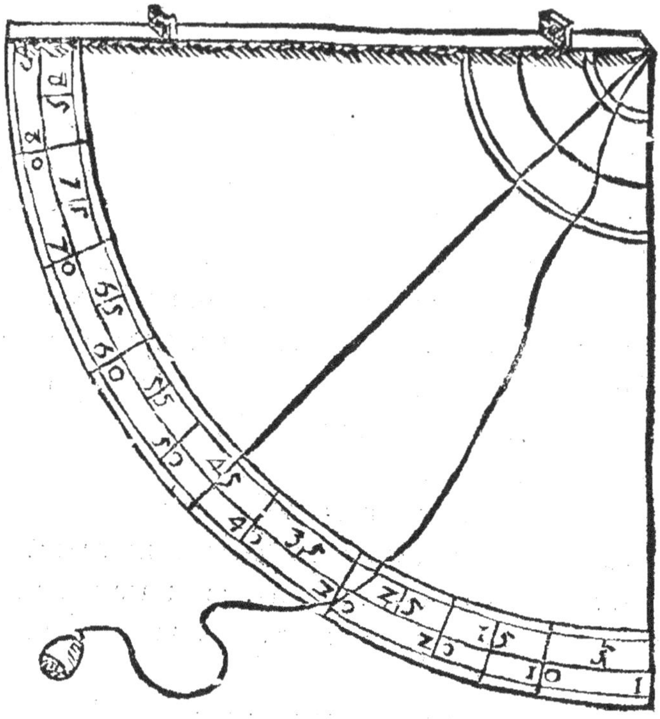

Illustration 4.12. Explanation of the quadrant for taking the height of the Sun and the stars, and its use. In: *Instrucción Náutica*, by Diego García de Palacio. Book I, Chapter III, p. 26, 1587 © Obras clásicas de náutica y navegación / José Ignacio González-Aller Hierro (comp.) Madrid: Fundación Histórica Tavera, D.L. 1998. Copy taken from the Fundación Mapfre.

> But not all of the cross-staffs which navigators use have the same measurement, sir [...] and even if they were all made with the same design and measurement and had the same graduation [...] inexperienced pilots may make false measurements [...] so that the good pilot must understand his cross-staff as well as a good harquebusier does his harquebus, for to hit the target he knows how to raise or lower his aim'.[63]

Diego García's *Instrucción Náutica* explained how to make a cross-staff, set down the rules for observing the Pole Star and gave detailed instructions

the calculations of the declination of the sun which the pilot gave him and noting them down in his navigation treatise, they continue with the dialogue, which deals with the height [of the heavenly bodies] and the things needed for and pertaining to the sound use and understanding of it', chapter, 'Taking the height of the north pole with the cross-staff', page 110.

63 Ibid., page 111.

Illustration 4.13. Drawing of the use of the astronomical staff or cross-staff. Sellers, John, 1672. Based on the book of Pedro de Medina.

on the use of the instrument at sea.[64] Among the main rules that had to be taken into account were, first, that the Pole Star, which ordinary people and sailors call the North Star, is not the true pole, since, if it were, it would always be fixed.[65] The second rule was that the Guardians of the Pole, or those stars which always accompany the Pole Star, indicate the side to which the star deviates, which enables one to make the needed corrections. García de Palacio offered the following instructions on the use of the cross-staff:

> When your worship has to take the height, wait until the front guard is on course with the North Star, in any of what are known as the eight [compass] winds. And if the weather allows it to be in one of two—Northeast, Southwest—because this position is less distant from the pole and sailors believe it is easier, then you will take your rod or cross-staff, and put the top of the staff just below your eye. And raise and lower it until the lower part of the hammer [main staff] moves and aligns with the Horizon; and

64 García de Palacio, Diego, 'Instrucción Náutica, para el buen [...]', op. cit., Book I, Chapter V, 'Manufacture of the cross-staff', pages 35–37.

65 The pole star is at a distance of 3½ degrees from the North Pole.

Altura del Norte con la Ballestilla.

Illustration 4.14. *On the height of the north.* In: *Regimiento de Navegación*, de Pedro de Medina. Book III, Folio XXXV, 1563 © Obras clásicas de náutica y navegación / José Ignacio González-Aller Hierro (comp.) Madrid: Fundación Histórica Tavera, D.L. 1998. Copy taken from the Fundación Mapfre.

doing it in this way, the other or top part of the hammer will join with the Star and if it does not reach it and is adjusted to it, you will lengthen that hammer until the lower part is on the Horizon and the upper part on the Star; and the degrees which the flat part of the hammer indicate on the staff will be those of the Star above the Horizon, and then you will see if you have to add or remove, in accordance with the rules we give, and the result will be the height or distance from the western line.[66]

66 García de Palacio, Diego, 'Instrucción Náutica, para el buen [...]', op. cit., Book I, Chapter VI, 'Instructions on the North Star', page 40.

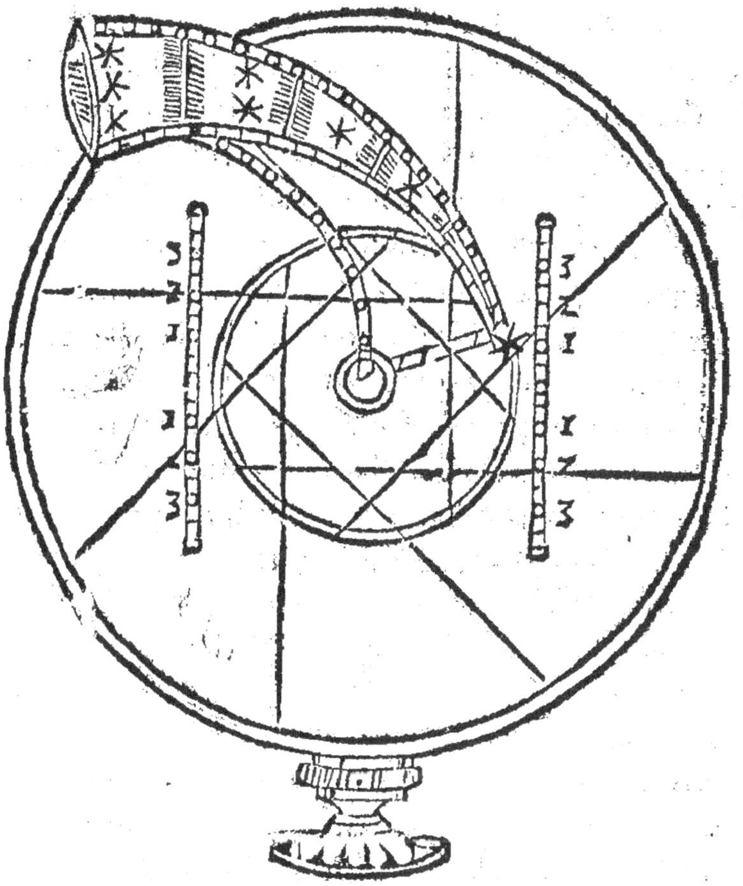

Illustration 4.15. Use of the cross-staff. In: *Instrucción Náutica*, by Diego García de Palacio. Book I, p. 39, 1587 © Obras clásicas de náutica y navegación / José Ignacio González-Aller Hierro (comp.) Madrid: Fundación Histórica Tavera, D.L. 1998. Copy taken from the Fundación Mapfre.

Measuring the height above the horizon, both of the North Pole Star and South Pole Star, is a simple method for determining the latitude. For Europeans familiar with the constellations visible in the northern hemisphere, the most common and best-known measurements were naturally done with the Pole Star and its Guards. However, once you cross the equator, measuring the latitude requires other astronomical references. Another important consequence of exploration beyond the Mediterranean and a notable occurrence in the history of science, or to be more precise, astronomy, was the appearance of a new sky. The most important referent in navigation south of the equatorial line would be the Southern Cross: 'What is known as the cross are four stars which are in the Antarctic pole, and the one closest to the pole, stands at thirty degrees from the axis'.[67]

### The mariner's compass

The function of the mariner's compass is to mark the direction of the ship—of the keel—in terms of the north–south line by means of a needle that points to magnetic north. In the absence of visual referents in the middle of the ocean, it would be impossible to fix a route and keep to the desired course without the support of compasses. The pilots on the high seas cannot maintain their course by locating visual referents on the horizon, so that it is indispensable to look toward the sky and make use of this small device on the ship, on the bridge, or in their own hands.

The authors of the manuals assigned a crucial importance to the compass for navigation. Escalante, for example, spoke of three basic resources for the sailor who wishes to keep to his course:

> The first guide is the compass, which always shows a single point which is level with the horizon and hence shows the course [...] 2. The height which he takes of the sun or the north [...] 3. The imagination or fantasy [creativity] of the good seaman, who traces and measures the day's run [the distance covered by a ship] and route, noting it down in his chart or pilot's manual.[68]

Hence, he continued, the compass 'is the most important and fundamental instrument in navigation to obtain certainty about one's course'.[69] For Chaves as well, the compass had an outstanding place among navigational instruments: 'this instrument is what guides us and shows us the place and part we wish to go to and from which we have come'.[70]

67 Ibid., Book I, Chapter VIII, 'Order and rule of the Southern Cross, in order to take its height, both at sea and on land', page 43.

68 Escalante de Mendoza, Juan de, 'Itinerario de Navegación de los [...]', op. cit., Book III, dialogue, 'dialogue which deals with the compass needle and its northeastern and northwestern variations; and other things pertaining to it which navigators must know to sail well and with certainty', chapter, 'Certainty of the sailing course which is chosen', page 204.

69 Ibid.

70 Chaves, Alonso de, 'Quatri Partitu en Cosmographia Practica [...]', op. cit., Book I, Second Treatise, Chapter I, 'which deals with the instrument known as the compass needle and its usefulness and advantages', page 104.

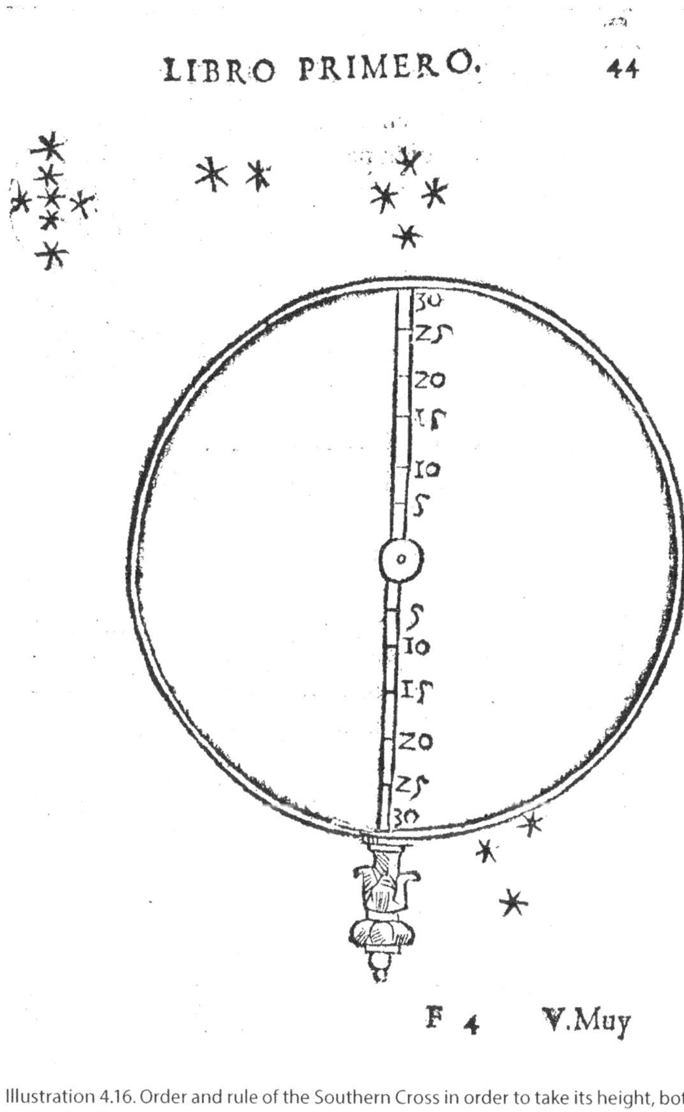


Illustration 4.16. Order and rule of the Southern Cross in order to take its height, both on sea and on land. In: *Instrucción Náutica*, de Diego García de Palacio. Book I, Chapter VIII, p. 44, 1587 © Obras clásicas de náutica y navegación / José Ignacio González-Aller Hierro (comp.) Madrid: Fundación Histórica Tavera, D.L. 1998. Copy taken from the Fundación Mapfre.

The compass is the wound up of many nautical and astronomical traditions but, for the purposes of this book, we need to explain its uses in sixteenth-century seamanship. There are many references to magnetic needles used in navigation in the second half of the 13$^{th}$ century, when it turned into a crucial instrument for sailing in the Mediterranean world. Its use is closely related to the handling of navigation or sea charts, which delineate the directions

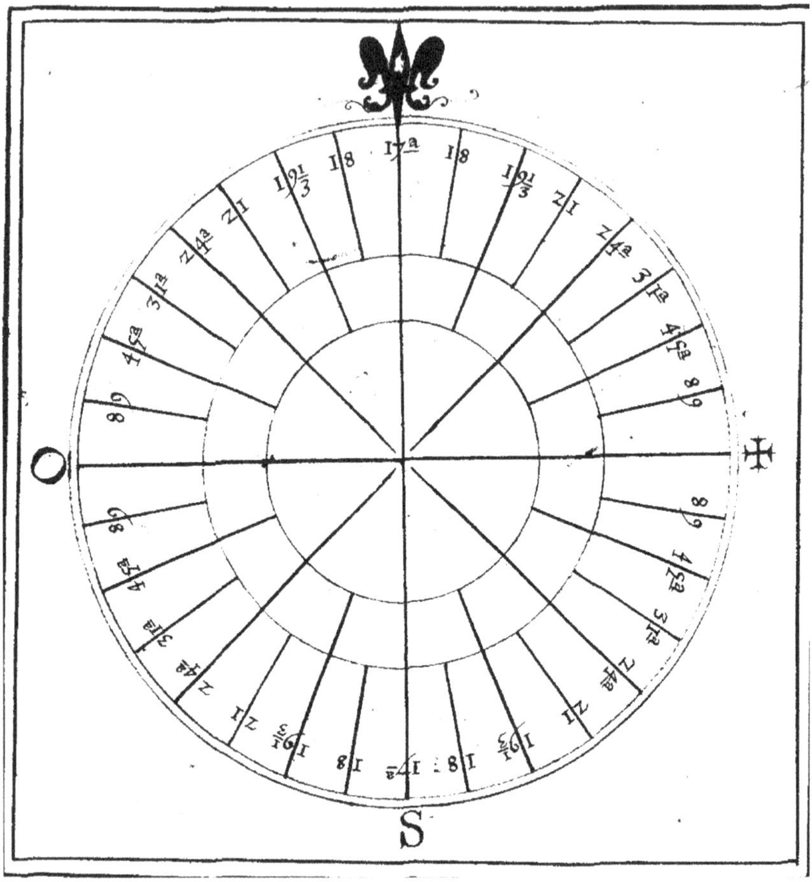


Illustration 4.17. Explanation of the mariner's compass. In: *Itinerario de Navegación*, de Juan Escalante de Mendoza, p. 112, 1575 © Obras clásicas de náutica y navegación / José Ignacio González-Aller Hierro (comp.) Madrid: Fundación Histórica Tavera, D.L. 1998. Copy taken from the Fundación Mapfre. While Escalante included 'a demonstration of the compass needle, which indicates the number of leagues which have to be sailed over for each of its thirty-two courses, in order to ascend or descend one degree of altitude', he did not fail to mention the importance of the sailor's 'sound experience' in avoiding miscalculations.

and courses that converge on the compass rose, a figure in the shape of a star with a symmetrical number of points—from eight in the simplest figures to 64 in the most complex—each of which marks a direction or a wind. The most visible point is usually decorated with a *flor* [*fleur*] *de lis* on the upper part of the maps. The star shape of this graphic representation, which has played a leading role in cartography and navigation, evidences the close relation between the fields of knowledge and astronomy.

The charts that dominated navigation in the Mediterranean world from the 12th century onwards belonged to the tradition of Portolan charts, navigational

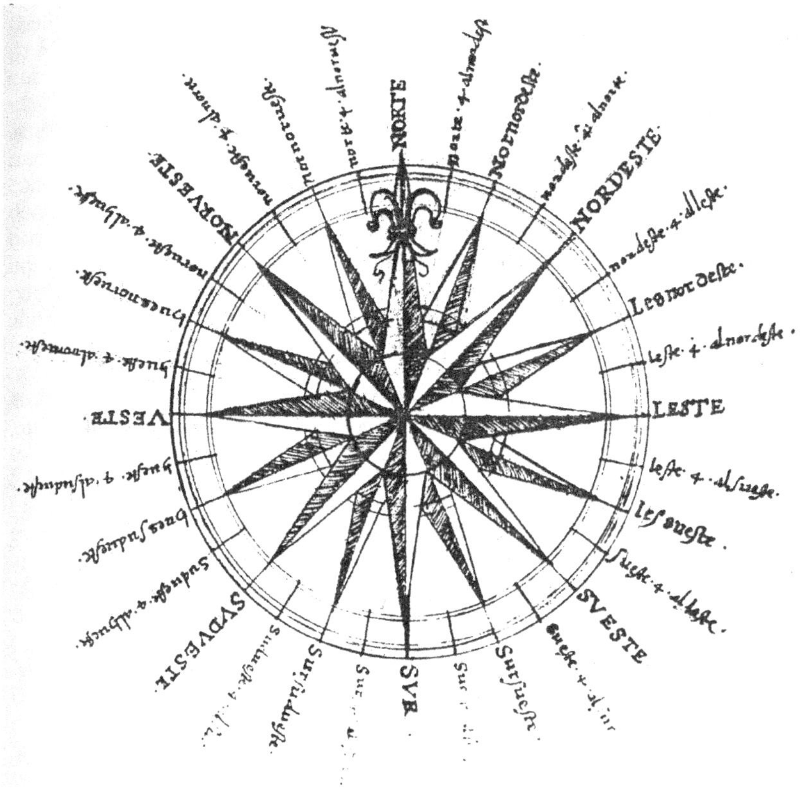


Illustration 4.18. Compass needle. In: *Quatri Partitu en Cosmographia Practica,* by Alonso de Chaves. Book I, Second Treatise, Chapter I, p. 107. © Obras clásicas de náutica y navegación / José Ignacio González-Aller Hierro (comp.) Madrid: Fundación Histórica Tavera, D.L. 1998. Copy taken from the Fundación Mapfre.

maps with descriptions of harbors and coasts, and one of its oldest prototypes was the *Carta Pisana* (Pisan Chart, c. 1275). The most notable characteristic of these maps is that they were crisscrossed by a mesh of lines that converged at central points so that those lines sketched out the courses. The outlines of the coasts of the Mediterranean Sea were precisely traced and the names of the most notable places were written, crosswise, on the coastline.

The nautical needle used by seamen in the Mediterranean showed the convergence of two traditions: on the one hand, knowledge of the winds and their names and directions, which were usually represented on the Portolan maps by the compass rose; and, on the other, the use of the magnetic needle, with its capacity to point to magnetic north or south. Thus, by superposing a magnetic needle upon a compass rose you join the two that gave rise to the modern compass. To begin with, the cardinal points continued to be

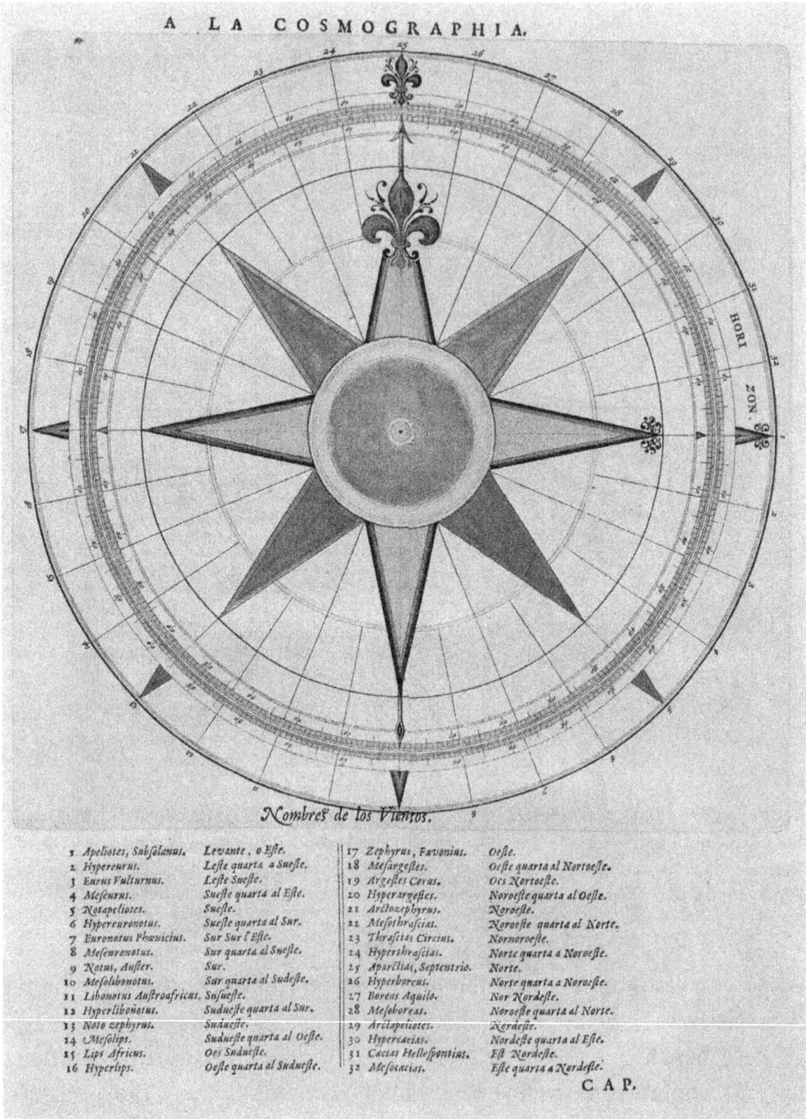


Illustration 4.19. Compass rose. In: *Instrumentos de navegación*, by Manuel Sellés, p. 17. The compass rose wound up being divided into thirty-two courses in the *Atlas Maior* of Blaeu, 1662.  Naval Museum, Madrid.

identified with the names of winds; the 'winds' signified the same as the direction or course and often had names that corresponded to the places from where they blew. However, as a result of the use of the fundamental mathematical notions found in Ptolemy's *Geography*, the terrestrial sphere was soon divided into 360 degrees and the courses stopped being the names

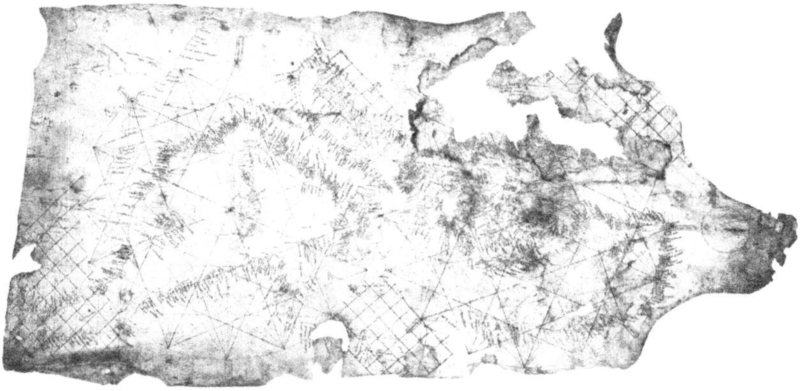

Illustration 4.20. Carta pisana (Pisan map), c. 1275. This is possibly the oldest Portolan chart that has been conserved and clearly shows the mesh of lines that mark out the routes.

of places and were instead named after the cardinal points: north, south, east, and west. Later, the courses would be expressed by degrees of circumference.

García de Palacio spoke of this transition:

> Leaving behind the opinion of the ancients and the names they gave to the winds and the number of them they had, I will speak of what the mariners of our days now use and will accordingly give you the names. Eight winds make up the whole, which are: The East, Northeast, North, Northwest, West, Southwest, South, Southeast. And between these they place another eight which they call the middle ones.[71]

The use of the magnetic needle, the new techniques of cartographical representation, the fact that there were no familiar destinations or winds when exploring the Atlantic, and surely the ever greater ability of ships to navigate in directions different from that of the wind, gradually transformed the tradition of using the names of winds for the courses and the adoption, instead, of the cardinal points and degrees on the celestial sphere.

As was usual with all instruments of measurement, the reliability of the needles depended on their proper calibration so that, in order to know whether a needle was accurate, one had to try it out before sailing. In addition, its capacity to point to the north varied depending on the observer's location on the Earth, a phenomenon known as magnetic declination or

71 García de Palacio, Diego, 'Instrucción Náutica, para el buen [...]', op. cit., Book I, Chapter II, 'Rule for understanding the tables of the declination of the Sun', page 23.

variation.[72] The difficulties that this caused were not so important within the Mediterranean, but they could have had serious consequences on long transatlantic routes.

It is thought that Columbus was the first to note the variations in the compass and that, in doing so, he was surely referring to this phenomenon. However, it is highly probable that other travelers had noticed the variation and that it was only the advance of astronomical navigation that made the changes in the compass evident.[73] Voyages to the Western Indies and India at the end of the 15th century and beginning of the 16th century yielded clear evidences of the changes in the direction of the needle. The magnetic declination on a point of the Earth is the angle between the local magnetic north and true north or geographical north. On a long transatlantic voyage a variation of a few degrees or even a few minutes might have led to variations in the course or deviations that, at times, may even have proved fatal. This is the reason why the problem was the subject of much debate among the Iberian sailors and cosmographers of the 16th century and, centuries later, among explorers all over the world. While a definitive conclusion was not reached, it was neverthless a widely discussed topic in the nautical manuals of that period. In his *Breve Compendio de la Sphera*, Martín Cortés tried to explain the variations in the needle on the supposition of a magnetic pole that was different to that of terrestrial one, a hypothesis that had an enormous influence on later debates about terrestrial magnetism.[74] On this subject, Escalante wrote: 'On this matter of the variation of the needle, there have been and there are, sir, many and diverse opinions, among men who are wise, cautious and learned, in which all disagree and almost no one hits the target, not because of any fault of theirs, but the experience they do not have or had'.[75] He also added: 'from which one concludes that that variation and the differences which the needles show do not derive from the pole, nor the Needle, nor the course you sail along'.[76] Without giving a convincing

72 Due to disturbances in the terrestrial magnetic field, the magnetic needle does not exactly point toward magnetic north at all points on Earth: its position varies over the course of time and in accordance with the geographical position of the observer.

73 Taylor, Eva G. R., 'The Haven-Finding Art: A […]' op. cit., page 165.

74 Cortés, Martín, 'Breve Compendio de la Sphera […]', op. cit., third part, Chapter V, Folio LXXI. On the importance of Cortés in the debate, see: López Piñero, José María, 'El arte de navegar en la […]', op. cit., pages 187–190.

75 Escalante de Mendoza, Juan de, 'Itinerario de Navegación de los […]', op. cit., Book III, dialogue, 'dialogue which deals with the compass needle and its northeastern and northwestern variations; and other things pertaining to it which navigators must know to sail well and with certainty', chapter, 'What the northwesting and northeasting of the compass needle is', page 207.

76 Ibid.

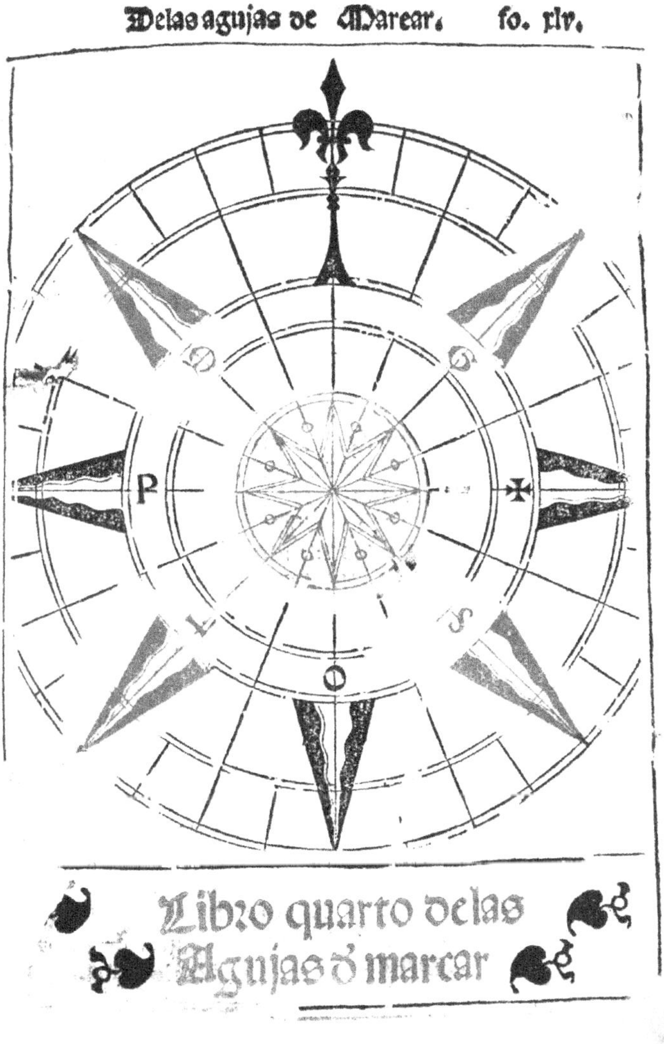


Illustration 4.21. On the mariner's compasses. In: *Regimiento de Navegación*, by Pedro de Medina. Book IV, Folio XIV, 1563 © Obras clásicas de náutica y navegación / José Ignacio González-Aller Hierro (comp.) Madrid: Fundación Histórica Tavera, D.L. 1998. Copy taken from the Fundación Mapfre.

explanation he made it clear that, instead of worrying about theories, the most important thing was to warn navigators about the variation.[77]

Diego García de Palacio devoted a chapter to the subject, titled 'On the north-easting and north-westing of the compass needle', which analyzed the

77 Ibid., page 208.

causes of and ways to correct such variations. Nevertheless, after considering several opinions, he admitted that the nature of this phenomenon was still uncertain. In his opinion, all sailors know about the variations in the compass, 'but they do not find any cause or reason for this northeasting and northwesting, nor know how much the difference is'.[78] Medina equally acknowledged that there were three things whose consequences were known but whose causes no one could explain: the winds, currents, and variation of the compass needle.[79]

For a long time, explaining and correcting these variations was a major scientific problem that was difficult to solve. It was already a subject of heated debate among the Iberian cosmographers of the 16th century, nor was it a concern limited to seamen, and soon became a topic that received considerable attention from learned elites. In fact, it turned into an outstanding subject in the history of modern science, closely related to the later studies of William Gilbert and Johannes Kepler, and in the end a subject of crucial importance in what is known as the Copernican revolution and the birth of modern physics.[80]

## Time and clocks

There were two basic ways to measure the passing of time in a sixteenth-century vessel: with sandglasses or with astronomical techniques for following the Sun and the stars. However, the difficulties at sea were obvious:

> Up to now, *señor* Tristan, no clock has been found or invented which can precisely verify the hour of the day at sea except for the astrolabe with which you measure the height of the sun [...] and all the other clocks which are invented for the sea are uncertain due to the continuous movement of the ship.[81]

78 Garcia de Palacio, Diego, 'Instrucción Náutica, para el buen [...]', op. cit., Book I, Chapter IX, 'The northeasting and northwesting of the Needle, with some doubts which are well explained and an instrument for understanding it better', pages 44–49.

79 Medina, Pedro de, 'Regimiento de Navegación. Contiene las [...]', op. cit., Folio XIVII.

80 See: Pumfrey, Stephen, *Latitud. La verdadera historia del descubridor del magnetismo terrestre,* Barcelona, Editorial Juventud, 2008.

81 Escalante de Mendoza, Juan de, 'Itinerario de Navegación de los [...]', op. cit., Book III, dialogue 'Tristán speaks to the pilot, having just finished taking the height of the sun, and the dialogue begins which deals with the clocks which are most commonly used and how they should be used when navigating the sea', chapter, 'The astrolabe is the most accurate clock at sea to determine the point of midday', page 223.

The first mechanical clocks were formidable, expensive, and heavy devices that were difficult to maintain. To work well, they needed firm and stable bases, which were of little benefit for the lightness inherent in the techniques of navigation. One of the best-known clocks of that time was the one built for the cathedral of Strasbourg in 1350, which had a moving calendar and an astrolabe that was suitable for showing the movement of the Moon and planets. Those enormous machines, which were installed in the towers of inner city churches, did not only cause important changes in the regulation of urban life, but also became symbols of prestige and power.[82] The earliest mechanical clocks were devices fixed into the structure of buildings; the creation of versions that could be transported presented a major technical challenge.[83]

The demand for more precise and portable versions of those devices came from navigators and cartographers. A reliable measurement of time was indispensable for navigating over long distances for three main reasons: first, because to determine the precise location of the ship on the globe required certainty about the local time; second, because the measurement of time was essential for measuring the speed of the ship and the distances it traveled; and, finally, the setting of regular and well-defined shifts was needed to ensure that the work of the crew would be disciplined and efficient. Sandglasses were an essential part of life at sea in the 16th century. And different sandglasses were used to measure different spans of time: an hour, half an hour, a quarter of an hour. Each vessel had to carry sandglasses to measure different spans of time, but the those most commonly used to mark the shifts were half hour sandglasses. Thanks to the hourglass, one knew the time that was passing during the day and could assign or divide up the guards and the sentinels. Life on board, which required discipline, routines, and control of the crew, largely depended on these small devices. When the shifts were of four hours, the time that marked the change of all activities on the ship, the ship (or cabin) boys were responsible for looking after the sandglasses and turning them over every half hour: this was also a way for them to prove that they had not fallen asleep and provided an opportunity for them to recite a prayer or verse[84] when they informed the others of the time. Such clocks governed

82 Cipolla, Carlo M., 'Las máquinas del tiempo y [...]', op. cit., page 15.

83 A complete history of the solution of the problem of longitude is found in: Sobel, Dava, *Longitude: The True Story of a Lone Genius Who Solved the Greatest Scientific Problem of His Time*, London, Penguin Books, 1995.

84 Martínez, José Luis, 'Pasajeros de Indias. Viajes [...]', op. cit., page 93.

life aboard ship and it would not have been possible to organize the crew and their tasks without them. Hence, sailors did not think of time in terms of hours but in terms of sandglasses and those shifts when they were on watch or guard—or, to be more precise, of shifts of eight turns of the sandglass.

The apparent simplicity of a sand clock should not lead us to underestimate the technical difficulties of building one. As García de Palacio pointed out: 'for the hour glass to be an accurate and true clock it must be made of very clear glass and the little hole which the sands falls through very round and not markedly small, and the sand must be very clean—what is called 'clock sand'—and very dry without any kind of moisture'.[85] The hourglasses had to be synchronized with good clocks on land and protected from moisture to ensure their precision, which was not an easy job at sea. The glass vials (*ampolletas*, another name for the sand clock itself) were fragile devices that were constantly handled in unstable conditions, therefore, they not only had to be treated with care but the ship also needed to carry some spare ones. Magellan, for example, had eighteen sandglasses aboard *La Capitana*.

Measuring the passage of time is also important for calculating the speed of the ship and the distances it has covered. The sandglasses were useful for calculating speed because they enabled a navigator to measure the time it took for an object on the surface of the water to move past the hull of the ship from the prow to the stern. With a sandglass, a knotted rope, and a reel, you could measure the speed of the ship. This measuring device was made of a small triangular piece of wood fastened to the logline held in the reel. Along the length of the rope there were knots, set at a distance close to fifteen meters from each other. When you wanted to calculate the ship's speed, you threw the piece of wood overboard, which was dragged along by the water as the reel loosened the line. Then, with the help of the sandglass, you would count the number of knots that passed in a given period of time.

However, to measure longer periods of time and calculate the local time you needed to resort to an analysis of the regularity of the position of the heavenly bodies. At night, the stars and their apparent movement from east to west were a useful referent: in the northern hemisphere, the Pole Star or Stella Maris and its guardians were the ones most used at sea. There is no star that exactly coincides with the North Pole, but the center of the Ursa

85 García de Palacio, Diego, 'Instrucción Náutica, para el buen [...]', op. cit., page 224. Also see: Chaves, Diego, 'Quatri Partitu en Cosmographia Practica [...]', op. cit., page 131.

Illustration 4.22. Sixteenth-century mariner's sandglass. In: *Instrumentos de Navegación*, by Manuel Sellés, p. 32.  Naval Museum, Madrid.

Minor constellation closely approaches it and at night it seems to be the point around which the whole of the celestial sphere revolves. You have to imagine that the compass rose or compass needle is centered on the North Star and that the stars around it revolve in a regular way at night. The brightest star in Ursa Minor is the Pole Star, the star which Kochab and Pherkad—the Guardian stars—revolve around every twenty-four hours: Columbus called

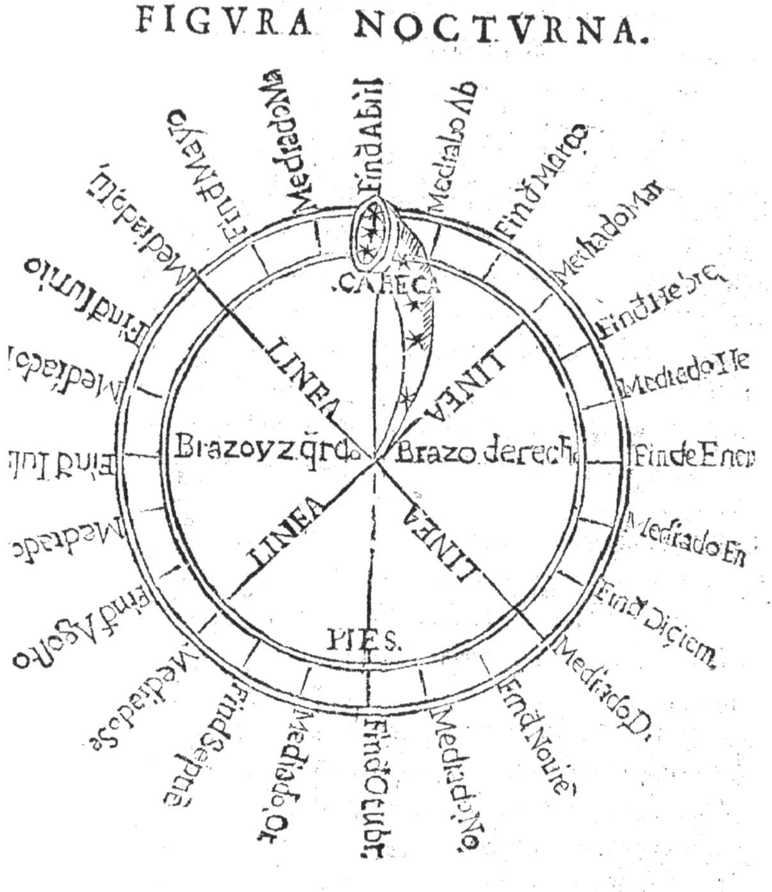


Illustration 4.23. Night chart to determine the hours with the North Star. In: *Instrucción Náutica*, by Diego de García de Palacio. Book I, Chapter VII, p. 42, 1587 © Obras clásicas de náutica y navegación / José Ignacio González-Aller Hierro (comp.) Madrid: Fundación Histórica Tavera, D.L. 1998. Copy taken from the Fundación Mapfre. Rules to find out what hour it is at night by the North Star, in accordance with the relevant calculations.

them the 'guardians of the sky'. Kochab works in a similar way to a modern clock and its position may indicate the time at night.

Diego García de Palacio wrote a long chapter on the 'rules for knowing what time it is at night from the North Star, in accordance with the new calculation'. It is also possible to follow the passage of time by day and calculate the local time by making observations of the Sun with instruments such the astrolabe and comparing these with those in the astronomical tables and manuals.

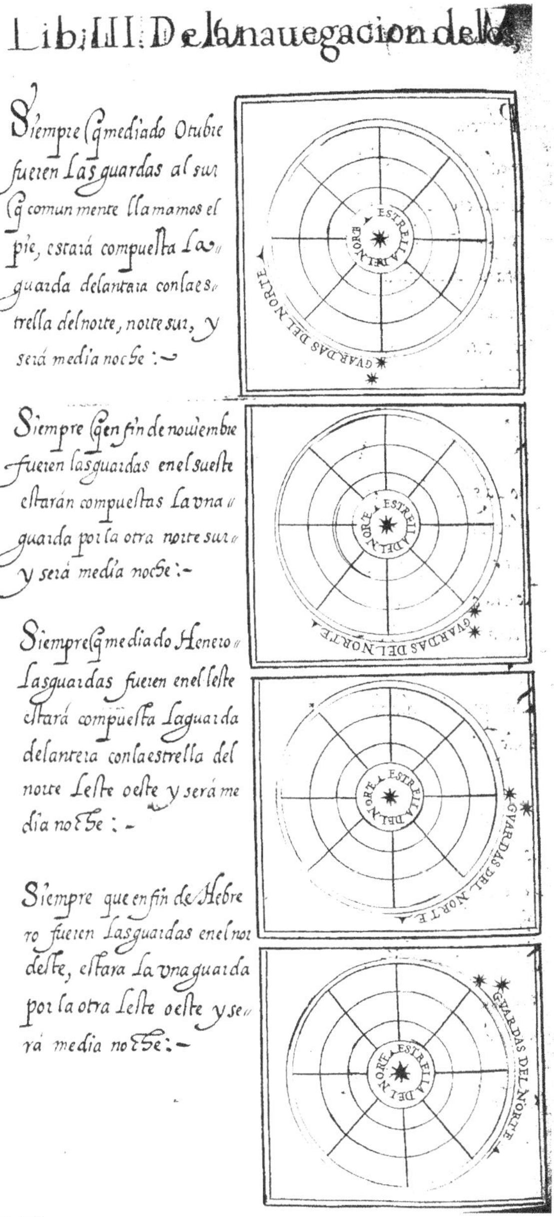
Lib. III. De la nauegacion del ...

Siempre (q mediado Otubre fueren las guardas al sur (q comun mente llamamos el pie, estará compuesta la guarda delantera con la estrella del norte, norte sur, y será media noche :—

Siempre (q en fin de nouiembre fueren las guardas en el sueste estarán compuestas la vna guarda por la otra norte sur y será media noche :-

Siempre (q mediado Henero las guardas fueren en el leste estará compuesta la guarda delantera con la estrella del norte leste oeste y será media noche :-

Siempre que en fin de Hebrero fueren las guardas en el nordeste, estara la vna guarda por la otra leste oeste y será media noche :-



Illustration 4.24. On sailing in the western seas and lands. In: *Itinerario de Navegación,* by Juan Escalante de Mendoza. Book III, p. 252, 1575 © Obras clásicas de náutica y navegación / José Ignacio González-Aller Hierro (comp.) Madrid: Fundación Histórica Tavera, D.L. 1998. Copy taken from the Fundación Mapfre. In these drawings Escalante showed the location of the Guardians of the North Star with respect to the North Star for each month of the year.

## The sounding/plumb line

The plumb line is needed to measure the depth of the sea on coasts, bays, and other difficult stretches of water. It is perhaps the oldest nautical instrument: Phoenician navigators were familiar with it, and for centuries it was essential for saving ships and the lives of sailors.[86] Simpler than the other devices mentioned in this section, the sounding line consists of a hemp cord to which a weight called a deep-sea lead (*escandallo*) is tied. The cord was tied around the narrow part of the lead, which was bell-shaped, while the thick part was smeared with a mass of soft and sticky grease or fat in order to extract samples from the seabed. In that way, one knew whether the ship was situated above a seabed made of sand, slime, or gravel and, consequently, whether it was a suitable place on which to cast the anchor.[87]

## The navigation charts

Although Chapter V, which is about cartography, deals at length with the navigation chart[88] as it was explained in the manuals on seamanship; it was one of the basic instruments of the new celestial navigation of the 16th century and, therefore, has an important place in this chapter. Its usefulness was acknowledged by the authors of Spanish texts on cosmography. Diego García de Palacio, for example, defined it as follows: 'the navigation chart is nothing more nor serves [for anything else] than to show the distance between some places and others, and the wind which blows along the course which the *nao* wants to follow'.[89]

In contrast with most navigational instruments of the 16th century, such as astrolabes, sandglasses, and magnetic needles—examples of which are conserved in museums—the sea charts that were used in sixteenth-century ships no longer exist and we only know of them through references and testimonies or what may be examples of charts published in manuals.[90] Drawn on paper and constantly modified to incorporate new findings in that period of exploration, the charts were, in a sense,

86 See: Taylor, Eva G. R., 'The Haven-Finding Art: A [...]', op. cit., page 35.

87 Chaves, Alonso de, 'Quatri Partitu en Cosmographia Practica [...]', op. cit., Book I, Second Treatise, Chapter VI, 'which deals with the lead line and its use', pages 129–130.

88 See: Illustration 5.4.

89 Garcia de Palacio, Diego, 'Instrucción Náutica, para el buen [...]', op. cit., Book III, Chapter II, 'On the Sea Chart', page 72.

90 Sandman, Alison, 'Spanish Nautical Cartography [...]', op. cit., page 1095.

perishable and disposable. It was not only the hard use given to them that caused their deterioration, since they would be marred by the points of the divider and lines drawn in ink, but also the difficult climatic conditions on board. In addition, the constant growth and adjustment of geographical knowledge meant that they were soon out of date and had to be replaced.

Even though the chart was the most fragile and perishable of the instruments required for navigation, it might be argued that it was the most powerful device available to a pilot. Navigation charts were not only an indispensable part of the equipment on any ship, they also represented an accumulation of many types of knowledge and practices related to the exploration and conquest of the world.

Alonso de Chaves wrote of the chart in the following terms: 'with this instrument, in a brief amount or space, we may understand everything which we would not be able to with very many volumes of books and writings [...] In a brief way, we explain what cannot be understood with [the use of] many words'.[91] A chart was so efficient that, laid on a work table, the whole world was available to its user: 'to look at the navigation chart, you spread it over a table, like tablecloths, and in the drawing and interweaving of its lines and winds we clearly see the work and figure of the world, just as some tablecloths show us their embroidered patterns'.[92] Diego García, for his part, acknowledged the obvious flaw of any map: 'they are imperfect because, since they are flat, they cannot accurately represent the round Earth'.[93] While the globe is the most faithful representation of the world, maps drawn on a flat surface are more practical and effective instruments. The fact that they are flat, two-dimensional, and drawn to scale makes them easy to transport and use: their 'imperfection' is thus their greatest advantage. Like all navigational instruments of the time, their first and main characteristic was that they were transportable. A device that is flat, made of paper or parchment, and of a suitable size for pilots to plan, follow, and predict their courses is a very powerful device. With the help of his other instruments and his experience at sea, the pilot can plot the precise location of the ship by consulting the global coordinates on the map. He can be sure or think he is sure where he is

91 Chaves, Alonso de, 'Quatri Partitu en Cosmographia Practica [...]', op. cit., Book I, Second treatise, Chapter III, 'which deals with the sea chart and the usefulness and advantages of this instrument', page 110.

92 Ibid., page 111.

93 García de Palacio, Diego, 'Instrucción Náutica, para el buen [...]', op. cit., Book III, Chapter II, 'On the Sea Chart', page 72.

and, with his finger, dividers, or a pen, can point to and mark out a spot on the map and say: 'we are here', 'we have covered this distance', or 'this is our destination'. A mariner with a notion of cosmology can then know or at least estimate the distance traveled and the course and stretch that still has to be sailed to reach a destination. With a map, you can dominate the ocean from the pilot's cabin or bridge. With a map stretched out on a worktable it is possible to travel through unknown places with a degree of security, predict what lies ahead, or define and plan the course; you can also measure the distance that has already been covered and the distance you still have to sail. With a good sea chart, the pilot holds the world in his hands, since he can master its immenseness at the glance of an eye. With a good navigational chart, the pilot can see places from a privileged point of view and on a convenient scale. Like maps, such sea charts are a striking example of the domination of space, in this case of the great ocean.

Pedro de Medina devoted a chapter to the use of charts: 'How One Should Prick a Point on the Chart in Accordance With the Height [of the Sun or Stars Above the Horizon] Which You Take in Order to Know the Exact Place Where the Ship Is'. He went on to explain that:

> For he is who sailing, it is a very convenient thing to know the place where his ship is, in order to set his course [...] So, to know that, the pilot marks a point on his chart [...] because there is a big difference between marking your place on land and marking it when you are sailing. That is because on land the chart is still and does not move'.[94]

The constant movement of the ship makes it much more difficult to carry out any activity at sea than on land. At sea, he continued, 'it as though someone is shaking the table on which the man who is writing has placed his paper, so that, due to that movement, he cannot arrange the words and the lines in an even way'.[95]

Pedro de Medina's *Regimiento* gave detailed instructions on how to mark a point on the navigational chart, that is, with the simultaneous use of two dividers, determine the course and latitude of a vessel at any point of the

94 Medina, Pedro de, 'Regimiento de Navegación. Contiene las [...]', op. cit., Book I, 'On the Sea Chart', Chapter IV, 'How One Should Prick a Point on the Chart in Accordance With the Height [of the Sun or Stars Above the Horizon] Which You Take in Order to Know the Exact Place Where the Ship Is', Folio XI.

95 Ibid.

globe. To understand how complex this procedure was, it is worth quoting an excerpt from that book:

> Thus, when the pilot or another person who is navigating wishes to mark the point on his chart, he takes hold of two compass dividers, and he places the point of one at the place he departed from and the other point of the same divider on the course where he has been sailing. And he takes the other divider and places one point on the degrees of altitude on which he has found that number on the graduated table of the chart. Where the two points of both dividers meet, the one he put on the place from which the ship left and the other at the altitude in which you find the point where those two points meet, that is the point where the *nao* is.[96]

The use of the chart is only possible insofar as it is linked with the sound use of other instruments. To measure distances, you needed two dividers and, naturally, trained pilots. Likewise, those instruments would have been useless without the chart. Once again, we can see how, in practice, some instruments depend on others and form a complex network of natural phenomena, devices, and people.

### Astronomical tables

A second instrument, or set of printed instruments, equally indispensable in navigation are the astronomical tables or *Regimientos de navegación*—the rules of navigation. The accurate astronomical calculations that are needed for navigation and cartography, the measurement of latitudes or longitudes, and a precise calculation of the hours in the day or night require reference data in order for them to be compared. The measurement of the height of a star or the Sun and the observation of the time when a phenomenon occurs are not useful unless you can count on tables and data to make the appropriate comparisons and calculations. Similarly, the *Regimientos* or astronomical tables are not useful for navigation without the observations and measurements taken *in situ* with the astrolabe, cross-staff, or quadrant. The data in the tables and the observations of the navigator have to be checked against each other before making the due calculations and determining the latitude in which the observer finds himself.

96 Ibid.

Año .i. despues de bisiesto. fo. xxvij.

| Dias | Enero. | | Febrero. | | Março. | | Abril. | |
|---|---|---|---|---|---|---|---|---|
| dl mes | Gra. | Mi. | Gra. | Mi. | Gra. | Mi. | Gra. | Mi. |
| 1 | 21 | 51 | 13 | 58 | 3 | 39 | 8 | 21 |
| 2 | 21 | 41 | 13 | 38 | 3 | 15 | 8 | 43 |
| 3 | 21 | 31 | 13 | 18 | 2 | 52 | 9 | 5 |
| 4 | 21 | 20 | 12 | 58 | 2 | 28 | 9 | 26 |
| 5 | 21 | 9 | 12 | 37 | 2 | 4 | 9 | 48 |
| 6 | 20 | 58 | 12 | 16 | 1 | 41 | 10 | 9 |
| 7 | 20 | 45 | 11 | 55 | 1 | 17 | 10 | 30 |
| 8 | 20 | 33 | 11 | 34 | 0 | 53 | 10 | 52 |
| 9 | 20 | 21 | 11 | 12 | 0 | 30 | 11 | 12 |
| 10 | 20 | 8 | 10 | 51 | 0 | 6 | 11 | 33 |
| 11 | 19 | 55 | 10 | 29 | 0 | 18 | 11 | 53 |
| 12 | 19 | 41 | 10 | 7 | 0 | 42 | 12 | 14 |
| 13 | 19 | 27 | 9 | 45 | 1 | 5 | 12 | 33 |
| 14 | 19 | 13 | 9 | 23 | 1 | 29 | 12 | 53 |
| 15 | 18 | 58 | 9 | 1 | 1 | 52 | 13 | 13 |
| 16 | 18 | 42 | 8 | 38 | 2 | 16 | 13 | 32 |
| 17 | 18 | 27 | 8 | 15 | 2 | 40 | 13 | 51 |
| 18 | 18 | 11 | 7 | 52 | 3 | 3 | 14 | 10 |
| 19 | 17 | 54 | 7 | 30 | 3 | 26 | 14 | 29 |
| 20 | 17 | 38 | 7 | 7 | 3 | 49 | 14 | 47 |
| 21 | 17 | 22 | 6 | 44 | 4 | 12 | 15 | 6 |
| 22 | 17 | 5 | 6 | 21 | 4 | 35 | 15 | 24 |
| 23 | 16 | 47 | 5 | 58 | 4 | 58 | 15 | 42 |
| 24 | 16 | 29 | 5 | 35 | 5 | 21 | 15 | 59 |
| 25 | 16 | 11 | 5 | 11 | 5 | 44 | 16 | 16 |
| 26 | 15 | 53 | 4 | 48 | 6 | 6 | 16 | 33 |
| 27 | 15 | 35 | 4 | 24 | 6 | 29 | 16 | 50 |
| 28 | 15 | 16 | 4 | 1 | 6 | 53 | 17 | 7 |
| 29 | 14 | 56 | | | 7 | 15 | 17 | 23 |
| 30 | 14 | 37 | | | 7 | 37 | 17 | 38 |
| 31 | 14 | 18 | | | 7 | 59 | | |

c iij

Illustration 4.25. Table of the Sun's declination. In, *Regimiento de Navegación*, by Pedro de Medina. Book III, Folio XXVII, 1563 © Obras clásicas de náutica y navegación / José Ignacio González-Aller Hierro (comp.) Madrid: Fundación Histórica Tavera, D.L. 1998. Copy taken from the Fundación Mapfre.

In the 16th century, the '*Regimiento*' was understood to be a set of tables and rules that a pilot needed in order to estimate his latitude. Among the most widely used were the *Regimiento* of the Sun, astrolabes, North, leagues, and the tables of the Moon, all of which formed what was known as the *Regimiento de navegación.*

The *Regimiento* of the Sun was perhaps the most widely used and most often dealt with in the manuals. Those tables list the solar declinations, that is, the degrees of the Sun's distance, north or south, from the equinoctial line—the equator—on each day of the year in each of the four years of the solar cycle. As we know, the apparent movement of the Sun around the Earth does not occur along the meridional line but its path above the globe deviates by 23 degrees and 26 minutes in relation to the equator, both to the north and south. The line or path of the sun over the globe is known as the ecliptic and the number of degrees in which the Sun deviates from the equator on each day of the year is known as the declination. To accurately measure latitude, you thus not only need to observe and measure the height of the Sun, but also know the declination that corresponds to that day of the year. The precise determination of the height of the Sun above the horizon is therefore a fundamental aspect of celestial navigation. As Alonso de Chaves explained: 'it is worth noting that, of all of the movements and effects of the heavenly bodies, understanding and precisely using the declination of the Sun and its elevation is the most advantageous and necessary'.[97] The diurnal measurements of the elevation of the Sun at noon must be adjusted with the data in the tables of solar declination. These are compilations of numerical data in degrees for the declination on each day of year, a number that has to be added or subtracted from the measurement one makes depending on the hemisphere and the time of the year when the measurements are made.

The *Regimiento* of the astrolabe included the rules for knowing when to add the declination to or subtract it from the meridional height of the Sun. The *Regimiento de Leguas* (Rules of Leagues) yielded the increase in latitude in accordance with the chosen course. The Regimiento del Norte provided the corrections that were needed to obtain latitudes by means of the height of the Pole Star.[98]

97 Chaves, Alonso de, 'Quatri Partitu en Cosmographia Practica [...]', op. cit., Book II, First Treatise, Chapter II, 'which deals with the true declination of the sun, general and perpetual, equaled to the meridian of Seville', page 157.

98 Escalante, de Mendoza, Juan, 'Itinerario de Navegación de los [...]', op. cit., footnote, page 99.

These tables were the result of the accumulation and refinement of an astronomical expertise acquired over many years after thousands of calculations and observations. In the end, they were the fruit of a long process of the systematization of observations: they circulated among pilots in the form of manuscripts. The first compendium of tables to be printed was that of Pedro García, which was published in Spanish in 1485 and, later, in French.[99] Most of the manuals of seamanship in the 16th century included these tables of astronomical observations[100] and some retained the term '*Regimiento*' in their titles. This was the case for those written by Pedro de Medina and Andrés García de Céspedes, titled *Regimientos de Navegación*, compendiums that, in addition to such tables, also included other navigational topics such as the use of instruments.

For most authors, the use of the tables was a subject they could not omit, and their recommendations were similar. As an example, we quote Alonso de Chaves' recommendations for measuring the height of the sun and determining the latitude:

> To know the true parallel or degree of latitude where we are, there are four things to take into consideration to start with. The first is that we look at the day on which we want to know this: if the Sun is in the northern or southern signs, which you will know from the tables of the movement of the sun. The second, to know how much the declination of the sun is on that day, which can be found out from the tables for the declinations of the sun, and to see if that declination is northern or southern. The third, we have to wait for when the sun, by its movement [*motu rapu*], will reach our meridian and then see what its elevation above the horizon is, which will be known with the astrolabe and quadrant. The fourth is to watch, when the sun reaches our meridian, where our shadows lie or any other thing which it does which tells us if they are falling to the north or to midday. And having made these four considerations, the rule is applied and it says the following: when both the sun and the shadows fall to the north or both to midday, in that case, the elevation of the sun having been taken, you will add the number of which it is short of ninety to the declination of the sun on

99 Ibid., page 99.

100 See, for example: Fernández de Enciso, Martín, 'Suma de Geographia que trata [...]', op. cit., Folios IX–XXI; García de Palacio, Diego, 'Instrucción Náutica, para el buen [...]', op. cit., Book I, Chapter II, 'Rules for understanding the declination tables', pages 13–23; and Chaves, Alonso de, 'Quatri Partitu en Cosmographia Practica [...]', op. cit., Book II, 'dealing with all celestial matters and touching on practical cosmology and the art of navigation', pages 147–207.

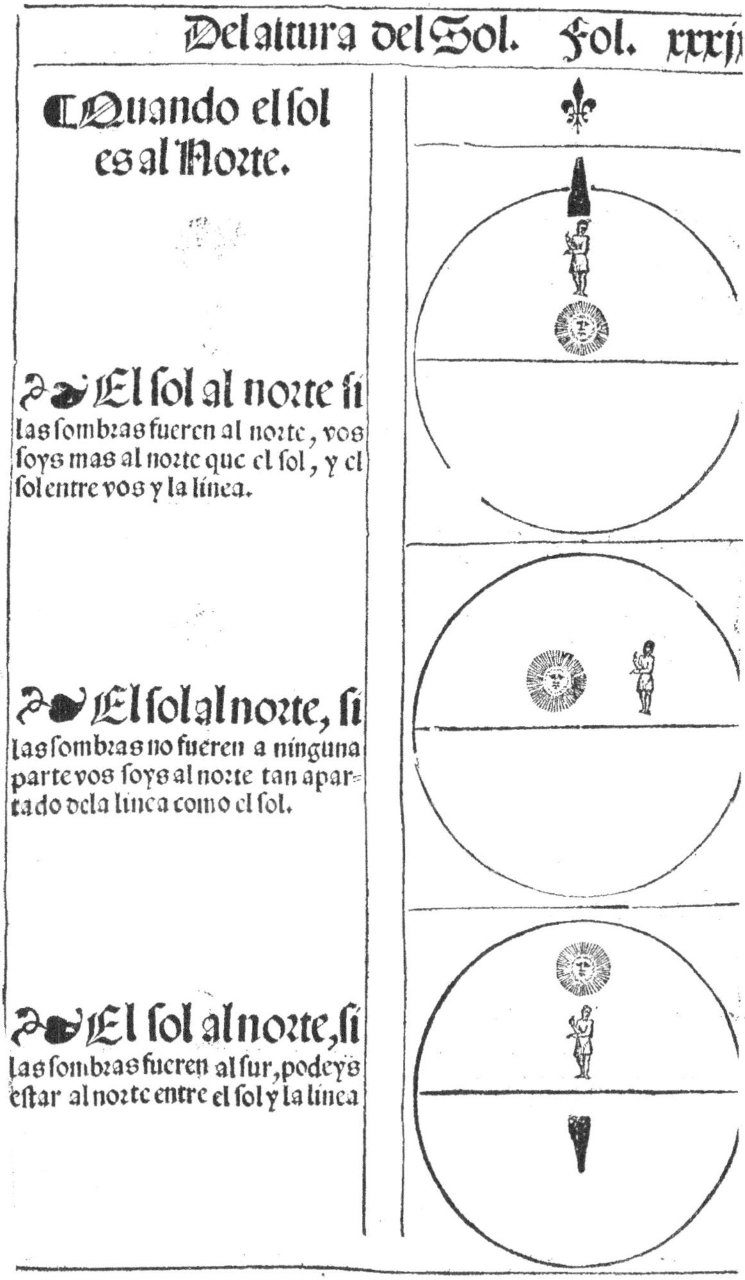

Del altura del Sol. Fol. xxxij

Quando el ſol es al Norte.

El ſol al norte ſi las ſombras fueren al norte, vos ſoys mas al norte que el ſol, y el ſol entre vos y la línea.

El ſol al norte, ſi las ſombras no fueren a ninguna parte vos ſoys al norte tan apartado dela linea como el ſol.

El ſol al norte, ſi las ſombras fueren al ſur, podeys eſtar al norte entre el ſol y la linea

Illustration 4.26. On the height of the Sun and the rules for navigating by it. In: *Arte de Navegar,* by Pedro de Medina. Book III, Chapter IIII, Folio XXXIX, 1545 © Obras clásicas de náutica y navegación / José Ignacio González-Aller Hierro (comp.) Madrid: Fundación Histórica Tavera, D.L. 1998. Copy taken from the Fundación Mapfre. This is one of many examples of the instructions found in manuals of seamanship: it shows the shape of the shadows cast by the position of the Sun at the meridian.

> that day and all of that together, in so many degrees and minutes, will be the latitude, distant from the equinoctial line at the part where the sun falls.

After the author explained all the possible situations, he concluded:

> From the canons [rules] and tables and the above manner of proceeding, you will be able to know [all this], generally and forever, and thus, both in maritime and terrestrial places, the true parallel and degree of latitude in which we are, by the elevation of the sun and its declination, as we have explained.[101]

The importance of the calculations of the Moon lies in the fact is that is the second most prominent body in the sky and, like the Sun, has regular cycles that are useful in nautical cosmography. The lunar cycles are important not so much for their usefulness in determining geographical coordinates on the globe as their relation to the behavior of the sea.

Escalante de Mendoza devoted a long section to the Moon and its effects on navigation:

> It falls to the pilot who has to be responsible for the sailing of any *nao* to first know the particularities of the Moon [...] because it is very necessary to understand the rising, moving and turning of neap tides and spring tides, because the movement of waters and rising of tides usually cause the changes and movements of winds and weather.[102]

And further on, he explained:

> A tide is the daily launching of the waters of the sea against its coasts and shores, which we understand to be in accordance with the movement which the moon makes from east to west as it is forced by the violence of the prime mover, by which movement the waters rise in some parts of the earth and fall in others.[103]

101 Chaves, Alonso de, 'Quatri Partitu en Cosmographia Practica [...]', op. cit., Book II, First Treatise, Chapter III, 'which deals with the canon for determining the true parallel or degree of latitude by the elevation of the sun and its declination', pages 159–160.

102 Escalante de Mendoza, Juan, 'Itinerario de Navegación de los [...]', op. cit., page 227; also see: García de Palacio, Diego, 'Instrucción Náutica, para el buen [...]', op. cit., Book III, Chapter III, 'The lunar table and its explanation', pages 76–77.

103 Ibid., page 233.

Libro quinto.

# Tabla perpetua para saber las conjunciones de la Luna.

| | Aureo nume, | Enero | Febre. | Março | Abril, | Mayo, | Iunio, | Iulio, | Agos, | Setie, | Octu, | Nouie | Dizie, |
|---|---|---|---|---|---|---|---|---|---|---|---|---|---|
| 1563 | 6 | 23 | 22 | 23 | 21 | 21 | 20 | 19 | 18 | 16 | 16 | 14 | 14 |
| | 7 | 13 | 11 | 12 | 10 | 10 | 8 | 8 | 6 | 5 | 4 | 3 | 3 |
| | 8 | 1.31 | 0 | 1.31 | 29 | 28 | 27 | 27 | 25 | 23 | 23 | 22 | 22 |
| | 9 | 20 | 19 | 20 | 19 | 18 | 16 | 16 | 14 | 13 | 12 | 11 | 11 |
| | 10 | 9 | 8 | 10 | 8 | 8 | 6 | 5 | 4 | 2 | 2 | 30 | 29 |
| | 11 | 28 | 27 | 28 | 26 | 26 | 24 | 23 | 22 | 20 | 20 | 18 | 18 |
| | 12 | 16 | 15 | 17 | 15 | 15 | 14 | 13 | 11 | 10 | 9 | 8 | 7 |
| | 13 | 6 | 4 | 6 | 4 | 4 | 3 | 2 | 1.30 | 29 | 28 | 27 | 26 |
| | 14 | 25 | 23 | 25 | 23 | 23 | 21 | 21 | 20 | 18 | 18 | 16 | 16 |
| | 15 | 14 | 13 | 13 | 12 | 11 | 11 | 9 | 8 | 6 | 6 | 5 | 4 |
| | 16 | 3 | 1 | 3 | 1 | 30 | 28 | 28 | 26 | 25 | 25 | 24 | 23 |
| | 17 | 22 | 20 | 22 | 20 | 19 | 18 | 17 | 16 | 14 | 14 | 13 | 12 |
| | 18 | 11 | 10 | 11 | 10 | 9 | 8 | 7 | 5 | 3 | 3 | 2 | 1.31 |
| | 19 | 30 | 29 | 29 | 28 | 27 | 26 | 25 | 23 | 21 | 21 | 20 | 19 |
| | 1 | 18 | 17 | 18 | 17 | 17 | 15 | 14 | 13 | 11 | 11 | 9 | 8 |
| | 2 | 7 | 6 | 7 | 6 | 6 | 4 | 4 | 2 | 1 | 30 | 28 | 28 |
| | 3 | 26 | 25 | 26 | 25 | 25 | 23 | 23 | 21 | 19 | 19 | 18 | 17 |
| | 4 | 16 | 14 | 15 | 13 | 13 | 11 | 11 | 9 | 8 | 8 | 6 | 6 |
| | 5 | 4 | 3 | 4 | 2 | 2 | 1 | 29 | 28 | 27 | 27 | 25 | 25 |

Esta cuenta del aureo numero como en esta tabla esta sirue hasta el año del nascimie[illegible] de .1581 que son .5. de Aureo numero. E luego el año [illegible] de .1582. boluer se ha al principio desta tabla que son .6. de aureo numero: y assi cada vez q̃ esta tabla se acabare que es de .19. en .19. años tornar se ha a començar del principio: y desta manera es perpetua.

Illustration 4.27. Perpetual table to find out lunar conjunctions. In: *Regimiento de Navegación*, by Pedro de Medina. Book V, Chapter I, Folio XIV, 1563 © Obras clásicas de náutica y navegación / José Ignacio González-Aller Hierro (comp.) Madrid: Fundación Histórica Tavera, D.L. 1998. Copy taken from the Fundación Mapfre.

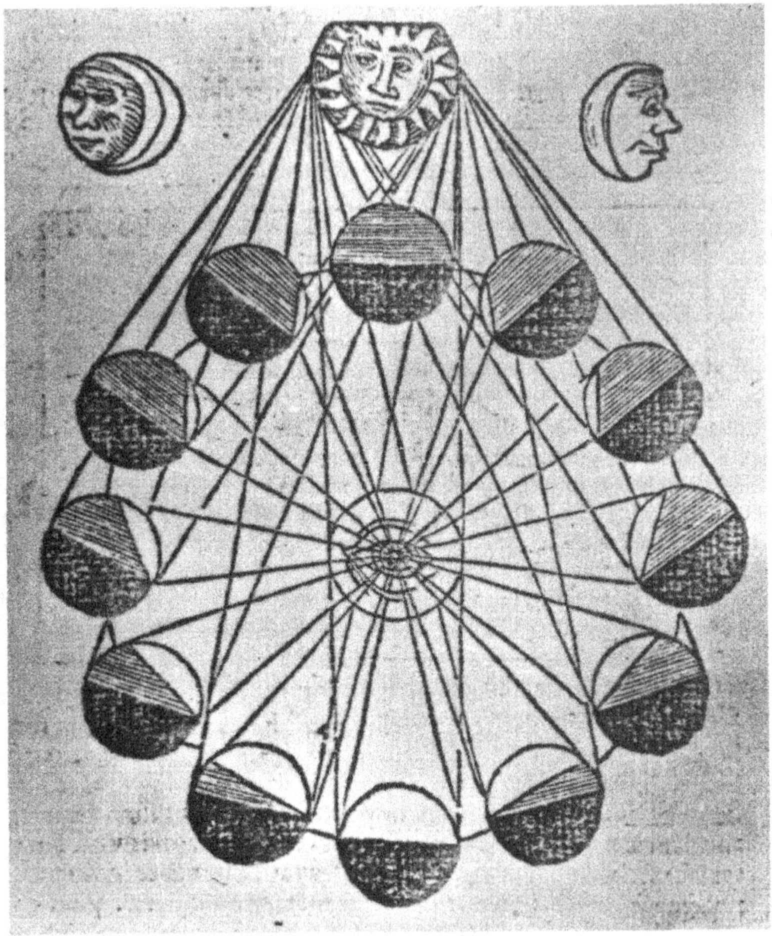

Fases de la Luna.

Illustration 4.28. Phases of the Moon. In: *Itinerario de Navegación*, by Juan Escalante de Mendoza, p. 228. © All rights reserved. Naval Museum Archive, Madrid.

## Instruments, measurements, precision, and standardization

The above description of the instruments of observation highlights the central problem of seamanship in the 16$^{\text{th}}$ century: precision. Before dealing with other aspects that are just as important for understanding the 'machines of empire', and as a way to round off this section, it is worth making some further comments on the main scientific challenge for Europeans in the 16$^{\text{th}}$ century: the transference, from remote places, of precise and reliable information.

Tabla para ſaber a que oras vienen las mareas. fo. liiij

| Dias ð luna | ¶ Rumbos. | Dias | Quartos. |
|---|---|---|---|
| 1 | Norte quarta al nordeſte. | 3 | 3 |
| 2 | Leſnordeſte. | 4 | 2 |
| 3 | Leſte quarta al nordeſte. | 5 | 1 |
| 4 | ¶ Leſte. | 6 | |
| 5 | Leſte quarta al ſueſte. | 6 | 3 |
| 6 | Leſſueſte. | 7 | 2 |
| 7 | Sueſte quarta al leſte. | 8 | 1 |
| 8 | ¶ Sueſte. | 9 | |
| 9 | Sueſte quarta al ſur. | 9 | 3 |
| 10 | Suſueſte. | 10 | 2 |
| 11 | Sur quarta al ſueſte. | 11 | 1 |
| 12 | ¶ Sur. | 12 | |
| 13 | Sur quarta al ſudueſte. | | 3 |
| 14 | Suſudueſte. | 1 | 2 |
| 15 | Sudueſte quarta al ſur. | 2 | 1 |
| 16 | Sudueſte quarta al oeſte. | 3 | 3 |
| 17 | Oeſſudueſte. | 4 | 2 |
| 18 | Oeſte quarta al ſudueſte. | 5 | 1 |
| 19 | ¶ Oeſte. | 6 | |
| 20 | Oeſte quarta al noroeſte. | 6 | 3 |
| 21 | Oeſnoroeſte. | 7 | 2 |
| 22 | Noroeſte quarta al oeſte. | 8 | 1 |
| 23 | ¶ Noroeſte. | 9 | |
| 24 | Noroeſte quarta al norte | 9 | 3 |
| 25 | Nornoroeſte. | 10 | 2 |
| 26 | Norte quarta al noroeſte. | 11 | 1 |
| | ¶ Norte. | 12 | |
| 28 | Norte quarta al nordeſte. | | 3 |
| 29 | Nornordeſte. | 1 | 2 |
| 30 | Nordeſte quarta al norte. | 2 | 1 |

Illustration 4.29. Tables to find out at what hours the tides come. In: *Regimiento de Navegación*, by Pedro de Medina. Book V, Chapter X, folio LIII, 1563 © Obras clásicas de náutica y navegación / José Ignacio González-Aller Hierro (comp.) Madrid: Fundación Histórica Tavera, D.L. 1998. Copy taken from the Fundación Mapfre.

Eugenio de Salazar, a monk who made the transatlantic crossing without any knowledge of experience of sailing, wrote a lively account of life on board a ship sailing the Atlantic in the 16th century: *La mar descrita por los mareados* (The Sea Described by the Seasick). Among other aspects of life on board, Salazar described the jobs of a pilot and, with a certain sense of humor, made fun of how the pilots tried to use their instruments without much success:

> These are times when you see the pilot take [the measurement] of the star, you see him take the cross-staff, and verify the north, and finally find 3000 or 4000 leagues from it; later you see him put the astrolabe in his hands at midday, raise his eyes to the sun, try to make it enter the sights of his astrolabe and, since he cannot, put it aside and you then see him look at his *Regimiento*; and finally use his own rough judgment about the height of the sun. And since at times it rises so much, it rises a thousand degrees above him. And other times, it falls so low, it would not reach that point in a thousand years; and above all, it tired me [to] see that he wanted to keep the degree or point and the leagues which they think the ship has sailed along its course; although later, when I understood the cause, which is that they see that they never get it right or understand it, I was patient, seeing that they are right not to reveal signs of their foolish aim; because the height they take is a matter of more or less, and the width of a pinhead in their instrument will lead to an error of 500 leagues in their judgment. What do you see? they ask each other. How many degrees has your honor measured? One says, sixteen; another, barely twenty; and another, thirteen and a half. Later, they ask, where are we in relation to the land, your honor? One says, I find myself forty leagues from the land. Another, one hundred and fifty. Another says: I had us at ninety-two leagues in the morning and whether it is three or three hundred, none of them agrees with another nor finds the truth.[104]

This testimony not only clearly shows the concrete difficulties of making precise measurements, but also the magnitude of the technical challenge that the *Casa de Contratación* faced in producing a precise body of geographical knowledge. Control at a distance turned into a problem of communication in which the data, observation, and measurement made in remote places,

104 Salazar, Eugenio de, 'Letter written to doctor Mirando de Ron, a special friend of the author, which describes a ship, and the life and tasks of the officers and sailors, and how those who voyage on the sea pass their time. It is useful for understanding the language of sailors'. *La mar descrita por los mareados*, 1573, in: Martínez, José Luis, 'Pasajeros de Indias. Viajes [...]', op. cit., APPENDIX 3. page 308.

very far from Seville and Madrid, had to be comparable and useful for those storehouses of information.

That implied, in turn, that the instruments, charts, or instructions produced in Madrid or Seville had to be equally useful and comparable in any part of the empire and the world. For a navigational chart made in Seville to be really useful on the other side of the Atlantic, a high degree of consensus and standardization of the methods of geographical representation was required. The success or failure of an enterprise like the *Padrón Real* (Model or Master Map) was essentially a problem of regulated and disciplined observations.

A modern reader, who may be familiar with satellite devices for location finding, will easily recognize the high degree of imprecision in the nautical science of the 16th century. Naval historians often mention the great uncertainty and lack of geographical precision of Columbus and those who followed him, but such anachronistic comparisons do not have much historical interest. What is true is that the use of astronomical instruments and new methods of celestial navigation was neither obvious nor easy in the 16th century. Precision was a clear and persistent purpose of the cosmographers, cartographers, and pilots in the age of discovery, but it was also an enterprise that was full of difficulties. Precision presupposes stability in the units of measurement, methods, and the instruments of observation and the latter requires that those who use the instruments are disciplined. In other words, precision is not possible without agreements about the rules of the game and synchronized, collective practices in which the devices play an essential role. The use of instruments and the search for precision were inseparable from the problem of establishing units of measurement, so that the creation of measures of distance and weight were a transcendental feature of seamanship and engineering. As we shall see in Chapter five, where we deal with cartography, the adjustment that was perhaps most urgently needed was to define the longitude of a degree on the globe, since the basic aspects of cartography, and in the end of navigation and imperial policy, would depend on that convention. Thus, it is not surprising that this was an indispensable subject in the manuals of the time.

## The crew

So far, we have described the ships and their parts, the complex procedures for building them, and the combination of bodies of knowledge and techniques that allowed them to function and sail. Such pieces of equipment and instruments were the components of the 'machines of empire' and, in turn, were part of a complex network in which human beings themselves were an essential part. Without the latter, artefacts would have been no more than lifeless pieces of wood, metal, or canvas. The human component of the 'machines of empire' is a subject that is equally complex and varied and, just like the artefacts used at sea, was a subject that the Spanish manuals on seamanship dealt with in detail.

Diego García spoke of the ships as complex machines made up of diverse inanimate elements, some of which were described in the previous section: the ship, its parts, the instruments, and the tools. But it was precisely in their interaction with humans that those artefacts became a powerful and indispensable tool of the Spanish Empire. Montañés, the personage in Diego García's *Instrucción Náutica*, reminded the reader that:

> One thing is clear: that although what we have discussed is the precise arrangement for navigating, which has to do with inanimate things, they themselves are useless if the man for whom they are made does not move them, and thus the leading person of the *nao* is its captain, and afterwards the *maestre* (shipmaster), *Piloto* (pilot), *contramaestre* (quartermaster), *guardián*, (boatswain), *despensero*, (steward), *carpintero* (carpenter), *calafate*, (caulker) *barbero*, (barber) *condestable* and *lombarderos* (gunners), common seaman, *grumetes* and *pajes* (ship's or cabin boys), a certain and sufficient number of [all of] whom is ordained by the ordinances of Seville.[105]

At the end of that chapter, the man from Biscay remarked: 'And since we have already raised the mainyards, with a good headroom, cordage, masts, sails and other details for any voyage, all that is lacking is to discuss the people who have to be governed, how many there are, what jobs they do'.[106]

Edwin Hutchins, a psychologist specializing in the field of cognition, has often called navigation a 'system of interactions' between humans and instruments,[107] and sociologists of science, such as John Law, have said that it is a network of associations and the sum of heterogeneous actors.[108] Although

105 García de Palacio, Diego, 'Instrucción Náutica, para el buen [...]', op. cit., Book IV, Chapter XIX, 'On tar, pitch, tow and other small items which are necessary for the ship and its stores', page 110.

106 Ibid., pages 110–111.

107 Writing about cognition, a field which seems very remote from the subject of this book, Edward Hutchins develops his thesis by writing about the cultural and material aspects of the cognitive processes used aboard a twenty-first-century warship. Hutchins, Edwin, *Cognition in the Wild*, Cambridge / London, The MIT Press, 1995.

108 Law, John, 'On Methods of Long-Distance [...]', op. cit.

phrased in a different way, similar ideas were found in sixteenth-century writings on navigation.

In the book cited above, *The Sea Described by the Seasick*, Eugenio de Salazar compared a ship with a town or city and, with his customary sense of humor, described life onboard as follows:

> In this town there are a multitude of people and inhabitants, where they have their callings and positions in accordance with their ranks and standings [...] Because the pilot is responsible for governing it, acting as the lieutenant of the wind, who is the governor and owner. The *maestre* guards the treasury; the *contramaestre* [is responsible for] loading and unloading; the mariners, for marinating the ship; cabin boys and *mozos* and *grumetes* help the sailors; the *pajes* serve the mariners; they mop, scrub and say the prayers and watch over the city. The *guardián* is not one of the Franciscan friars, but he guards the skiff and takes care to guard what might be stolen from the passengers and orders water to be brought; the *despensero* [steward] guards the supplies, and the shipwright [caulker] is the engineer who fortifies the ship and closes the holes where the enemy might enter. In this town there is a barber/surgeon to scrape off the hair from the nape of the neck of the seamen and bleed them if it is necessary.[109]

The governance of a ship can be compared to that of a city: it can be regarded as a complex social organization, just as Eugenio de Salazar described it. Ships may also be thought of, as Diego García did, as complex entities, organisms with a body and soul.

This section began by stating that the total of the devices and instruments that make up a ship do not have any role, power, or meaning if they do not interact with humans, but it is also true that without their instruments and tools humans would not have any significance in this story. It might be said that pilots and captains such as Vasco de Gama, Columbus, or Magellan were considered powerful figures in the history of the Christian conquest of a large part of the world only to the extent that they formed part of those mighty imperial machines. In other words, pilots and sailors were central actors in as much as their interactions with the navigation charts, astrolabes, quadrants, compass needles, or sandglasses in their hands. Additionally, those navigators were on the deck of powerful and complex vessels, and on

109 Salazar, Eugenio de, 'La mar descrita por los [...]', in: Martínez, José Luis, 'Pasajeros de Indias. Viajes [...]', op. cit., page 298.

the bridges of ships equipped with the most sophisticated technology and surrounded by a diligent crew.

Perhaps the image of the helpless and powerless shipwrecked sailor who has lost his ship and his life or who survives in the solitude of some remote place is a striking symbol of the fact that the nature of human power is only real when men form part of complex networks, that is, when they join with other humans, devices, and natural agents and operate together to achieve a common aim. When a person interacts with other humans and with tools in an organized system, he acquires cognitive skills and capacities that are radically different from those the same person would have outside of that context.[110] On dry land, the bravest and most skillful sailor, the boldest captain, and the most talented pilot is no more than a human being who is as defenseless as any other.

The human component of seamanship is a complex subject in itself and deserves to be described in some detail. However, before describing the different callings carried out on board a ship, it is worth asking ourselves who the sailors of the Indies Run were and what place they occupied in sixteenth-century society.

Speaking of the difficulties of the sea, Friar Antonio Guevara remarked:

> The sea is not such a hospitable place that anyone dares to enter it by his own choice, but necessity, because I say and declare that the man who sails, if it is not to ease his conscience, or defend his honor, or make a living, is foolish or bored or stark raving mad. The sea is delightful to look at but very dangerous to travel on.[111]

What is certain is that, for the sailors of the 16th century, life at sea was a means of survival and rarely a voluntary choice or a matter of preference. Their motivations varied, but both ordinary seamen and officers chose a life at sea in a world where there were few opportunities for survival or social ascent. Even some of the officers and pilots, as might have happened with Columbus, Magellan, and other renowned seamen, were trying to escape from a reality that offered them few opportunities for social mobility and they thus chose the adventurous profession of maritime exploration, an activity full of risks and difficulties. In the years immediately following Columbus' first voyage, and with the aim of encouraging men to enlist on

110 Ibid., page XVI.

111 Guevara, Fray Antonio de, 'De muchos trabajos que pasan en las galeras', in: Martínez, José Luis, 'Pasajeros de Indias. Viajes [...]', op. cit., APPENDIX 1, pages 245–246.

the ships sailing to the Indies, the Kings of Spain decreed that those who embarked for the island of Hispaniola would be exonerated of their crimes; the island was also used as a place of exile for criminals. The number of ships and sailors rapidly rose in the early years of the century. In 1511, anyone who wished to go to the Indies, the islands, and the Spanish Main was allowed to, without any need to ask for permission: they only had to register their names. However, when the conquests spread and the rumors of enormous wealth grew, Spain became concerned with exerting a stricter control of foreigners and heretics.[112] The sanctions for foreigners without a license were very severe. Those found on the ships could be executed without any need to consult higher authorities.

The seamen of the 15th and 16th centuries usually lacked any education beyond experience. They often began to work at a very young age and earned their living in ports and ships. Nevertheless, this situation gradually changed, and the ships soon needed pilots educated in the theoretical fields pertaining to celestial navigation. As we shall see in some detail below, life at sea was tough and full of discomforts and keeping ship-operating men who were trained in specialized callings was a necessity; they were constantly on the job and needed strong and healthy bodies. While such training usually began at the age of ten or twelve years, when they started as ship's boys (*'pages'*), the ideal age for being a good sailor was between 20 and 40. But, in addition to enjoying good health and physical strength, the sailor had to obey rules, receive constant training, and submit to harsh discipline.

At the end of the 16th century a distinction was made between two kinds of sailors: those who sailed *'costa y derrota'* (coastal routes)—for instance Biscay to Flanders, France, England, and Andalusia without going far from the coasts—and the new class of 'high seas' sailors, who traveled to the East or Western Indies, or sailed round the world, and navigated by *'altura y escuadra'* (measuring the height of the sun or stars with an astrolabe or cross-staff). This new breed of seadogs, formed in the Iberian Peninsula, are the subject of the rest of this chapter.

Since the caravels and galleons were driven by the wind, they needed a smaller crew than the large oar-driven galleys, perhaps a tenth for ships of the same capacity.[113] On long crossings, nevertheless, running a sailing ship still meant that the crew had to be working all the time and that they had to carry out highly specialized functions. A ship of 100 tons needed twelve ordinary seamen, eight *grumetes*, and three *pajes*, while one of 500

112 Martínez, José Luis, 'Pasajeros de Indias. Viajes [...]', op. cit., page 32.

113 Morison, Samuel Eliot, 'El Almirante de la Mar [...]', op. cit., page 28.

tons needed twice that number and one of a thousand tons more than 95 persons on board, without counting the shipmaster, pilot, scribe, and other officers.[114] For an understanding of how such vessels were sailed, however, the size of the crew matters less than the complex manner in which the performance of different duties activated those 'technological systems' discussed in the previous chapter.

In addition to the pilot and marines, all the ships needed a specialized crew and organized team of officers arranged in a strict hierarchical order. A description of the different ranks and jobs undertaken by the seamen on board a vessel was an unvarying topic in the manuals on seamanship and almost of them speak of the same categories: the captain, pilot, *maestre* (shipmaster), *contramestre* (quartermaster), *escribano* (scribe), *despensero* (steward), *carpintero* (carpenter), *calafate* (caulker), *marineros* (ordinary seamen), *grumetes*, *pajes* (ship's boys), *lombarderos* (gunners), and *cirujano* (surgeon) being among the most important.

As we shall see there were marked differences in social standing among the crew, thus the captain or pilot did not belong to the same social class as an ordinary seaman or gunner. To get an idea of the hierarchies in a ship, it is worth mentioning the salaries each post received. In 1571, the following annual salaries in maravedis (Spanish coins)[115] were recorded: *paje* (ship's boy) 9000; *grumete* 12,000; *marinero* (ordinary seaman) 18,000; *artillero* (gunner) 22,500; *notario* or *veedor* (notary or inspector), *vigía, alguacil del agua, despensero* (steward) carpenter, and *contramestre* (quartermaster) 27,000; and, finally, the captain, 36,000. Another form of paying the sailors was to assign some percentage of the freight charges to the crew, which was also shared out in accordance with their rank. As Alonso de Chaves remarked, the distribution of those charges had to be done in a way that 'three *pajes* (ship's boys) earn as much as a seaman and three *grumetes* (midshipmen) as much as two *marineros* (seaman), and the *contramaestre* (quartermaster) and *escribano* (scribe) and *despensero* (steward) and carpenter and *calafate* (caulker), each as much as two seamen. The *maestre* (shipmaster) and the pilot, each as much as four *marineros* (seamen)'.[116] While the differences

114 Martínez, José Luis, 'Pasajeros de Indias. Viajes [...]', op. cit., page 73.

115 Martínez, José Luis, 'Cruzar el Atlántico', op. cit.; 'Pasajeros de Indias. Viajes [...]', op. cit., page 21; and Chaves, Alonso de, 'Quatri Partitu en Cosmographia Practica [...]', op. cit., Book III, Third Treatise, Chapter II, 'which deals with the people and the provisions which the ship should have and the arms and munitions it should carry to sail well prepared and well-armed', pages 226–227.

116 Chaves, Alonso de, 'Quatri Partitu en Cosmographia Practica [...]', op. cit., Book III, Third Treatise, Chapter II, 'which deals with the people and provisions which the ship should have

and ranks were obvious, coexisting in a tight space for weeks on end, with limited provisions and many shared responsibilities, meant that many of the social differences found on land were forcibly worn down at sea.

To better understand the complexity of this social organization and the human dimension of these 'machines of empire', it is useful to describe each of these posts and callings separately and, in the end, try to get a grasp of the collective dynamics of life aboard a ship.

### The captain/admiral

The overall head of the crew was the captain, who was often the owner or part owner of the vessel. The captain was paid for the services he rendered and was responsible for choosing and hiring the pilot, as well as the other officers and crew members. He set the course without necessarily taking part in navigation. However, often due to his class and education, he was one of the few crew members who had a knowledge of navigation and astronomy or some training in reading sea charts and the use of instruments.

Before weighing anchor, the captain had to visit his ships; inspect the crew, artillery, equipment, and munitions; and confirm that everything necessary for navigation and combat was in order. The manuals insisted on the need for captains to have special moral and human virtues and a suitable temperament for leading and controlling the crew. For that reason, they had to be wellborn, sober, knowledgeable, and good Christians.[117] The sailing of regular fleets established the rank of captain general for the man who was in charge of the lead ship, that is, the one whose ship the others had to follow. When there was no captain general in the fleet, the admiral had to carry out the same functions and therefore, had to meet the same requirements of education and moral virtues as the captain.

### The pilot

The pilot was responsible for sailing the ship to the destination chosen by its owner or the captain, and therefore had to have a technical knowledge of navigation and know how to mark the sea charts and use the other

and the weapons and munitions it must carry in order to sail well prepared and well-armed', page 227.

117 García de Palacio, Diego, 'Instrucción Náutica, para el buen [...]', op. cit., Book IV, Chapters XX–XXXII, 'The captain, shipmaster, pilot, quartermaster, guard [...] and other persons and officers of the ship', pages 111–120.

navigational instruments: the astrolabe, the compass needle, the quadrant, the sandglass, and the plumb line. The success or failure of the crossing depended on his expertise, which explains why the training of pilots was a matter of the utmost importance. The Chair of the Art of Navigation and Cosmography was founded in Seville in 1508, and the post of pilot major was created with the main aim of teaching and testing pilots.

Without exception, the manuals stressed the importance of the post of pilot above all others. To a large extent, the purpose of the sixteenth-century Spanish texts on seamanship was to train pilots since, as we have already pointed out, the royal authorities were concerned about the shortage of pilots who had the requisite experience and knowledge 'to safely carry so many souls, currency and things which are entrusted to them'.[118] Pedro de Medina highlighted the grave responsibility that the post of the pilots entailed, men who had to have a sound knowledge of everything to do with good sailing and, at the same time, were obliged to protect many human lives, in some cases the lives of princes and kings, as well as valuable cargoes of gold, silver, and other precious merchandise. The pilots had to be Spaniards, so that foreign pilots needed a special permit both to test their skills and to obtain maps of the New World.

There were pilots for sandbars or rivers and sea pilots; the former knew the rivers and sailed the ships out into the sea, while the latter sailed long distances on the open sea. The reason for this difference was that a knowledge of the currents and beds of rivers, which was indispensable for keeping ships far from shallows, was regarded as a permanent calling and linked to those places of which the pilots had a detailed knowledge. Sea pilots, by contrast, had a mastery of a very different kind of knowledge, one that was not necessary for river or coastal navigation. For the purposes of this section, it would be interesting to describe the skills and tasks of a deep-sea pilot.

Escalante summed up the main bodies of knowledge that a pilot had to have in five points: first, he had to understand the compass needle and its variations; second, he had to know how to read his sea chart in order to keep to his course, take the height of the heavenly bodies with his astrolabe, measure the elevation of the North Star with his cross-staff, and use the *Regimiento* and determine the declination of the Sun every day; third, he had to have a knowledge of the weather and winds, and the order and course of the tides and their relation with the Moon; fourth, he had to know how to set the sails in accordance with the wind; and, finally, he had to have a

118 Ibid., page 111.

knowledge of the land he was going to and the signs and markers that would orient him along the way.[119]

The character and skills required of the pilot clearly revealed the above-mentioned tension between practical and theoretical bodies of knowledge. On the one hand, he needed to have an expertise in the aspects of astronomy and mathematics that were essential for celestial navigation and the compilation of geographical data that would be useful for the *Padrón Real* and, on the other, a long, real-life experience at sea. Escalante's *Itinerario de Navegación* included a lengthy discussion of the subject:[120]

> Those who have understood and learned the sciences of cosmography and astrology, *señor* Tristán, know in detail many rules, calculations and the differences between them. And that to those among them who have not sailed, nor are practiced or experienced in the manners and secrets of the sea, it seems that the mariner who does not know rules as long and extensive as they do cannot be good pilots. In that, they are deceiving themselves somewhat [...] And therefore the calculations and rules which, in any form, are made and ordered for navigators should be easy and clear and not very subtle or complicated, but certain and true, without much astrology.[121]

However, as to whether pilots or captains should be learned men, Escalante remarked:

> and so if all the pilots and shipmasters and other leaders [*caudillos*] who sail on this and similar voyages could first be graduate masters in arts, or at least cosmographers and astrologers, it is undeniable that this would be a great and good advantage which is general and universal for all those who have some interest in the sea, but, speaking here from our point of view, it is nearly impossible because most of the mariners who navigate

119 Escalante de Mendoza, Juan de, 'Itinerario de Navegación de los [...]', op. cit., Book I, dialogue, 'After the interlocutors hear Mass on Sunday, one in Sanlúcar and the other in Bonanza and return to embark on the admiral's ship, the dialogue between them begins, which deals with the work of any sea pilot and what he must know for his navigation, and the way he must set his sails and leave any port or sandbar similar to that of Sanlúcar de Barrameda where he was anchored', chapter, 'The five main things which a good sailor should know well and understand in order to exercise the office of pilot', page 65.

120 Ibid., dialogue, 'dialogue which deals with what any good sailor should know about taking the altitude, and the use and practice of it, in order to navigate well and accurately by means of calculations and reasoning', chapter, 'The difference in pilots', page 96.

121 Ibid., page 98.

> the sea are of two [different] kinds. The first are all those who begin to sail as their way of life, as poor men and the sons of poor parents, and that was the most suitable calling that they found to support themselves, especially since they are born in places where there are ports and seashores [...] The other kind of men who become mariners are those who were born with a yen for the art of navigation and the calling of the militia [...] nor can good pilots and shipmasters and great sailors study and be graduates or masters in arts, nor cosmographers or great astrologers, before learning the callings of the sea, because it is advisable to enter into each learned profession and profession of the sea from childhood.[122]

The sailor who set out to become a pilot had to be as much a lover of books as of the sea, but in the 16th century, it was difficult to find these aptitudes in the same person. In a chapter devoted to the pilot, García de Palacio dealt with this problem and wrote that, while there were manuals to instruct the pilot, he had to be of 'a good age and much experience and have had fortunate experiences at sea and if it were possible for him to know about astrology, mathematics and cosmography, it will be very advantageous even though he does not realize it'.[123] In any case, he must 'be skilled in measuring the height [of the heavenly bodies] with the astrolabe, cross-staff and quadrant, [know about] the moons and tides and the land and the plumb line; and imaginative about pricking his navigation chart'.[124]

In the text mentioned above, Eugenio de Salazar drew the following picture of the pilot in action: 'Look at the pilot, the lieutenant of the wind [...] turned into a Neptune, he sets out to command the sea and its waves'.[125] He also singled out the unquestioned authority with which his orders were followed: 'I have not seen a gentleman who is so well served, nor have I seen rogues who serve so well and deserve their wages so much as these sailors'.[126]

The pilot's job was a full-time one and thus other members of the crew, such as the shipmaster, had to be ready to replace him and carry out his functions. At night, they had to be alert to the sky and make the needed astronomical

122 Ibid., dialogue, 'dialogue which deals with taking the altitude and the things which are necessary and important for the sound use and understanding of it', chapter, 'It is very difficult for a doctor and graduate in the arts to be a pilot', pages 115–116.

123 García de Palacio, Diego, 'Instruccion Náutica, para el buen [...]', op. cit., Book IV, Chapter XXII, 'The Pilot', page 112.

124 Ibid., page 113.

125 Salazar, Eugenio de, 'La mar vista por los [...]', in Martínez, José Luis, 'Pasajeros de Indias. Viajes...', op. cit., page 299.

126 Ibid.

observations and therefore it was better that their hours of rest should be from dawn to midday, when they had to measure the height of the Sun.

### The shipmaster (*maestre*) and quartermaster (*contramaestre*)

In order of importance, the pilot was followed by the shipmaster. He had to be able to carry out the same tasks as the captain or pilot when they were absent. The specific responsibilities of the shipmaster were to prepare and equip the *nao*, supplying it with all the equipment and things it needed in order to sail. Everything that was loaded or unloaded in the ship had to be authorized by the shipmaster since he was responsible for keeping an account of all the persons and merchandise put on board. García de Palacio wrote that the shipmaster 'has to load the merchandise well and supply the cargo', but he also needed to 'be a good sailor and know about measuring the height of the heavenly bodies, from experience and science'.[127]

The *contramaestre* (quartermaster) occupied fourth place in the hierarchy of the ship's officers, just behind the captain, the pilot, and the shipmaster. Among his duties, he had to replace them when they were off duty and, like all of them, know how to read and write in order to keep the logbook or ship's log (*Libro de abordo*). He was likewise responsible for looking after the cargo and equipping the ship, always had to be and live in it[128] and, in collaboration with the shipmaster, ensure that the sails and other fittings remained in a good state and protect them from moisture and mice.

### The boatswain (*guardián*)

The fifth position in the hierarchy was that of the boatswain (*guardián*), who commanded the midshipmen and ship's boys and thus had to have a strict temperament. In addition to replacing the *contramaestre* when necessary, his rank entailed more specific duties, such as fastening and loosening the cables:[129]

> going to the sail of the foremast, so that from there he can order the sails to be filled with wind quickly and diligently; get all of the tackle of the

127 García de Palacio, Diego, 'Instrucción Náutica, para el buen [...]', op. cit., Book IV, Chapter XXI, 'The Shipmaster', page 112.

128 Ibid., Book IV, Chapter XXIII, 'The Quartermaster, page 113.

129 Escalante de Mendoza, Juan de, 'Itinerario de Navegación de los [...]', op. cit., Book I, dialogue, 'After the interlocutors have eaten, they begin the dialogue which discusses the measurements of tons and topmasts and lateen yards and other riggings, maritime instruments and people which any ship, in accordance with its size, must carry in it, in order to sail well', page 46.

> prow ready every evening; and when it is necessary, punish those in his charge and visit them at night to make sure that there is no fire in the stove and all of the lamps and sails are ready and in order.[130]

### The ordinary seamen (*marineros*)

The sailors were the most numerous members of the crew and had to carry out a wide variety of jobs. A total of 50 sailors were needed for a ship of 500 to 600 tons; an average of 35 sailors were needed for a ship of 300 to 500 tons, and around 20 sailors for a ship of 100 to 300 tons. Sailors had many skills, but their main job was to keep the ship operating, following the orders of their superiors. They had to be: 'good helmsmen, skillful at handling any rigging or sails, taking and letting go of the sheets'.[131] García de Palacio even wrote that it would be convenient for them to know 'how to prick the chart, take the height [of the heavenly bodies] with the quadrant, astrolabe and cross-staff, find the course by the Guardians of the Pole, etc.',[132] though their most important duty was to 'be diligent in obeying [orders] and going to the tackle and rigging to where they were ordered and be lively at their tasks'.[133]

### Midshipmen (*grumetes*) and cabin boys (*pajes*)

The *grumetes* (midshipmen) were between sixteen and twenty years-old, since they had to attain three years' experience in sailing and serve under sailors and officers such as the captain, *maestre* (shipmaster), *contramaestre* (quartermaster), and *guardián* before rising to a higher rank. The total number of *grumetes* a ship had to have was about two thirds the number of sailors. One of their most important duties was to look after the hourglasses, which had to be turned over as soon as the sand had fallen into the lower bulb. The corresponding announcement of the time was probably done by their recitation of a prayer; thus, those young sailors were the machinery that moved the only clock on board. They also had other responsibilities in the handling of the ship, such as climbing the mainmast to adjust the sails, rowing in the small boat or skiff, and, when the ship was in in combat, helping the gunners.[134]

130 García de Palacio, Diego, 'Instrucción Náutica, para el buen […]', op. cit., Book IV, Chapter XXIIII. 'On the quartermaster', page 115.

131 Ibid.

132 Ibid., Book IV, Chapter XXX, 'On the Sailors', page 119.

133 Ibid.

134 Ibid., Book IV, Chapter XXXI, 'On the Midshipmen', page 119; and Escalante de Mendoza, Juan de, 'Itinerario de Navegación de los […]', op. cit., Book I, dialogue, 'After the interlocutors', page 48.

The number of cabin boys had to be about a tenth of the number of sailors, in addition to those who served the captain, shipmaster, and pilot. Boys aged twelve and sixteen years did all the jobs they were ordered to do. There were different kinds: the cabin boy of the captain was usually a relative of the captain or the son of a friend, while the cabin boy of the shipmaster would be his son or nephew; the same was true of the pilot and quartermaster. Cabin boys did various kinds of manual jobs: mopping the ship and keeping it clean, announcing and serving at meals, taking bread out of the storeroom, attending to the summons of any sailor, looking after their hourglass and praying.[135]

## The carpenter, steward, cooper, and cook

Ships had to be constantly repaired and the responsibility of keeping them in good order often fell to the carpenter. The carpenter had to be a ship's carpenter, since repairing a vessel required experience in that trade. He also had to have good tools, a long list of which can be found in the manuals of seamanship: 'saws of different sizes, axes, chisels, gouges, hammers, files, saw sets, planes, trowels, red ochre and wool yarn for signals'.[136] Very close to the jobs a carpenter did were those of the caulker, who was responsible for closing the joints in the ship's timbers with tow and tar to stop water from seeping. One of the responsibilities of the caulker was to maintain and work the bilge pump and watch over the hold and deck 'so that if some water appears, he removes it'.[137] By the same token, the sails needed to be constantly repaired, so that ships had their own specialists in mending their canvases.

The feeding of the crew and the passengers entailed several jobs. The steward was burdened with enormous responsibilities since, as the majordomo of the shipmaster, he had to care for the bread, wine, water, and other provisions that sustained the people on board. The majordomo kept all such supplies under lock and key, exclusively handled the keys to the hatches, ensured that the oldest food was consumed first, and guarded the rations throughout the voyage. The manuals stressed that he had to be a man of a restrained and moderate character. He also had to be literate, in order to keep a record of the supplies in his charge: ship's biscuits, beverages, meat, fish, vegetables, oil, vinegar, and everything to do with the food and drink for the people on his *nao*.[138] An even more specific responsibility

135 Ibid., Book IV, Chapter XXXII, 'On the Ship's Boys', page 120.
136 Ibid., Book IV, Chapter XXVI, 'On the Carpenter', page 117.
137 Ibid., Book IV, Chapter XXVII, 'On the Caulker', page 117.
138 Ibid., Book IV, Chapter XVI, 'The Steward', pages 115–116.

for conserving the provisions on board fell to the cooper, who cared for the barrels of water and wine. Finally, preparing the meals was the direct responsibility of the cook: cooking for several hundred men in a tight and ill-equipped space was a difficult and at times impossible job.

### The scribe, master-at-arms, and overseer

The ships required the presence of other callings that were not necessarily related to sailing but, rather, to official matters of a commercial or legal nature that were very important, as in the case of the scribe and master-at-arms.

The scribe had to keep a book in which he recorded and measured each article that was put on board, and also served as a notary who legalized the Crown's taking possession of newly discovered lands. The master-at-arms was responsible for punishing criminal offences on board and the overseer for keeping accounts of expenses and guarding the percentage of gold that belonged to the Crown, the 'Royal Fifth'.

In addition to the crew members we have already mentioned, the ships carried passengers and often cargoes of slaves as well.

### The cannoneer

So far we have only detailed the crew members of a commercial ship, and so now we should go on to detail those onboard a warship. Naturally, some vessels combined both functions since the large majority of those on the Indies Run carried some kind of armament. Such weapons and the handling of these weapons were another facet and a further example of the complexity of the equipment and functions of a ship.[139]

While there were no differences in the general structure of a merchant ship and a warship, what happened on deck was different since, if there was a battle, the sailors also had to act as soldiers. Such warfare thus required the performance of other specialized jobs, such as that of the *lombardero*, the cannoneer who handled a lombard, a sixteenth-century cannon. In line with his responsibility for the artillery and munitions, the *lombardero* had to care for the gunpowder, grenades, pineapple grenades, and darts. As was mentioned in the section on ships, the manuals had a long list of weapons for attacking an enemy.[140] Keeping the artillery clean and in working order was an important feature of this calling, as was a constant attention to the

139 Ibid., Book IV, Chapter XXIX, 'On the Constable and Gunners', page 117.

140 See: Chapter 4, page XX.

gunpowder; this was especially crucial due to the danger of a fire, which could cause any gunpowder onboard to explode and thereby cause a shipwreck.

### The barber/surgeon

Conditions of health on the ships were very poor and there were many health dangers for the crew. Not all of the vessels had a doctor on board, but the need to have one became increasingly necessary. García de Palacio describes the barber and surgeon as officers who usually ate at the same table as the captain, shipmaster, and pilot. They mostly had to deal with the common health hazards onboard: fevers, seasickness, and injury. They had to be equipped with the specialized instruments and medicines of their profession. Any long voyage entailed constant problems of health, but it was when the ship was engaged in combat that the surgeons played the crucial and difficult role of dealing with the consequent injuries and wounds. García de Palacio speaks of this as follows: 'the surgeon should be below decks, in a safe place, with a lighted brazier and instruments, surgical cotton, eggs, turpentine, cotton cloth, and with two assistants who are least needed [in battle], so that he may heal the wounds there'.[141]

### The priest

While the priest was not part of the crew, a ship rarely departed for the New World without one or two priests on board. As we shall see below, such priests played a decisive role in life on board: they officiated masses, led prayers, ensured that the required religious festivals were held, and acted as confessors and spiritual guides for sailors and passengers.

## Life on board

*'La mar es muy deleitosa de mirar y muy peligrosa de pasear'.*[142]

We have seen how the organization of the tasks on a ship also implied a hierarchical differentiation that was essential for the maintenance of

141 Ibid., Book IV, Chapter XXVIII, 'On the War Ship', page 124.
142 'The sea is very delightful to look at and very dangerous to cross'. Guevara, Fray Antonio de, 'De muchos trabajos que pasan [...]', in: Martínez, José Luis, 'Pasajeros de Indias. Viajes [...]', op. cit., pages 245–246.

discipline and order. The officers had to be good at ordering their inferiors around and have a suitably tough character to do so, while the sailors and the rest of the crew had to submit to their orders and be diligent. However, the crowded conditions, tight space, and the interactions that the activities on board necessitated meant that the traditional social ranks on land were modified by life at sea. Speaking ironically, Friar Antonio de Guevara mentioned that despite, the many difficulties of life on a ship, he was struck by the fact that the strict hierarchical relations among men and the rigid conventions that obtained on land seemed to vanish aboard a ship: 'As much of an honored, rich and swollen-headed gentleman a passenger who goes on a ship may be, he has to call its captain 'sir' and address its master as a relative, its boatswain as a friend, its workers as brothers and its oarsmen as companions, and the reason for this is that just as a galley slave is not free, everyone there has to'.[143]

As is evident from our description of the different jobs aboard sixteenth-century ships, they were far from being places where one could rest. To maintain the course and safety of a ship was a task that required permanent attention. The crew had to be in constant activity and the officers, without pause, had to give orders to guide the multiple tasks being carried out. The *nao* imposed a new organization on daily life. The crew would change guard every four hours: at midnight and at four, eight, and twelve a.m., as well as at four and eight p.m. Every four hours a ship's boy or cabin boy turned the hourglass over and marked the half hour by reciting the different ditties or prayers that corresponded to each hour of the day and night. The songs the ship's boy sang invariably invoked the protection of God, the Virgin Mary, or the saints for a safe voyage.[144]

## The argot of the sailors

One way to approach and picture the world that characterized life at sea is to scrutinize the complex and special language used by the sailors. Eugenio de Salazar was 'captivated when I looked at this city and the activities of

143 Ibid., page 235.

144 Alonso Chaves, in the chapter of his book titled 'The Form and Manner Which Should Be Followed When Sailing and the Care and Watchfulness Which Everyone Must Observe in His Calling' gave a detailed description of the routines on a ship, from the handling of the vessel, shifts and responsibilities of the different officers and sea men to the times when prayers were said. Chaves, Alonso de, op. cit., Book III, Third Treatise, Chapter III, pages 229–233. Also see: Morison, Samuel Eliot, 'El almirante de la Mar [...]', op. cit., Chapter XII, pages 261–278.

the people in it and was amazed to hear the language of the sailors, which, like the ebb and flow of the sea, only sounded like a meaningless bellowing to me'.[145]

Friar Antonio Guevara wrote about this complex sailors' language in the Chapter six of his book, titled 'Of the Many Jobs Which are Done on the Galleys', saying that 'their way of speaking is as strange as their way of living'. He went over a long list of expressions and words that seemed incomprehensible to him and finished by saying: 'Such is the jargon which they speak on the galley: if we had to list all their strange terms here, we would never finish'.[146]

These are the impressions of two friars having their first experience of life on a ship, so it is not strange that they found the language of the sea barbaric. From the point of the view of the captain, pilot, and ordinary seamen, on the other hand, using a precise language was a matter of necessity. Throughout the history of sailing and up to the present day, seamanship has had its own jargon, known as the *parla marinera*, and it seems to have a word for each piece of equipment on the ship and a verb for each activity carried out onboard. While the impressions and accounts of these passengers may have been exaggerated, they do give us a good idea of the complexity of a sailor's language and also explain why a particular jargon was essential for the sound handling of the ships themselves.

Even though this language seems to be intricate, it was not capricious nor the result of a lack of education. Very much on the contrary, what one should consider when contemplating the strange nature of nautical language was the complex world of a sailing ship. Part of the reason for the existence of this specialized language was the urgent need to give and follow orders and instructions in a precise way, which required a language without any ambiguity. This maritime jargon was therefore a technical necessity in which it was vital to assign a specific name to all the parts of a ship and the maneuvers it carried out. If it was not understood, an order about handling the sails or the rudder might have had fatal consequences. This language was a further example of the urgent need for standardization, which every complex set of technical practices requires and, therefore, it was to be expected that the manuals would deal with the subject in specific detail. In fact, some had glossaries, which were a basic feature of the education of

145 Salazar, Eugenio de, 'La mar vista por los [...]', in Martínez, José Luis, 'Pasajeros de Indias. Viajes [...]', op. cit., page 301.

146 Guevara, Fray Antonio de, 'De muchos trabajos que pasan [...]', in: Martínez, José Luis, 'Pasajeros de Indias. Viajes [...]', op. cit., pages 244–245.

navigators.[147] Diego García de Palacios, for example, included a '[v]ocabulary of the names which the people of the sea use for all which pertains to their art, in alphabetical order':[148] it comprised 510 entries.

## Overcrowding

One of the most serious problems on transatlantic crossings was a lack of space. The smallest ships carried a minimum crew of thirty men and a similar number of passengers. Thus, it had to accommodate nearly sixty persons in a space smaller than twenty-five meters in length and eight meters in width wherein there were constant activities, and which was often already full of the ship's equipment and cargo.[149] In the 16th century, the weight of provisions per person might reach 900 kilos when the ship embarked, that is, nearly a ton per person. It is worth clarifying that this number only applied to provisions because the weight of a person's baggage and bedding had to be added to it.[150] With a single cabin for the captain and, in some cases, for the officers too, the places where a person could rest or sleep were few and uncomfortable, either in the open air or below deck with ventilation.

For the passengers who had no responsibility for handling the ship and, in most cases, no experience of the sea, the voyage was no less uncomfortable. There are many testimonies that spoke of the hardships one suffered on the ships. For the passenger of today, the overcrowding, terrible hygienic conditions, poor food, and a scarcity of drinking water at sea would be an unbearable prospect. Friar Tomás de la Torre described the sufferings as soon as the ship embarked:

> In short, the sea made us understand that a habitat for men was not to be found there and all of our spirits sank as if we were dead [...] there was no choice but to force ourselves to eat something, although we felt nauseous, but we were not short of thirst; you cannot imagine a lodging hospital more dirty and full of groans than this one: some went below decks, where they were boiled alive, other stayed on deck, where they were

147 Chaves, Alonso de, 'Quatri Partitu en Cosmographia Practica [...]', op. cit., Book III, Third Treatise, Chapter I, 'which deals with the ship and its parts and the terms used in navigation in Spain', pages 210–223.

148 García de Palacio, Diego, 'Instrucción Náutica, para el buen [...]', op. cit., Book IV, 'vocabulary of the names which seamen use, in all which pertains to their craft, in alphabetical order', pages 129–157.

149 Martínez, José Luis, 'Pasajeros de Indias. Viajes [...]', op. cit., page 80.

150 Ibid., page 98.

> roasted by the sun, throwing themselves on the ground, where they were stepped on and so trampled and dirty there are no words to explain it.[151]

The consequences of overcrowding and poor hygienic conditions made life on board a harsh and often intolerable experience. Friar Antonio de Guevara described it in the following way:

> every passenger who wishes to purge his stomach and go to the toilet is forced to go to the latrines in the prow or approach a loophole, and the shamelessness of it is unspeakable, and more so when it is done so publicly [when] everyone can see you sitting down to do what you must do as if they were seeing you eat at the table.[152]

For his part, Friar Tomás de la Torre described the poor treatments and humiliations suffered on board:

> there are an infinite number of lice who eat men alive, and you cannot wash your clothes because the seawater won't allow it; there is a bad smell, especially below decks, which is intolerable in the whole ship when the pump is working. We felt a great heat on those two days which I cannot describe. We felt it a great deal [...] because the tar of the ship was burning, and many people were crowded there.[153]

Fray Tomás summed up his impressions of life on a ship by comparing it with a prison writing 'the ship is a very tight and very strong prison from which no one can escape even though there are no bars or chains and so cruel that there is no difference between the prisoners'.[154]

Eugene de Salazar gave a similar account of mortifying situations: 'Men, women, boys and old men, dirty and clean: all are turned into a stew and pudding, stuck to each other, and so one standing next to the other, one person belches, another vomits, another farts, another loosens his bowels'.[155]

151 Torre, Fray Tomás de la, *Diario del Viaje de Salamanca a Ciudad real, 1544–1545*, in: Martínez, José Luis, 'Pasajeros de Indias. Viajes [...]', op. cit., page 263.

152 Guevara, Fray Antonio de, 'De muchos trabajos que pasan [...]', in: Martínez, José Luis, 'Pasajeros de Indias. Viajes [...]', op. cit., page 240.

153 Torre, Fray Tomás de la, 'Diario del Viaje de Salamanca [...]', in: Martínez, José Luis, 'Pasajeros de Indias. Viajes [...]', op. cit., page 263.

154 Ibid., page 264.

155 Salazar, Eugenio de, 'La mar vista por los [...]', in: Martínez, José Luis, 'Pasajeros de Indias. Viajes [...]', op. cit., page 104.

It would take centuries before ships became places for relaxation and comfort. The technical resources of the 16th century were not able to deal with such problems, but it is worth mentioning one of the innovations, which came from the New World, that made transatlantic crossings less unbearable: the hammock. The generalized use of that kind of bedding was a response to the shortage of space on European vessels.

## Food and health

Perhaps the biggest problem in the exploration of the high seas was the supply of drinking water for the crew and passengers crossing could take several weeks during which there was no access to dry land or sources of sweet water. Just as the storms and high winds were a threat to the life of sailors, long calms and an absence of winds posed another mortal trap since they lengthened the voyage and aggravated the problems caused by the scarcity of food and drinking water. To avoid a slow and painful death from thirst or hunger, both the choice and care of the provisions and a knowledge of the winds and right seasons of the year for undertaking a voyage were vital. Friar Tomás mentioned these dangers in his diary: 'because everyone said that the calms which occur at that time of the year are a very dangerous thing and it thus seemed that we were doomed to die of thirst and hunger at sea'.[156] After several days of calm, he described the situation: 'the sea was as still as milk and the ship did not move back and forth; the boards and rigging burned with the great heat, and with the fish we ate, our thirst grew a great deal, and seeing we could not move, we cut the ration of water'.[157]

The sixteenth-century manuals of seamanship almost always included instructions about the types, amounts, and storage of provisions. To get an idea of what a ship carried on a transatlantic voyage, and therefore the food that was usually eaten, it is worth looking at the list of provisions, which, according to Chaves, were necessary for a month's crossing on a ship of 200 tons, manned by 50 sailors:

> biscuit bread, fourteen *quintales* [hundredweights of 112 pounds]; wine, at least five *pipas* [barrels or casks]; oil, three *arrobas* [roughly, 25 pounds each]; garlic and onions, four strings of each; provisions for the sick, what they wish; firewood for cooking, a batch; vinegar, two *arrobas*;

156 Torre, Fray Tomás de la, 'Diario del Viaje de Salamanca [...]', in: Martínez, José Luis, 'Pasajeros de India. Viaje [...]', op. cit., page 260.

157 Ibid., page 272.

> sweet water, two barrels; salted meat, a *bota* [a large barrel]; dried fish, a *bota*; spicy sardines, a *bota*; broad beans and chickpeas, a *bota*; salt, half a bushel; cheese, a dozen small ones; tallow candles, thirty pounds; stoves, one for meat and the other for fish; barrels to bring water, half a dozen; funnels to refill them, two; plates and bowls, two dozen; cooking utensils, what suffices.[158]

Preparing meals on a ship was obviously difficult. How could one cook for a hundred or several hundred men in such a confined space? The ships only had one stove, located in the forecastle: it consisted of a metal plate with sand on top and wood for cooking the meal. It could not be lit in a rough sea or bad weather, and handling it required a great deal of care since a fire was one of the most frequent causes of accidents and shipwrecks. On the high seas, dinner was always eaten by day, and when it was possible to do so, since there was only one hot meal per day.

On some ships the officers ate in a special place: 'the *contramaestre* (quartermaster) sat down at the head of the table and after him, at his sides, the officers of the *nao* and after them, the ordinary seamen and after them, the *grumetes* (midshipmen), and the cabin boys would serve the meal'.[159] The captain shared the table with the pilot and shipmaster, and, at times, the surgeon, but the rest of the crew and the passengers had to carry their own plates and eat in any available place on the ship.

During the 16th century, the diet of sailors seemed to be very regulated: 'if it is lunch, a little sea biscuit [hardtack], some garlic teeth, a few sardines or pieces of cheese, and two servings of wine to everyone, and meat only on Sunday and Thursday and fish and vegetables for the rest of the week'.[160] Speaking of the food, Fray Tomás remarked:

> You suffered with the food, because there was usually very little. I think that was partly the reason why many complained about it: a little bacon in the morning and at midday a bit of cured meat that was cooked and a little cheese: it was the same at night. You'd be lucky to get a couple of

158 Chaves, Alonso de, 'Quatri Partitu en Cosmographia Practica [...]', op. cit., Third Treatise, Chapter II, 'which deals with the people and provisions which the ship should have and the weapons and munitions it must carry in order to sail well prepared and well-armed', pages 225–226. Also see: García de Palacio, Diego, 'Instrucción Náutica, para el buen [...]', op. cit., Book IV, Chapter XVI, 'On the Provisions', page 109; and Haring, Clarence H., 'Trade and Navigation between Spain [...]', op. cit., page 346.

159 García de Palacio, Diego, 'Instrucción Náutica, para el buen [...]', op. cit., page 115.

160 Ibid., Book IV, Chapter XVI, 'The Steward', page 116.

> eggs at a meal. The thirst you suffer is incredible. We drank much more than the ration, even though the cupfulls were measured out. What did the others do?[161]

For his part, Fray Antonio Guevara wrote about the food on the ships in the following terms: 'all who go on them have to eat the common ship's biscuit, on the condition that it be covered with spider's webs and black with maggots, eaten by rats, little and poorly soaked'.[162]

The worst of these inconveniences was the shortage of water and consequent thirst. Remembering that suffering, Eugenio de Salazar wrote: 'because in the ship nothing is more desired and less abundant than water'.[163] The wooden barrels did not always provide the best conditions for conserving the water or wine: 'even with the water, you need to lose your sense of taste and smell and sight in order to drink it and not be affected',[164] and 'with the wine you drink at sea you could eat lettuces on land'.[165]

Not all the voyages were so harsh but to get an idea of the extreme scarcity of provisions that might occur on the transatlantic crossings in the 16th century, it is useful to read Antonio Pigafetta's narrative of his journey around the world. Once they reached the Pacific, an ocean much larger and more difficult to cross than the Atlantic, they suffered from hunger and illness:

> We ate ship's biscuit, but it was no longer a biscuit but powder, with handfuls of maggots, because they had eaten the best of them; it fiendishly smelled of rat's urine. And we drank yellowish water, putrid for many days by then, completing our meal with the strips of ox leather from the platform of the mainmast which protected the rigging from the dew: the hides were extremely hardened by the sun, rain and wind. We soaked them in the sea for four or five days, then dried them a little on the embers. They weren't so bad to eat, better than the sawdust which we didn't despise either.[166]

161 Torre, Fray Tomás de la, 'Diario del Viaje de Salamanca [...]', in Martínez, José Luis, 'Pasajeros de Indias. Viajes [...]', op. cit., page 272.
162 Guevara, Fray Antonio de, 'On the many jobs which are done', in: Martínez, José Luis, 'Pasajeros de Indias. Viajes [...]', op. cit., page 235.
163 Salazar, Eugenio de, 'La mar vista por los [...]', in: Martínez, José Luis, 'Pasajeros de Indias. Viajes [...]', op. cit., page 104.
164 Ibid.
165 Ibid.
166 Pigafetta, Antonio, 'El primer viaje alrededor del [...]', op. cit., page 70.

The conditions of life aboard ship, the lack of a healthy diet, and the arrival at climates that were often alien to Europeans meant that maintaining good health was a major challenge on the voyages of exploration. A whole chapter in the history of medicine has to do with the lessons learned at sea and on the voyages of exploration. It soon became common for vessels to carry a surgeon on board, who had several responsibilities. The poor hygienic conditions were a threat to the health of all the sailors and the holds of the ship favored the proliferation of rats and parasites like ticks. The daily work on the ship entailed a high risk of accidents, so that the surgeon had to deal with any problem of health that might arise. In addition to this, attacks by pirates or enemy ships and the battles that resulted were a frequent cause of wounds and mutilations.

The poor food might have been the cause of many illnesses, but the most notorious illness on long crossings was scurvy.[167] For Antonio de Pigafetta, it was the worst of the sufferings at sea: 'but above all our trials, it was the worst: that the gums of some grew over their teeth, both the upper and lower in the mouth, so that there was no way they could eat: they died of that disease'.[168]

The extreme conditions we have just described confirm the idea that those who chose a life at sea either had few other opportunities to survive or else were moved by powerful reasons of a social, economic, and religious nature.

## Men of the sea and men of God

*'Si querés saber orar, aprended a* navegar'[169]

Samuel Morison says in his study about Columbus, with sound reasons, that sailors were the most religious of all the workers on sea or on land.[170] The religious aspect is crucial in the history of sixteenth-century Spain. In

167 A disease characterized by bleeding in the gums, joints and nails, usually accompanied by tiredness, irritability, and a loss of appetite. It is caused by a lack of Vitamin C/ascorbic acid, a substance that enables the body to produce collagen. Collagen has the same role in the human body as steel structures have for cement. In its absence, blood vessels lose their strength and drip and the bones and conjunctive tissues (cartilages) weaken. Although it has nearly been eradicated today, this disease was common among sailors who made long voyages without eating citrus fruits or fresh vegetables.
168 Pigafetta, Antonio, 'El primer viaje alrededor del [...]', op. cit., pages 70–71.
169 'If you want to know how to pray, learn to sail'. Fernandez de Oviedo, Gonzalo, 'Historia general y natural de [...]', op. cit., vol. 5, Book I, Chapter I, 'On the Father and Son Who Floated in the Sea on a Plank until the Father Died, and How the Son Escaped', page 308.
170 Ibid

fact, it was an important motivation in every naval enterprise and helped men to bear the difficulties of life aboard ships, so that it is essential for an understanding of how the 'machines of empire' worked. Instead of trying to give a social or cultural explanation for the role religion played in seamanship, this book wishes to explain how religion played an active part in the configuration of 16th century's social.

To understand the religious life on the ships, two aspects will have to be taken into account: first, that the very objective of the explorations they made had a strong and explicit religious motivation, and second, that the sailors were men of faith and life on board was a permanent affirmation of religious experiences.

To begin with, it is worth recalling the religious significance of the missions that explored the New World in the 16th century. The scientific output of sixteenth-century Spain, that is, the navigation manuals, cosmological treatises, natural and moral histories, and the accounts of the chroniclers explicitly mentioned both their political importance in serving the Empire and their religious one in serving God. The Iberian exploration and conquest of the Atlantic was a mission to expand the Christian religion and the authority of the Catholic Church. The monarchs, members of the clergy, pilots, chroniclers, conquistadors, and sailors regarded the imperial ventures of the Habsburgs as a spiritual mission. Even though this religious fervor is evident, in a constant and obvious way, in all the available testimonies, historians usually ignore the subject or speak of it as a rhetorical matter, an anachronism for which there is no justification. As we have already remarked, historians usually use the religious aspect of imperial history as a pretext to justify the preeminence of more 'tangible' or 'real' motives, such as economic, political, or commercial ones. The references to Christianity that are found in the Spanish writers of the 16th century should not be regarded as mere rhetorical flourishes but as a crucial part of their writings. The problem, of course, is that spirituality lies outside of the secular conventions of the social sciences, which are forced to subordinate it to other explanations of a 'social' nature.

Throughout this book, we have encountered examples of how the history of the conquest of the Indies in the 16th century was thought of as the working of divine providence. We have also seen how the manuals wrote of their imperial mission as something that was inseparable from their religious mission, and the Biblical and religious references in the writings on the Conquest fill pages and pages. As we have also pointed out, the nautical manuals were not an exception. By way of an example, it would be useful to recall the language in which those books were written. Escalante dedicated his book to King Philip II in the following words: 'To show us the evidence, our Lord and Catholic King, of the great importance of safe navigation, since by means of it God fought for and conserved the relics and the second progenitors of the human gender from that universal flood, with the help of the Patriarch Noah, in the Holy Ark'.[171] Later, Escalante explained that navigation is essential for the Catholic mission to 'extirpate idolatries and the deceits of the devil, with the enlightenment of our Holy Catholic Faith', as well as for 'the universal expansion of this, your monarchy of Spain'.[172] In the introduction to his book, Alonso de Chaves shared the same ideas and expressed them in a maxim that might serve as a summary of the religious sentiments of the sixteenth-century Spaniard: 'that the main foundation of our wisdom is to love and fear God'.[173]

This religious sentiment was not only found in such dedications and explanations of the purpose of the imperial mission; they also formed part of the concrete activities of daily life on the Indies Run, and life on board the ships that sailed it would be incomprehensible without taking this spiritual dimension into account. As follows, we present some examples and testimonies of the daily life of those sailors.

In the first place, no one could embark without attending Mass first and confessing and taking communion:

> It is wise advice that any man who wishes to go out to sea prays, be it in a *nao* or a *galera*. One confesses and takes communion and entrusts oneself to God, like a good and faithful Christian, because the life the sailor leads is as hazardous as he who enters a long battle.[174]

Confronting the dangers of the sea and the adventure of a journey to unknown places encouraged a sincere religious attitude. To start with, it is worth recalling the feeling of helplessness and fear in the face of what is strange and the possibility of death:

> Thus, a man, finding himself on a solitary ship, without seeing the land but only a sky and water that are not calm, walks through those dark-green

171 Escalante de Mendoza, Juan de, 'Itinerario de Navegación de los [...]', op. cit., 'Sumario', page 19.
172 Ibid., page 20.
173 Chaves, Alonso de, 'Quatri Partitu en Cosmographia Practica [...]', op. cit., Book I, First Treatise, Chapter I, page 79.
174 Guevara, Fray Antonio de, 'De muchos trabajos que pasan [...]', in: Martínez, José Luis, 'Pasajeros de Indias. Viajes [...]', op. cit., page 247.

> realms on a dark and frightful ground, without realizing that the ship is moving and leaves a wake, since he always seems to be surrounded by the same horizon and sees today's night the same as yesterday's, without noticing anything which offers variety.[175]

The travel diary of Fray Tomas de la Torre described what it was like to experience a storm on board a ship; these occurrences that always awakened a religious sentiment:

> At midnight on the day of Our Lady, a great storm arose which seemed to want to smash the ship into pieces and with a great devotion, we entrusted ourselves to God [...] we thought that our lives would end there, although we also had a great trust in God that our death would not give pleasure to the evil ones [...] we also had a great faith in the Masses and prayers which we knew our very dear brothers and friends said for us. With all this, we were afraid and confessed, awaiting what God would do.[176]

Invoking divine protection was a very obvious need at sea.[177] For Christians in that era, death was a time of repentance, and there were frequent references to making an act of contrition and asking forgiveness of the Creator. Like many other testimonies, the one of Friar Tomás clearly showed a belief in the permanent presence of the divine and the role of God in such situations:

> Saturday night was the worst up to then because the rain was terrible, and the waves seemed to want to reach the sky. They had already smashed the ship, and many washed over the stern, so we thought our last hour had come. The wind was so fierce it broke the mast they call the foremast. Some of the priests modestly entrusted themselves to God, others shouted, calling out the name of Our Lord Jesus Christ. The old, saintly one conjured the sea and in the name of Our Lord Jesus Christ ordered it to be quiet and stop roaring and shouted to the people to be silent and not be afraid, that God was with us, and we could not perish. That consoled us and we were no longer so afraid of dying and thought that if we were to die, the mercy of God would save us and so we began to sing hymns for a long

175 Salazar, Eugenio de, 'El mar visto por los [...]', in: Martínez, José Luis, 'Pasajeros de Indias. Viajes [...]', op. cit., page 102.

176 Torre, Fray Tomás de la, 'Diario del Viaje de Salamanca [...]', in: Martínez, José Luis, 'Pasajeros de Indias. Viajes [...]', op. cit., page 282.

177 See: Morison, Samuel Eliot, 'El Almirante de la Mar [...]', op. cit., chapter 'A Day at Sea'.

> while and while we were singing, a sailor came and said, 'Fathers, the storm has stopped' and it was truly as if an angel had told us and at once we began to sing the *Te Deum laudamus* and the storm was followed by a very great calm [...] I am not inclined to attribute things to a miracle when a natural cause can be found, and I have told you what happened: you can explain it any way you like. We gave thanks to God, certain that He had saved us.[178]

The travel literature of the 16th century was full of references to divine causes and the intervention of spiritual forces, and in that respect the will of the saints, demons, and God was always part of the story. The final section of Oviedo's *General and Natural History of the Indies* dealt with 'shipwrecks and misfortunes' and constantly mentioned the work of the divine will. After Oviedo spoke of the 'miracle' that the linen sails and timbers the ships were made of, which changed the history of the world, he believed, and just before he went on to retell many accounts of misfortunes and shipwrecks, he cited a proverb, 'If you want to know how to pray, learn to sail', and explained: 'because there is no doubt that the attention Christians pay to such disasters and shipwrecks is such that they entrust themselves to God and His Glorious Mother, and thus, it seems that He miraculously hears and helps them, as you shall see and it appears in the following examples and chapters'.[179] That was clearly the main subject of Oviedo's chapter: the permanent intervention of the divine, the will of God, and the faith of sailors. Oviedo recounted more than thirty stories of tragedies at sea and in all of them, without exception, the intervention of God was evident. Like all the narratives of shipwrecks and dangers at sea, that Chapter was full of assertions where the chronicler acknowledged divine will as an undeniable historical cause: 'God miraculously saved us and gave us life', 'which God permitted', 'it was provided by God', 'where God wished to guide us', and 'God willed that'. Phrases such as these and the constant references to the divine providence, the will of God, and divine mercy cannot be ignored or be reduced to mere stylistic convention.[180] To say that the references they made to God were only rhetorical would not be fair to the writers of the

178 Torre, Fray Tomás de la, 'Diario del Viaje de Salamanca [...]', in: Martínez, José Luis, 'Pasajeros de Indias. Viajes[...]', op. cit., page 283.

179 Fernández de Oviedo, Gonzalo, 'Historia general y natural de [...]', op. cit., vol. 5, Book I, Chapter I, page 308.

180 Ibid., vol. 5, Book I, Chapter IV, page 310; Chapter V, page 314; Chapter VIII, page 319; and Chapter IX, page 321.

16th century, because for Oviedo and the other chroniclers of his age, the first and most powerful of all historical causes was the will of the Creator.

Not only God, but saints and devils also form part of this story. It is worth recalling one such account, where Oviedo spoke of a fight among devils who were not able to sink the ship because it carried a figurine of the Virgin of Guadalupe.

> And they saw very fierce and horrible devils set on the bow and the stern of the ship, and in the air heard one of them say: *Put it off course.* Another responded: *I can't.* And a while later they heard another voice, saying: *Send it to the bottom; drown them.* Another voice answered, saying: *I can't, I can't.* And the one who seemed to be their leader replied in turn: *And why can't you?* And that cursed voice said: *I can't because the Virgin of Guadalupe is here* [...] And thus, miraculously, God saved us from the danger of the sea and the devil, by the intercession of his glorious mother.[181]

To set off on a ship on a long voyage, at a long distance from the place one came from, implied the rupture of a certain social order. Thus, the daily routines pertaining to the cities and towns on land ceased being part of life on board. The voyage, in fact, was a rupture in which one lost contact with one way of life and adopted another. This had a particular meaning in a treatise for navigators and discoverers who were sailing far away from the Christian world, so that, while they might be a great distance, in terms of space and time, from the centers of Christianity, they would not lose their relationship to and synchrony with the Christian order. This is the reason why an adequate management of the religious calendar was so important: it enabled them to strictly follow the dates and celebrations characteristic of a Christian people, no matter where on the planet they were.

Like many others, Chaves paid special attention to the religious celebrations on the ships and was concerned with the need to follow an accurate calendar in order to strictly comply with the dates of religious celebrations: 'And, therefore, before any other thing, it seemed to be that the principle and foundation of the present work should be to discuss and state the rules and ordinances which the Holy Mother Church has laid down for keeping and celebrating its festivities'.[182] The fact that the first chapter of his treatise dealt with the Roman calendar and the precise dates of the religious celebrations

181 Ibid., vol. 5, Book I, Chapter IX, page 321.

182 Chaves, Alonso de, 'Quatri Partitu en Cosmographia pratica [...]', op. cit., page 79.

is very significant. There is a thorough linkage or 'naturalization' of the calendar and Christian cosmology to the enterprise of conquest.

Thus, the first chapter included twelve tables with the religious celebrations for every day of the year. The second and third chapters dealt with the calendar from a more technical angle, referring to the 'golden lunar circle' and 'the solar circle and the Sunday lesson'. Here, the rigor of astronomy and of religion were linked in a way that naturalized Christian cosmology.

Looking once again at Fray Tomas de la Torre's travel diaries, we clearly see that, despite the adverse conditions on board, the religious ceremonies formed part of the life at sea.

> On Sundays and holidays, we sang Masses and there was a sermon for the whole ship and every night we sang the *Salve*. The day of Our Father [then August 3$^{rd}$], we held a great celebration and the whole ship was joyful. We shot off many rounds of artillery and supplicated Our Lord with much consolation. There was also a big fiesta the following day, because the ship is called *San Salvador*.[183]

Speaking of the consolation he felt, the friar described several of these celebrations:

> We felt very sorry to see that Christmas caught us at sea and we could not celebrate the birth of our redeemer as we would have liked to and so on the eve of Christmas, we sang the vespers with great devotion and in the evening the father vicar read the lesson and preached, and gave us a general absolution and we built an altar in the stern on that little deck or shelf I spoke of and we got out the Christ Child we had taken and wrapped him in some hay that was there and kept a vigil all night, lighting white candles and praying and singing songs of joy and Our Lord brought us copious tears and devotion. Just after dark, we sang many hymns and at midnight we sang the matins and celebrated the midnight Mass and continued until dawn, with all the solemnity we were capable of, without getting tired: with the fair weather at sea and our songs the sailors went to sleep, which would have been a great danger to us if God had not miraculously saved us.[184]

183 Torre, Fray Tomás de la, 'Diario del Viaje de Salamanca [...]', in: Martínez, José Luis, 'Pasajeros de Indias. Viajes [...]', op. cit., page 271.
184 Ibid., page 283.

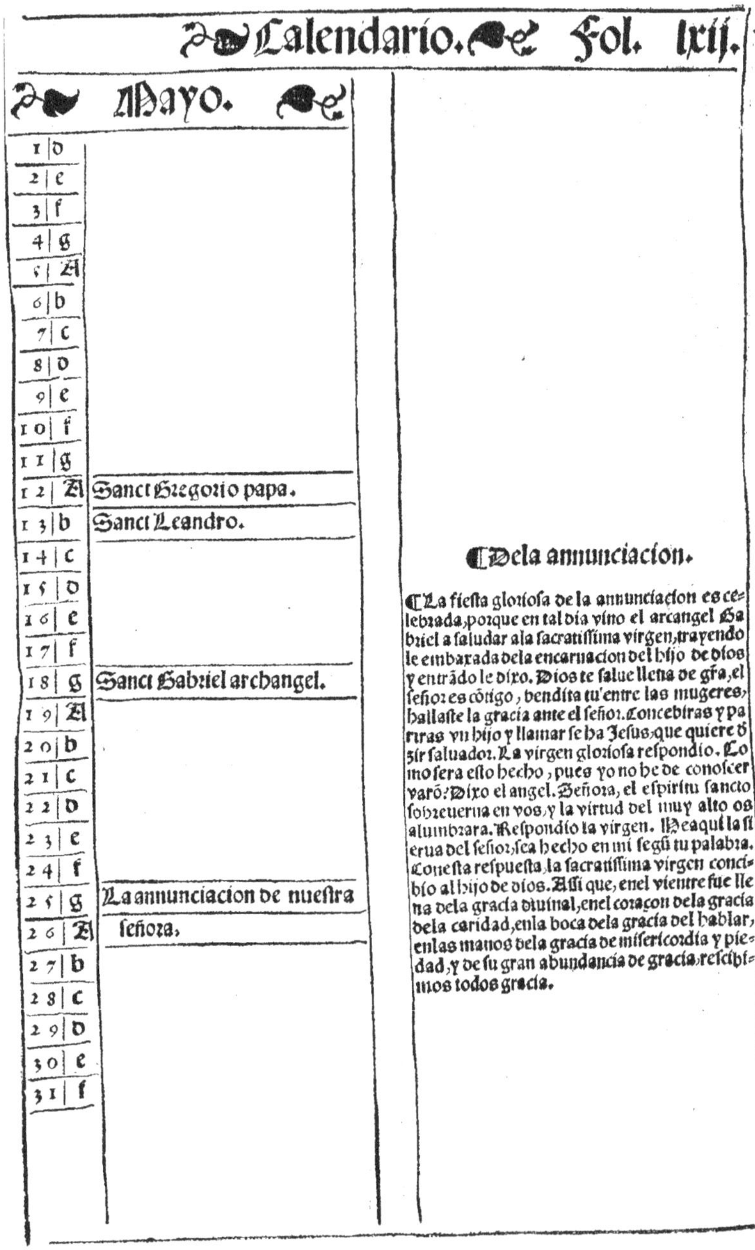
Calendario. Fol. lxij.

Mayo.

| | | |
|---|---|---|
| 1 | d | |
| 2 | e | |
| 3 | f | |
| 4 | g | |
| 5 | A | |
| 6 | b | |
| 7 | c | |
| 8 | d | |
| 9 | e | |
| 10 | f | |
| 11 | g | |
| 12 | A | Sanct Gregorio papa. |
| 13 | b | Sanct Leandro. |
| 14 | c | |
| 15 | d | |
| 16 | e | |
| 17 | f | |
| 18 | g | Sanct Gabriel archangel. |
| 19 | A | |
| 20 | b | |
| 21 | c | |
| 22 | d | |
| 23 | e | |
| 24 | f | |
| 25 | g | La annunciacion de nuestra |
| 26 | A | señora. |
| 27 | b | |
| 28 | c | |
| 29 | d | |
| 30 | e | |
| 31 | f | |

¶ Dela annunciacion.

¶ La fiesta gloriosa de la annunciacion es celebrada, porque en tal dia vino el arcangel Gabriel a saludar ala sacratissima virgen, trayendo le embaxada dela encarnacion del hijo de dios y entrãdo le dixo. Dios te salue llena de gra, el señor es cõtigo, bendita tu entre las mugeres, hallaste la gracia ante el señor. Concebiras y pariras vn hijo y llamar se ha Jesus, que quiere dezir saluador. La virgen gloriosa respondio. Como sera esto hecho, pues yo no he de conoscer varõ? Dixo el angel. Señora, el espiritu sancto sobreuerna en vos, y la virtud del muy alto os alumbrara. Respondio la virgen. He aqui la sierua del señor, sea hecho en mi segũ tu palabra. Con esta respuesta, la sacratissima virgen concibio al hijo de dios. Assi que, enel vientre fue llena dela gracia diuinal, enel coraçon dela gracia dela caridad, enla boca dela gracia del hablar, enlas manos dela gracia de misericordia y piedad, y de su gran abundancia de gracia, rescibimos todos gracia.

Illustration 4.30. Calendar. In: *Arte de Navegar*, by Pedro de Medina, Folio LXII, 1545 © Obras clásicas de náutica y navegación / José Ignacio González-Aller Hierro (comp.) Madrid: Fundación Histórica Tavera, D.L. 1998. Copy taken from the Fundación Mapfre. This is an example of the calendars which appeared in the manuals of navigation. The calendar sets out the dates of the festivals for the month of May in the Christian calendar and explains why the Feast of the Annunciation (Lady Day) is celebrated.

But, in addition to the masses and celebrations that Christians held on land, the routines of the sailors themselves also formed part of these religious practices. In his account of the ships of Columbus, Morison describes a number of prayers that were ritually repeated on those voyages. For example, at dusk the Virgin was hailed with an ancient canticle, the *Salve Regina*.[185] Eugenio de Salazar described how, at dusk, 'two cabin boys come out and recite the Christian doctrine and the *Pater Noster, Ave Maria, Credo,* and *Salve Regina* prayers'.[186] Such prayers were part of the daily routine throughout the voyages and, as we have mentioned when speaking of the use of hourglasses on board,[187] every time that one was turned over—that is, every half hour—the ship's boy in charge had to recite a prayer. Morison cites some of the verses that the boys recited when they turned over the hourglasses or when there was a change in the sailors' shifts:

*'Blessed be the hour when God was born*
*Santa Maria who gave birth to him,*
*San Juan who baptized him. The guard has changed,*
*the sand in the glass is falling,*
*we shall have a good voyage,*
*if God wills it'.*[188]

In that manner, the ship's boys led the crews in the recital of what was called the 'Christian doctrine'. Everyone recited the *Pater Noster, Ave Maria,* and the *Credo,* and they also sang the *Salve Regina.* On the first Saturday of the month, the prayers were recited collectively before an altar that had been previously set up with religious images and lighted candles.[189]

185 See: Morison, Samuel Eliot, 'El Almirante de la Mar [...]', op. cit., page 266.
186 Martínez, José Luis, 'Cruzar el Atlántico', op. cit., page 66.
187 Ibid.
188 Cited in: Morison, Samuel Eliot, 'El Almirante de la Mar [...]', op. cit., page 275.
189 Martínez, José Luis, 'Cruzar el Atlántico', op. cit., page 67.

# 5. The Master Map (*Padrón Real*) and the cartography of the New World

**Abstract**
This chapter presents one of the technological products that most explicitly represents European power over the New World: maps. In particular, the chapter focuses on the Spanish project of a *Padrón Real*, a great map of the entire world, and explains the leadership role of *Casa de Contratación* de Sevilla in the construction of a new picture of the Earth.

**Key words:** Maps, Cartography, Navigational charts, Geography, Cosmography

*'The great man is a little man looking at a good map'.*[1]

The uses of maps are varied: they may serve as navigational aids, charters of ownership, symbolic forms of appropriation, or tools for political administration. They are also useful for the defense of diplomatic interests or serving military or commercial purposes, or may be valuable possessions and saleable goods in their own right. Likewise, they may be appreciated as works of art since they are often objects with an undeniable aesthetic appeal, and because mapmaking rests on complex technical skills.

The Spanish cartography of the New World has been studied in depth and is well documented. Some recent investigations offer a fairly complete account of their history.[2] This chapter deals with a scientific project of

1 Latour, Bruno, 'Drawing things together [...]', op. cit., page 56.

2 Cerezo Martínez, Ricardo. *La cartografía náutica española en los siglos xiv, xv y xvi*, Madrid, Consejo Superior de Investigaciones Científicas, 1994; Martín-Merás, María Luisa, *Cartografía marítima hispana. La imagen de América*, Madrid, Lunwerg, 1993; Sandman, Alison, 'Spanish Nautical Cartography [...]' op. cit.; Castro y Bravo, Federico de, *Las naos españolas en la carrera*

Nieto Olarte, M., *Exploration, Religion, and Empire in the Sixteenth-century Ibero-Atlantic World: A New Perspective on the History of Modern Science*. Taylor & Francis Group, 2022
DOI 10.5117/9789463725316_CH05

great scope, closely linked to the *Casa de Contratación*, navigation, and the post of the pilot major: the drafting of a new map of the world or a Master Map. As indicated by its name in Spanish, it was a singular cartographic representation that was meant to serve as a point of comparison for all maps and as a model for all navigational charts. The idea was that, either in Seville or Madrid, the Crown would have a register and a faithful representation of the New World.

When the Crown appointed Amerigo Vespucci as the first pilot major, despite his lack of experience as a navigator,[3] it ordered him to:

> to make a master map and so as to make it more faithful, we order our officials at the *Casa de Contratación* of Seville to assemble all our pilots, the most skillful that may be found on land at that time, and in the presence of yourself, the said Amerigo Vespucci, our pilot major, a register of all the lands and islands of the Indies which have been discovered to date is to be arranged and made.[4]

Starting in 1508, the pilots authorized by the *Casa de Contratación* in Seville had to systematically provide information for the improvement of the Master Map. The pilot major was to be responsible for assembling the new information and making the master drawing. However, putting the project into practice entailed many difficulties. The standardization of the information depended on the adequate use of calibrated instruments and training the pilots in the art of cartography, an indispensable skill: they would also have to be skillful at making astronomical measurements.[5] The regulations governing the maps, and the care taken to keep their circulation limited to

*de las Indias. Armadas y flotas en la segunda mitad del siglo xvi*, Madrid, Voluntad, 1927; and Fernández-Armesto, Felipe, 'Maps and Exploration in the Sixteenth and Early Seventeenth Centuries', in: Woodward, David, *History of Cartography. Cartography in the European Renaissance*, vol. 3, Chicago / London, University of Chicago Press, 2007, pages 738–773.

3 Edward Collins remarks that 'There is uncertainty surrounding the abilities of Vespucci as Piloto Mayor, though it is certain that he was not an experienced navigator', in: 'Interactions of Portuguese artisanal culture in the maritime enterprise of 16th-century Seville', page 205.

4 'Título de Piloto Mayor para Américo Vespucio', in: Pulido Rubio, José, 'El piloto mayor de la Casa [...]', op. cit., page 462.

5 Fernández de Oviedo, Gonzalo, 'Historia general y natural de [...]', op. cit., vol. 4, Book XXXIX, Chapter III: 'Continuing with the geography of the coast of the Spanish Main in the southern sea, from the gulf and port of la Posesión, which is in the governorship of Nicaragua, following the route to the west until the river of Sancti, which up to now is the farthest point to the west of New Spain noted on the navigational chart, you turn to the North as will be specifically explained in this chapter, in accordance with the drawing of the modern chart of the cosmographer Alonso de Chaves', page 346.

those who were officially authorized to use them, caused problems since updating the Master Map lagged behind exploration, and because making the official maps of the New World seemed to be an endless task.

At one and the same time, the pilots were users of and sources of information for drafting and correcting the Master Map. Dealing with many hazards of the sea did not only require trained pilots but trustworthy maps as well. This difficulty was explicitly mentioned in the royal decree of December 21, 1512, which commissioned Juan Solis and Juan Vespucci to make the Master Map:

> [Since] with the help of our Lord we expect to issue orders for the discovery of other lands, it is necessary that there are more expert and better trained persons who know about the things needed for such voyages, so that those who may be under your governance and pilotage may travel more safely and those such pilots may have and be familiar with the chart insofar as they may know how to undertake the voyages and navigations which they might make, and because I have learned that there are many plans of such charts made in diverse manners and by diverse 'masters' (*maestros*), and they have placed and established the lands of the Indies and those of the mainland of the ocean sea which belong to us [on them], and those we have newly ordered to be discovered for ourselves and the navigation of some of them is very different from that of others, in both the course to and the settlement of those lands, which may cause and does cause many inconveniences for the pilots, hence, they should be ruled and governed so that it be orderly and so that those pilots may have a map for their exclusive use and ease wherever they follow and undertake their voyages, it is my order and will that their Master Map be made, for the reasons I have given.[6]

As a summary of the findings of the authorities on the subject that it was, made by persons trained in cartographic techniques and based on the largest possible amount of information, the Master Map counted on the endorsement of the most reputable cosmographers and could only be periodically revised by meetings of a broad spectrum of experts.[7]

Cartographic knowledge was necessary for the exploration, conquest, and administration of new territories, and for this reason the maps were often regarded as top-secret materials. In the hands of enemies, they could

6 Pulido Rubio, José, 'El piloto mayor de la Casa [...]', op. cit., pages 467–469.

7 Sandman, Alison, 'Spanish Nautical Cartography [...]', op. cit., page 1088 and APPENDIX 40.3.

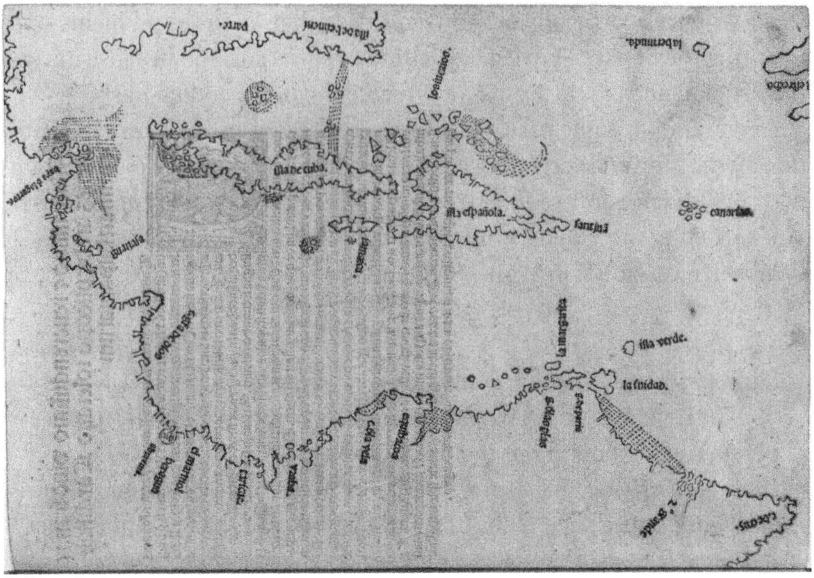


Illustration 5.1. Map of the recently discovered New World. In: *Décadas del Nuevo Mundo*, by Pedro Mártir de Angleria, 1511.

be dangerous for the security of the State. One copy of the Master Map was kept in a 'chest with three keys'[8] and another in the Council of Indies in the King's court. The cartography of the New World was valuable information, indispensable for Spanish navigators and of interest to the enemies and rivals of Spain. Thus, despite the measures that were taken to keep the contents of the maps secret, many manuscript copies—or partial versions of them—had to be circulated, licensed or not, within and beyond Spain.

The tradition of sixteenth-century Spanish cartography was essentially one of manuscripts. There are few examples of maps that reached the printing presses, partly because printing them was technically difficult and cost a great deal, and partly because of the strategic and secret nature of the geographical information they showed. Few maps of America were printed in Spain in the 16th century. Some of the most important examples are: the Map of the New World, which appears in the *Décadas* of Pedro Martir de Angleria (Seville, Jacobo Cronberger, 1511); the navigational chart of the New World in the *Arte de Navegar* of Pedro de Medina (Valladolid, Francisco Fernández de Córdoba, 1545); the Map of the New World in the *Libro de Grandezas y Cosas Memorables de España* of Pedro de Medina (Seville, Dominico Robertis, 1549); the Navigational Chart in the *Breve*

8 Portuondo, María M., 'Secret Science. Spanish cartography [...]', op. cit., page 99.

Illustration 5.2. Map of the New World. In: Crónica del Perú, by Pedro Cieza de León, 1554. Taken from the book by Francisco Vindel, *Mapas de América en los libros españoles entre el siglo XVI al XVIII*, 1955, p. 43. 

*Compendio de la Sphera y de la Arte de Navegar* of Martín Cortés (Seville, Antonio Álvarez, 1551); the map of the New World in the *Crónica de Perú* of Pedro de Cieza de León (Antwerp, Juan Lacío, 1554); the Map of the Universe of Jerónimo Girava (Milan, Juan Antonio Castellán and Cristóbal Cacón, 1556); the Map of the New World in *Relaciones Universales del Mundo*, of Juan Botero Benes (Vallalodid, Diego Fernández de Córdoba, 1599); the

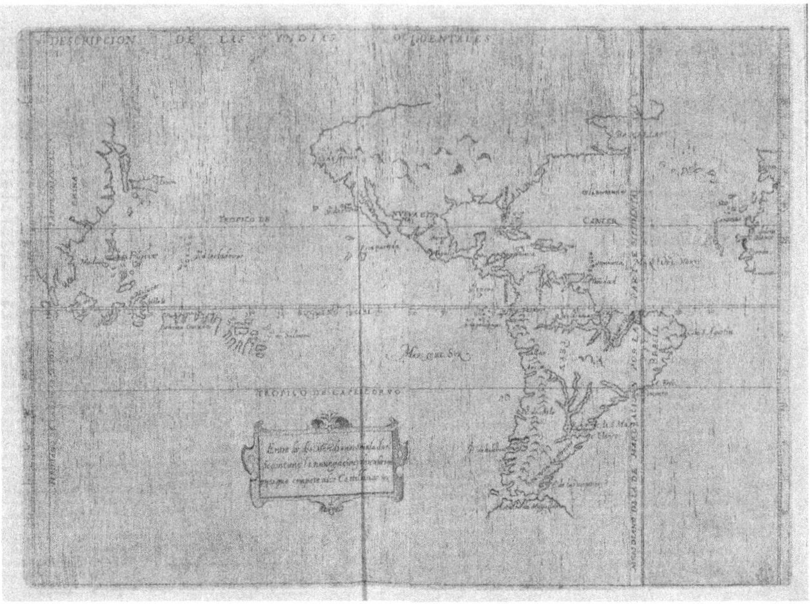

Illustration 5.3. Maps of different parts of America. In: *Historia general de los hechos de los castellanos en las islas y tierra firme del mar océano*, by Antonio de Herrera, Madrid (1601–1615). Taken from the book by Francisco Vindel, *Mapas de América en los libros españoles entre el siglo XVI al XVIII*, 1955, p. 43. © All rights reserved. National Library of Spain.

map of different parts of America in the *Historia general de los hechos castellanos en las islas y tierra firme del mar océano*, better known as the *Décadas*, of Antonio de Herrera, which also included another fourteen maps of the different regions of America (Madrid, Juan Flamenco and Juan de la Cuesta, 1601–1615); and the sea chart or map of the universe in the *Regimiento de Navegación* of Andrés García de Céspedes (Madrid, Juan de la Cuesta, 1606).[9]

Like any other form of knowledge, maps must be compiled on the basis of prior bodies of knowledge and follow standardized codes: they are always partial or complete copies of other maps. However, the maps of places that have never been explored before must incorporate hitherto unknown data and observations. The job of making a cartographic representation of the New World continuously brought together new elements, but the project of the *Padrón Real* combined two efforts that were difficult to reconcile: having a single chart and incorporating new observations.

9 A complete compilation of these printed maps is found in: Vindel, Francisco (ed.), *Mapas de América en los libros españoles de los siglos XVI al XVIII (1503–1798)*, Madrid, Ministerio de Asuntos Exteriores, Dirección General de Relaciones Culturales y Científicas, 1991.

## Nautical charts and how they were made

Since the section devoted to navigational instruments explained the use of nautical or sea charts,[10] we now turn to how these charts were made and their relation to the project of the Master Map, whose main purpose was to serve as a model for the sea charts. As we mentioned above, while many of the maps of the New World have been published and reprinted, the large majority of the sea charts that once existed have not been conserved.[11] Most of the maps of the New World that were reproduced in the 16$^{th}$ century were large; they were expensive and elegant representations that were difficult or impossible to acquire; they were also of little practical use in navigation. The finely wrought maps that circulated in that era and that are well-known today were probably made as gifts or exhibited as a tangible demonstration of the possessions of the Empire. By contrast, the charts that the pilots bought and used on their voyages could not have been expensive, nor did they have the detail, decoration, nor care that was taken in making the large maps of the period that are still known today.[12] It is difficult to precisely determine the extent to which the charts used by sailors were faithful copies of the *Padrón Real*; it is also certain that the Map was constantly modified in the course of the 16$^{th}$ century.

Fernández-Armesto has asked whether these charts were really used and questions their importance for navigation in the 16$^{th}$ century. While one must avoid the anachronism of singling out their great imprecision compared with later or current maps, it is undeniable that maps that would have been useful for crossing the Atlantic could not have existed at the start of the 16$^{th}$ century. In the early stage of the discovery and conquest, Europeans did not have the capacity to make maps that would appear to be of some use from our point of view today. Columbus and those who followed him sailed with maps that did not show the continent of America.

The use of maps as a guide for travelers or as instruments for voyages or navigation now seems normal. However, this particular use of maps is a modern practice.[13] The importance of nautical charts in the history of navigation and science is not directly related with their usefulness on the earliest voyages of exploration, because, in the end, there was no real

10 See: Chapter 4, page xx.

11 Sandman, Alison, 'Spanish Nautical Cartography [...]', op. cit., page 1095.

12 A thorough study of the cartography of the New World in the 16$^{th}$ century is found in: Fernández-Armesto, Felipe, 'Maps and Exploration in the [...]', op. cit.

13 Akerman, James R. (ed.), *Cartographies of Travel and Navigation*, Chicago, University of Chicago Press, 2006.

knowledge of that part of the world. On the contrary, the most interesting aspect of that point in the history of navigation was precisely the process and the practices that transformed sea charts into indispensable and powerful instruments. The sea charts and the cartography that were employed in the activities of exploration should be understood as part of the process of inventing and developing modern techniques of navigation.[14] It was in the 16th century that a set of practices that encouraged collaboration among people, and that even required the forming of institutions to produce maps that gave an account of a new geographical reality, were truly set into motion.

For the rest, the fact that we no longer have examples of the charts that were used back then is also understandable. In contrast with artefacts made of wood or metal, such as the astrolabe, cross-staff, or compass, charts were made of perishable materials such as paper; even more significantly, the sea charts were disposable instruments. Since the pilot would draw additions or corrections on them, it is reasonable to assume that each chart was only used for a single voyage.[15] Furthermore, the information contained in a chart was of a secret nature and its circulation was prohibited, so that its possession and manufacture were subject to control and its circulation limited. Such charts were a valuable booty for pirates or explorers supported by Spain's enemies, so that when there was a battle at sea, with the possibility that they would be seized, the charts were destroyed.

Fernández-Armesto's skepticism about the exaggerated attention that historians gave to an object for which, in his view, there is no evidence[16] does not seem very convincing, since there are many testimonies about the men who made them and their costs and uses. The same storehouses and merchants who sold instruments provided the maps to the pilots who were authorized to use them, and those maps were an important aspect of the scientific output that revolved around the Indies Run.[17]

All of the navigational manuals of the 16th century necessarily paid a great deal of attention to such charts, hence some included detailed instructions about their manufacture and use.[18] Some examples and models of nautical

14 Ibid., page 6. To find out more about this subject, see: Taylor, Eva G. R, 'The Haven-Finding Art: A [...]', op. cit.

15 Portuondo, María M., 'Secret Science. Spanish cartography [...]', op. cit., pages 8–9.

16 Fernández-Armesto, Felipe, 'Maps and Exploration in the [...]', op. cit.

17 Sandman, Alison, 'Spanish Nautical Cartography [...]', op. cit., page 1130.

18 García de Palacio, Diego, 'Instrucción Náutica, para el buen [...]', op. cit., Book III, Chapter II, 'On the nautical chart, its drafting and form', pages 71–76; Medina, Pedro de, 'Regimiento de Navegación. Contiene las [...]', op. cit., Book I, 'On the rules of navigation of the sea where the sea chart explains the things about sailing in good faith'; Zamorano, Rodrigo, 'Compendio de

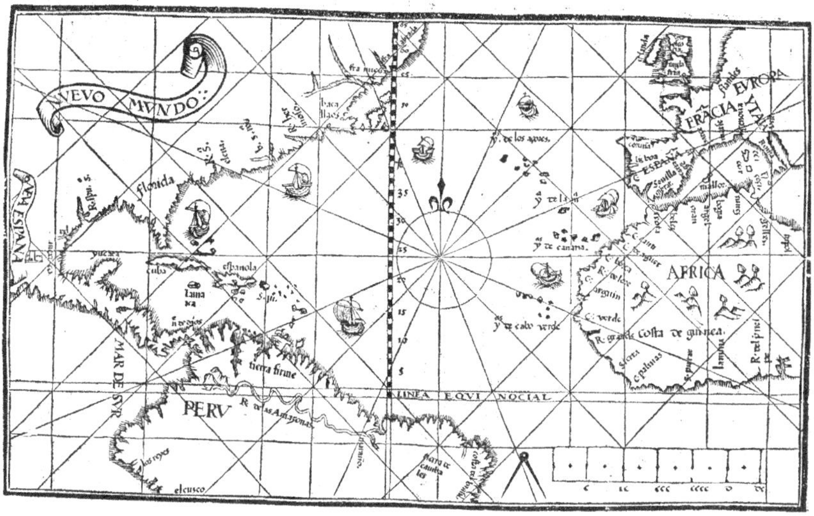


Illustration 5.4. Navigational chart containing the navigation of most of Europe, Africa, the Indies or New World, and their respective altitudes and degrees In: Arte de Navegar, by Pedro de Medina. Book III, Folio XXII, 1545 © Obras clásicas de náutica y navegación / José Ignacio González-Aller Hierro (comp.) Madrid: Fundación Histórica Tavera, D.L. 1998. Copy taken from the Fundación Mapfre.

charts were printed in those manuals. The most widely known of those charts was the one published by Pedro de Medina in his *Arte de Navegar.*

The manuals of both Enciso and Chaves must have included a chart, but none have been conserved. It is nearly certain that this was not a simple oversight, but a necessary precaution since such information was highly valuable.

## The making of a chart

Alonso de Chaves and García de Palacio included sections on drafting charts as well as detailed instructions on how to draw them. On the basis of the teachings of those authors, we can recreate the steps required for the manufacture of a chart.[19] The first step, following the basic notions of Ptolemy's *Geography,* was always to lay down a grid of astronomical

la arte de…', op. cit., 'On the Sea Chart', Chapters XVII–XX, pages 37–39; and Cortés y Albarcar, Martín, 'Breve Compendio de la Sphera […]', op. cit., Third Part, Chapter II, 'On the composition of the nautical chart', Folio LXI.

19 García de Palacio, Diego, 'Instrucción Náutica, para el buen […]', op. cit., Book iii, Chapter II, 'On the Sea Chart', pages 71–72; and Chaves, Alonso de, 'Quatri Partitu en Cosmographia Practica […]', op. cit., Book I, Second Treatise, Chapter II, 'which deals with the sea chart and

coordinates, on which the geographical information would be placed, and divide to the globe into regions: the 'torrid and very hot zone', 'the middling and temperate regions', and the 'frigid regions and furthest and uninhabitable ones'. Once the 'drawn guideline and graduation' was on the chart, the mapmaker turned to the locations and descriptions of the places on Earth.[20] Some mapmakers usually drew a small circle in color and painted the compass rose on it: a fleur-de-lis indicated the north; a cross, the east; an 'S', the south; and the upper part of the map corresponded to the north of the globe.

For García de Palacio, the starting point was to define the courses in the manner of a portolan chart:

> Your worship should take a parchment or paper and make two lines on it, one perpendicular, along the length of the parchment or paper, which will be East–West, and the other along its width, which will be North–South, so that they form right angles, with black ink. This point [where they cross] will be the center of a circle which is divided into 32 equal parts, the winds. Only the eight main winds are in black ink. The eight middle winds are in green and those in the middle, which are called quarters, are red-colored.[21]

Having done that, the mapmaker then had to locate and indicate all the coasts, ports, islands, capes, cities, rivers, and shallows, among other places; placing these on the chart was done in two possible ways: with the use of faithful, authorized maps with verified altitudes, or by means of measurements that the same mapmaker undertook on the basis of the differences of height shown by the astrolabe and cross-staff. To incorporate geographical information, he had to adopt a place that was well known, famous, or familiar as a starting point, a river or cape whose global coordinates he was completely certain of, especially its latitude, and then find another nearby reference point. Once their respective distance was marked on the paper, he then proceeded to make a fully detailed drawing of the coastline that joined them. This was followed by the job of putting down the names of the capes, rivers, ports, and other places, and highlighting the most important

the usefulness and advantages of this instrument and the drafting of this instrument known as the sea chart', pages 110–114.

20 Chaves, Alonso de, 'Quatri Partitu en Cosmographia pratica [...]', op. cit., page 113.

21 García de Palacio, Diego, 'Instrucción Náutica, para el buen [...]', op. cit., Book III, Chapter II, 'On the Sea Chart', page 72.

ones according to their size or color. The next step was to put down a third reference point and then follow the same procedure until the chart was finished.

Among the conventions that were used, sandbanks were indicated by many dots and underwater hazards with crosses. Likewise, compass roses with thirty two points had to be drawn, with a larger one in the center and the names of ancient and modern seas indicated in accordance with the regions and provinces that bordered them.[22] The use of color was also important—almost always in black and red—to better distinguish some places from others.

Another key aspect of these instructions covered toponymy. Once the coasts were described, it was necessary to write down the names of all the known places with a fine pen, using different kinds of handwriting to indicate their degree of importance. All the names had to go in the same direction, for reasons of aesthetics and clarity. Subsequently, the mapmaker would go on to include inland places.[23] If the mapmaker wished, he could adorn the chart with gold and other colors, painting miniature representations of ships, cities, or other things, although he would naturally have to be careful not to hide some city, port, or other place necessary for navigation.[24]

In addition to being powerful devices in which space is condensed into two dimensions on a human scale maps are closely related to the act of naming, which, like baptism, was a form of integrating places into the Christian world. As Adam did in paradise, the Christian conquistadores assumed the right, granted by God and the Crown, to name or rename places, animals, plants, and even the native inhabitants of the Western Indies.

On October 12, 1492, Columbus baptized the first island he found as San Salvador, to honor Christ the Savior. The second island he discovered was given the name of Santa María de la Concepción; the third, Fernandina (after the King of Spain); the fourth was named after the queen, Isabella; and the fifth after Juana, the heiress to the throne. Most of the names had Christian connotations in praise of the Holy Trinity, Christ, the Holy Spirit, and, above all, the Virgin Mary. There were also references to religious festivities, such as Christmas (the Navidad or Nativity) or allusions to the saint of the day on which the place was named. A few were descriptive, such as Cabo Bueno

22 Chaves, Alonso de, 'Quatri Partitu en Cosmographia Practica [...]', op. cit., Book I, Second Treatise, Chapter II, 'which deals with the sea chart and the usefulness and advantages of this instrument', pages 110–114.

23 Ibid.

24 Ibid.

(Good Cape), Cabo Lindo (Beautiful Cape), Puerto Bello (Beautiful Port), the Isla de Tortuga (Tortoise Island), or the Bahía del Caracol (Seashell Bay). The names that referred to royalty or places in Spain—famous, for example, in the case of La Española (Hispaniola)—explicitly showed that the Crown had taken possession of the land; that practice indicated the firm aim of making those places part of the realm of Spain. The act of naming was part of European exploration and giving a name to a place was, in a certain way, an act of translation.[25] The new names immediately inserted the New World into the known world and made the unknown familiar. That is why it was always an act of possession, domestication, and dominion, and, at the same time, a denial of the local.

The instructions for making a map showed the importance of Euclidian geometry as the fundamental framework for the representation of space, as well as the reach of the principles of Ptolemaic cartography. It is important to recall that these instructions basically motivated a judicious adherence to conventions, so that the very idea of teaching someone to make or read a chart implied the mastery of a common language, that is, the collective observation of identical linguistic codes. The starting point for assembling the charts were the classical astronomical conventions because by following them on a plane and making use of simple geometrical rules it was possible to obtain a complete representation of the terrestrial sphere. Thus, the fundamental lines of a map were and still are the equatorial or meridian line; the Tropics of Cancer and Capricorn, with their respective geographical zones; and the conventions that indicate the north at the top and the east at the right.

Another fundamental feature of maps is their scale. In fact, the power of maps lies in the very possibility of representing space on a human scale. Only in that way was the pilot of a ship able to possess and have a control of space from his own vessel. Diego García gave precise instructions about the adequate use of a map's scale. This feature—also known as 'graduation'—could be traced back to the Master Map, since, with the use of a compass, it was possible to reproduce the scales of the Master Map on a visible part of the map being drawn.[26]

The fundamental aspect was the trustworthiness and precision of the information, be it from a known cartographical source or new testimonies.

25 See: Callon, Michel, 'Some elements of a Sociology of Translation. Domestication of the Scallops and Fishermen of St Brieuc Bay', in: Law, John (ed.), 'Power, Action, and Belief: A [...]', op. cit., pages 196–223.

26 García de Palacio, Diego, 'Instrucción Náutica, para el buen [...]', op. cit., Book I, Chapter II, 'On the Sea Chart', pages 73–74.

As is seen in the history of cartography, the sources that a map drew on were mainly other maps. Martín Cortés gave another description of their manufacture, this time stressing the need to trace other maps. Nevertheless, as we have said, in the context of exploration and the making of maps of unknown territories, the most important sources were written descriptions or itineraries and the observations and measurements made on the terrain. The 16$^{th}$ century again revealed the tension between the classical authorities and the new geographical shape of the world based on the experience of explorers.

## The charts of *tierra firme*: the earliest maps of the New World

Charts were also used on land, where they served to plan or display the routes that were to be followed or to test the feasibility of future voyages, as occurred when the plans of Columbus and Magellan were examined.[27] Even more important, the charts were useful devices when it came to defining territorial borders and claims to possession, and they became essential for diplomatic treaties, since the most urgent geographical and political issue of the time was to define the line that divided the possessions of Portugal and those of Spain.

In that ambit of an unprecedented territorial expansion, a precise knowledge of the location of the new territories had important geopolitical implications. Even after the unification of the Crowns of Spain and Portugal, the debate on the precise location of the demarcation line established by the Pope in 1493, and revised by the Treaty of Tordesillas in 1494, continued. The definition of that line went far beyond the tension between those two States. In fact, it became even more problematical for other European States, which had been excluded when the Pope shared out the New World between the two Crowns, above all when news of the vast size and potential wealth of the new territories reached the whole of Europe. The French, English, and later the Dutch, as well as many pirates and corsairs, were not going to turn their backs on the new commercial possibilities of the Atlantic.[28] The precise geographical location of the territories of the world was therefore a pressing political problem, partly because of the abovementioned needs of navigation, but also, in a very obvious way, for the clarity required by Europe's sharing out of the new conquests.

27 Sandman, Alison, 'Spanish Nautical Cartography [...]', op. cit., page 1097.

28 Portuondo, María M., 'Secret Science. Spanish cartography [...]', op. cit., pages 5–6.

In 1493 the bull *Inter Caetera*, issued by Pope Alexander VI, decreed that the lands discovered to the west of a demarcation line one hundred leagues west of the Azores Islands would belong to the Spain and the lands east of the same line would belong to Portugal. In the Treaty of Tordesillas (1494), the signatories, Spain and Portugal, agreed that the line would be 370 leagues west of the islands of Azores and Cape Verde. It very soon became obvious that the agreement would turn into the cause of a long technical and political dispute due to the enormous difficulty of measuring leagues at sea and converting leagues into degrees of longitude at a time when the precise size of the globe was still unknown. We must not forget that the measurement of longitudes on the globe was one of the greatest scientific challenges faced by cosmography in the 16th century.

The controversy became more heated when the Spanish proclaimed their possession of territories in the Pacific.[29] Portugal's discovery and claim to possession of the Moluccas, found by Magellan in 1511, and the subsequent counter-claim made by Spain in 1521, were evidence of the urgent need to clearly and precisely define the demarcation line. That need turned cosmography into a field of intense technical debates and implied a battle—no longer military but technical—to determine who had the authority to define a single way to measure distances and represent territories on maps.[30]

The Treaty of Tordesillas and in general the precise definition of borders, is the best example of the convergence of a technical and political question. The idea that such a controversial debate would have a technical solution, when there was no clarity about the relation between leagues and degrees, was evidence of an exceptional challenge for the establishment of a body of knowledge that would be universally accepted: the standardization of common units, measurements, and guidelines.

From the discovery of the New World onwards, determining the extent and location of the new domains of the Empire was a major concern of the Crown. Sailing to the New World, as well as sailing between its islands and the coasts of the mainland, was an enterprise full of dangers in which it was difficult for ships to orient themselves, but the precise location of the new lands was also a critical political and diplomatic problem. For that enterprise to be successful, the Crown had to be able to know the global coordinates of the new places in terms of latitude and longitude. In theory, the principles behind such measurements were simple, but in practice and

29 Sandman, Alison, 'Spanish Nautical Cartography [...]', op. cit., page 1109.

30 Portuondo, María M., 'Secret Science. Spanish cartography [...]', op. cit., pages 66–67.

at sea, the job was full of difficulties, especially measuring longitude, a technical problem that would require centuries before it reached a certain degree of precision and consensus.[31]

The theory of measuring longitude appears to be clear: eclipses or other astronomical phenomena may serve as events for synchronizing global time. An observer in one part of the globe registers the exact local time when the eclipse occurs and, if he compares it with the local time at which the same phenomenon is observed in another place on Earth, he will find a difference in hours and minutes; this difference can easily be converted into degrees of longitude on the globe since an hour of time is equal to fifteen degrees on the surface of the Earth. While this appeared to be simple, complex and unusual technical solutions were needed to observe the phenomenon; complex because it presupposed the use of synchronized clocks in different places, and unusual because they depended on heavenly events that are infrequent and that can only be observed with sophisticated equipment.[32] In the Iberian context, most cosmographers studied the problem. Among them, there stood out Alonso de Santa Cruz (1505–1567), whose *Libro de Longitudes* (Book of Longitudes) explained twelve different methods for measuring longitudes, including the observation of eclipses.

Alonso de Santa Cruz was responsible for an impressive collection of cosmographic and cartographic studies, which included 338 maps found in two atlases: the *Islario General*, a map and document describing the world's islands; and another atlas of 169 maps that have since been lost. Different chronicles, geographical descriptions, and cosmographical treatises gradually formed a source of information that turned into an obligatory reference for later chroniclers and cosmographers of the *Consejo de Indias*, although, in principle, their publication was prohibited by royal decree.[33] The aim of Santa Cruz was to compile a geography of the world in the manner of Ptolemy, as he explained in the final part of the *Libro de Longitudes*, 'I wish to describe something of all the parts of the world in tables, placing [on them] their respective provinces, cities, places, rivers, mountains, and other notable things which there are in each one, and I will do the same for the Western Indies, newly discovered now, in many parts of which I have been'.[34] His project therefore consisted of two different parts, in accordance

31 Ibid., page 224.

32 Sobel, Dava, 'Longitude: The True Story of [...]', op. cit.

33 Ibid., page 71.

34 Santa Cruz, Alonso, 'El libro de las Longitudes', in: *Obra cosmográfica*, 1:273, cited by: Portuondo, María M., 'Secret Science. Spanish cartography [...]', op. cit., page 72.

with his sources and methods: in the first place, a description of the known world—that is, Europe, Asia, and Africa—following the classical traditions; and, in the second, a description of the New World whose main source of data was the direct testimonies and experiences of those who had traveled there. The main problem of a global cartographic endeavor, such as the one of Santa Cruz and many other sixteenth-century cosmographers, was precisely the determination of longitudes.

The method for measuring longitude on the basis of the comparative observation of the time at which eclipses occur had been known since antiquity, and has since spread to different parts of Europe along with the geographical methods of Ptolemy. It was one of the most important and controversial concerns of sixteenth-century cosmography, but no one was able to find a satisfactory and relatively consensual solution until the much later invention of mechanical clocks that worked without major errors on sea voyages.[35] Measuring longitude also ran into the problem of determining the distance of one degree on the globe.

If the Earth were a perfect sphere, all of the degrees of latitude and longitude would be equal—we now know that the degree of latitude is somewhat larger. The basic convention was already part of the classical tradition: the sphere divided into 360 degrees, each located at 60 minutes from one another. For Columbus, one degree corresponded to fourteen ⅙ leagues, perhaps because he thought that the Earth was smaller than it really is and was. Enciso enlarged that distance a little, assigning a value of sixteen ½ leagues to each degree, although that distance is still somewhat short.[36] Escalante defended the idea that each degree measures seventeen ½ leagues but admitted that there was 'a diversity of opinions' about it.[37]

The problem of standardizing distances did not end with an agreement how many leagues there are to a single degree. Just as important and complex was determining the distance of a league. Diego García explained it in the following way: 'because these leagues have different lengths, I would say that it has to be understood in this manner: four barley grains make a finger [finger's breadth or digit]; four fingers make a palm; four palms make

35 Ibid.

36 Fernández de Encisco, Martín, 'Suma de Geographia que trata [...]', op. cit., page 39.

37 Escalante de Mendoza, Juan de, 'Itinerario de Navegación de los [...]', op. cit., Book II, dialogue, 'which deals with the height [of the heavenly bodies]', chapter, 'The number of leagues which have to be sailed over for each of its thirty-two courses, in order to ascend or descend one degree of altitude', page 114; and García de Palacio, Diego, 'Instrucción Náutica, para el buen [...]', op. cit., Book II, Chapter IX, 'The rules for knowing how many leagues are sailed for each of the seven quarters', pages 63–65.

Libro primero.

## Las leguas que ſe cuentan por grado en cada rumbo de la nauegacion.

| NORTE. | xvij. me. | SUR. |
|---|---|---|
| Norte. 4. al nordeſte. | xviij. | Sur. 4. al ſudueſte. |
| Nornordeſte. | xix. y ter. | Suſudueſte. |
| Nordeſte. 4. al norte. | xxj. | Sudueſte. 4. al ſur. |
| NORDESTE | xxiiij. y me. | SUDUESTE |
| Nordeſte. 4. al leſte. | xxxj. y me. | Sudueſte. 4. al oeſte. |
| Leſnordeſte. | xlv. | Oeſudueſte. |
| Leſte. 4. al nordeſte. | lxxxix. | Oeſte. 4. al ſudueſte. |
| LESTE. | | OESTE. |
| Leſte. 4. al ſueſte. | lxxxix. | Oeſte. 4. al noroeſte. |
| Leſueſte. | xlv. | Oeſnoroeſte. |
| Sueſte. 4. al leſte. | xxxj. y me. | Noroeſte. 4. al oeſte. |
| SUESTE | xxxiiij. y me | NOROESTE |
| Sueſte. 4. al ſur. | xxj. | Noroeſte. 4. al norte. |
| Suſueſte. | xix. y ter. | Nornoroeſte. |
| Sur. 4. al ſueſte. | xviij. | Norte. 4. al noroeſte. |

DE Leſte Oeſte no ſe da altura ningũa porque el polo (yendo contino por eſta via) ni ſe alça ni ſe abaxa mas en vna parte que en otra: y aſſi el que Leſte Oeſte nauegare la cuenta de ſu camino es a ſu arbitrio cõſiderando lo que el nauio puede andar ſegun el tiempo y nauio y la carga fueren.

Illustration 5.5. The leagues calculated by degree in each course of the voyage. In: *Regimiento de Navegación*, by Pedro de Medina. First book, Chapter VIII, Folio XIII, 1563 © Obras clásicas de náutica y navegación / José Ignacio González-Aller Hierro (comp.) Madrid: Fundación Histórica Tavera, D.L. 1998. Copy taken from the Fundación Mapfre.

a foot, five feet make a geometric step; a league, 15,000 feet'.[38] In a similar manner, Escalante tried to define the measurement of a league: 'the width of four barley grains make a finger and four fingers, a hand, and four hands the length of a foot, five feet a geometric step or two simple steps and 1000 geometric steps, a mile and three miles a league'.[39]

It is also important to note that the problem of standardizing units and measures was not a pressing need for only cosmography and cartography. As we have seen in the previous chapters, it was a challenge in other fields of navigation, such as shipbuilding, determining the correct amount of ballast and cargo, and, of course, the control and oversight of trade.

## Three early maps of the New World

Sixteenth-century cartography has been widely studied, but a quick look at some of the most important maps made in the early stages of the Spanish exploration of the New World, closely connected with the *Casa de Contratación*, as well as personages such as Juan de la Cosa, Martin Waldseemüller, and Diego Ribero, will be sufficient to show the role of cartography as a form and appropriation and dominion.

### Juan de la Cosa (1500)

The first representation of the New World as part of a map of the known world was the famous map of Juan de la Cosa (1500). Most of the chart showed the Old World and its three main parts: Europe, Asia, and Africa. The novel feature of this parchment was an enormous green mass that appears in the upper part of the map. Not even a decade had passed since the return of Columbus from America, and the world was already markedly different.

On two hides glued together to form a rectangle, measuring 183 centimeters high and 96 centimeters wide, Juan de la Cosa assembled a map of the world that showed the cartographic knowledge of the 'old world' in detail and, for the first time, presented the American continent using the traditional parameters of portolan charts. Its importance lay in the incorporation of the geographical

38 García de Palacio, Diego, 'Instrucción náutica para el buen [...]', op, cit., Book II, Chapter IX, 'The rules for knowing how many leagues are sailed for each of the seven quarters', pages 63–65.

39 Escalante de Mendoza, Juan de, 'Itinerario de Navegación de los [...]', op. cit., Book II, dialogue, 'que trata de la altura [...]', chapter, 'the leagues to be navigated through each of the 32 courses that the needle represents to raise or lower a degree of height in navigation', page 114.

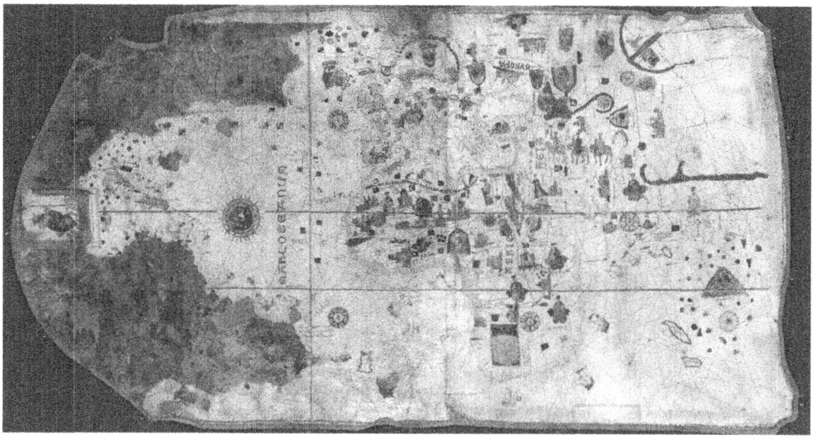

Illustration 5.6. Juan de la Cosa's Map, 1500.  Naval Museum, Madrid.

information that the most recent voyages of exploration had yielded and the fact that the already familiar depiction of the world was supplemented by another new and enormous part of the planet: a fourth continent.

One obvious purpose of this map was to determine the line that divides the world in two parts, from south to north, just as the 1494 Treaty of Tordesillas had stipulated. Since this map was a large title deed for the ownership of the world, it is not surprising that it was full of *naos* flying the flags of Castile and pennants[40] of Portugal.

The Ptolemaic paradigm for attaining a representation of the whole world was an effective method for incorporating new and unknown lands into familiar parameters. The chart of Juan de la Cosa and the maps that followed clearly expressed the process of the Christian incorporation and appropriation of the world.

On the chart, the New World was not only incorporated into a spatial frame of reference, but also as part of a political and religious order. As was characteristic of the Christian tradition, there were many Biblical references on the map. In the middle of the ocean, on the western part of the map, the key decoration was a compass rose adorned with a beautiful woodcut of the Virgin Mary, whose resemblance to the queen was noted by some, and the Christ Child.

Juan de la Cosa's map was usually reproduced and read in accordance with modern cartographic conventions, with the north on top, but most of the inscriptions were designed to be read with the map in a vertical position and the north on the right. Hence, the map was clearly in the shape of a

40 Triangular flags on the masts which indicate to whom the ship belongs.

Illustration 5.7. Compass rose with a picture of the Virgin Mary and Jesus at the center of the map by Juan de la Cosa, 1500.  Naval Museum, Madrid.

cross whose center was the compass rose in the Atlantic. In the upper part of the parchment, there was a depiction of Saint Christopher walking on the water, the three Wise Men or Magi, and the Tower of Babel, among other references to the Christian religion.

## Waldseemüller (1507)

Among the variety of world maps published in the 16$^{th}$ century, the planisphere of Martin Waldseemüller (1507) stands out. Closely related to the work of Amerigo Vespucci, the map was part of a pamphlet titled *Cosmographiae Introductio*.[41] The text that accompanied the map mentioned

41 See: Lester, Toby, *The Fourth Part of the World: The Race to the Ends of the Earth, and the Epic Story of the Map that Gave America its Name*, New York, Free Press, 2009.

Illustration 5.8. In the middle of the New World the figure of Saint Christopher crosses the seas, suggesting the idea of a passage to the Orient. Detail of the map by Juan de la Cosa, 1500. © All rights reserved. Naval Museum, Madrid.

the traditional division of the Earth into three parts—Europe, Asia, and Africa—but it also noted that recent explorations had brought news of a fourth part, and that this part was called the Land of Amerigo, or rather America, in line with the female names of the other three parts. Therefore, it was the first cartographic document that brandished the name of America and showed the new lands as a single geographical entity, independently

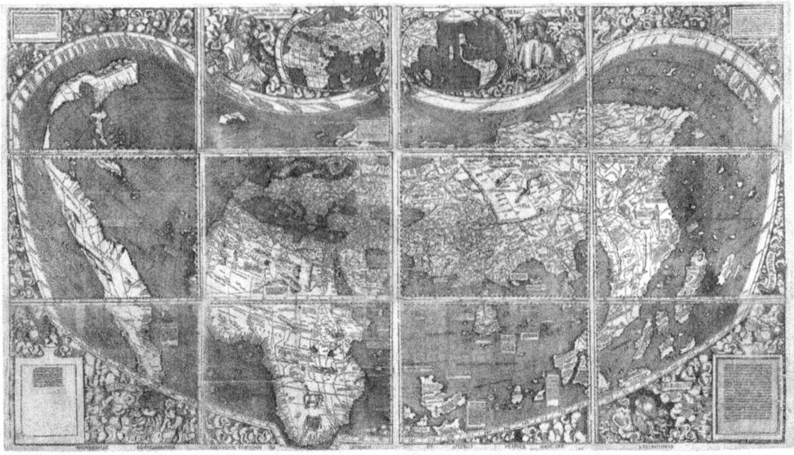

Illustration 5.9. *Universalis Cosmographie descriptio in plano*. In: *Introductio*, by Martin Waldseemüller, 1507.

of whether there was or was not a strait of sea between its northern and southern parts.

The *Terra Incognita* was thus termed America and incorporated both into the terraqueous globe and the history of the world, so that this map is not only a geographical but also an historical representation of the New World. In the upper part, the figures of Ptolemy, the great geographical authority of antiquity, and Vespucci, an allegorical figure of the New World, faced each other. This *mappa mundi* was therefore both a narrative and a celebration of the historic moment that marked the start of a new era, a new cosmos, and a new political order.

The text reminded its readers that the purpose of the publication was to present a spatial and political ordering of the world, demarcate the dominions of the Earth, and facilitate its exploration and conquest.

> The purpose of this little book is to write a description of the map of the world which we have designed as a globe and projection. The globe which I have designed is on a small scale, the map on a larger one. Just as farmers mark out and divide their lands with border lines, our purpose has been to mark out the world's main countries with the emblems of their rulers (and beginning with our continent) in the middle of Europe we have placed the eagles of the Roman Empire (governed by the monarchs of Europe) and with the key (which is the symbol of the Holy Father), we have framed almost all of Europe, which acknowledges the Roman Church. We have distinguished most of Africa and a part of the Asia with crescent moons,

> which are the emblems of the Sultan of Babylon, the prince of Egypt and part of Asia. We have surrounded the part of Asia which is known as Asia Minor with a saffron cross joined to a branding iron, which is the symbol of the Turkish sultans, and finally, on the fourth division of the earth discovered by the kings of Castile and Portugal, we have placed the emblems of those sovereigns.[42]

But, in addition to declaring dominion, the map was a navigational instrument: 'and what must be considered, we have marked the shallow places in the sea where there is a danger of shipwreck with crosses'.[43]

The texts and illustrations that accompanied these maps gave the territories they represented a place in time,[44] incorporating those without a history into the history of Europe. The inscription in the lower left corner read:

> A general delineation of the different lands and islands, some not mentioned by the ancients, later discovered between 1497 and 1504 on four sea voyages, two for Ferdinand of Castile, two for Manuel of Portugal, most serene monarchs, with Amerigo Vespucci as one of their navigators and navy officers.[45]

The introduction of Waldseemüller's map of the world made the power of a cartographic representation clear:

> [Those who are interested] will be able to satisfy their needs and be grateful to us for our labor when they see nearly the whole of what has been discovered or recently explored here and there, carefully and clearly put in a single place so that it may observed at a single glance.[46]

42 Waldseemüller, Martin, *Cosmographiae Introductio* [1507], in: Facsimile Followed by the Four Voyages of Amerigo Vespucci, Charles George Herbermann (ed.), New York, United States Catholic Historical Society, 1907.

43 Ibid.

44 The etymology of the word 'geography' is a reminder that that science is the daughter of writing. When we think of cartography today, we refer to maps, but we should not forget that cartography is an accessory and product of geographical literature. Maps are like visual images of verbal descriptions, so that they are not an independent and autonomous form of the geographical discourse but a complement or part of it. See: Zamora, Margarita, *Reading Columbus*, Berkeley, University of California Press, 1993, page 102.

45 Waldseemüller, Martin, 'Cosmographiae Introductio, in facsimile […]', op. cit.

46 Ibid.

One of the reasons why this map has been given so much importance is that it used the word 'America' to name this fourth part of the world for the first time. The new continent needed a name that would be in harmony with the other continents. Many names were used, but variety and ambiguity had to be eliminated to reach a consensus on a single name that would be 'universally' recognized. As the same pamphlet explained, the natives of Hispaniola called the land to the south of the island 'Bohio', while the Portuguese used the names 'Vera Cruz' and 'Terra de Santa Cruz', as initially mentioned by Cabral. Some cartographers used the name 'Brasil' but that led it to be confused with the name of an imaginary island in the Atlantic. 'Terra dei Pappagalli' (Land of Parrots) was a local name and only used for part of the continent. 'Parias' was a native name for a region near Trinidad. 'Nueva India' (New India) was regarded as inadequate, and 'Mundus Novus' (the New World) or 'Terra Incognita' (Unknown Land) were more descriptions than names. Therefore, it was very significant that, with the publication and later circulation of this cartographic material, the unknown world, this *terra incognita*, would be transformed into a world known as America, the land of Amerigo.[47]

There was a strong demand for this small book and its respective map, and it was widely circulated. In 1507 two editions were printed and more than a thousand copies were sold. Both the map and the texts were frequently reprinted, which helped the name of America to take root.[48]

## Diego Ribero (1520)

Today, not a single map with the name of the 'Padrón Real' is known. However, a map like the one of Diego Ribero might be an example of the charts that drew on the data compiled in the *Casa de Contratación* in Madrid. The purpose of this chart was evident from its title, which had the explicit aim of representing the whole of the world and how it was shared out between the Crowns of Spain and Portugal: Universal Chart, which contains all of the world that has been discovered up to now. Made by Diego Ribero, cosmographer of his majesty. Year 1520, Seville, which is divided into two parts in accordance with the capitulation [agreement] made by the catholic Kings of Spain and the king Don Juan of Portugal in Tordesillas, year 1494, Seville, 1529.

Although by 1520 the world was very different from that of the 15th century, such maps conserved the convention of portolan cartography. On

47 A detailed study of this map and its history is found in: Lester, Toby, 'The Fourth Part of the World [...]', op. cit.

48 Brown, Lloyd A., *The Story of Maps*, New York, Dover Publications, 1977, page 157.

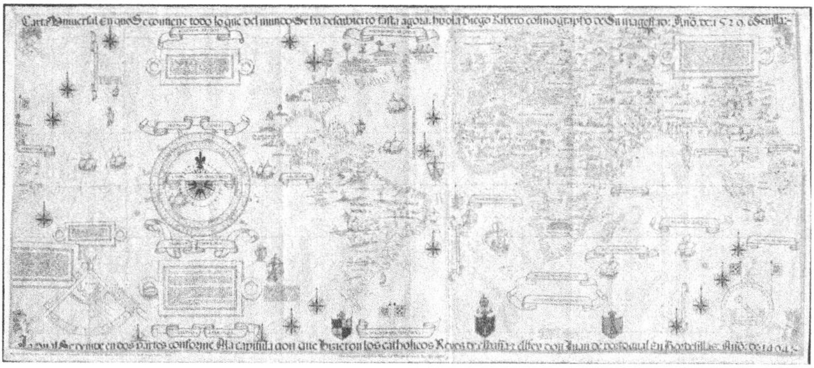

Illustration 5.10. Universal map containing all of the world that has been discovered up to now. Diego Ribero, Seville, 1529. The map is divided into two parts in accordance with Treaty of Tordesillas, 1594, in which the Catholic monarchs of Spain and King don Juan of Portugal divided the newly discovered lands outside of Europe between the two Kingdoms.  Naval Museum, Madrid.

the compass rose of Diego Ribero's map, the lines of sea routes converged in the manner of networks that seemed to cover the whole world, and the main purpose of the chart was still to define the demarcation line between the dominions of Spain and Portugal. The use of the classical astronomical coordinates was also plain: north in the upper part, the lines indicating the equator, and the Tropic of Cancer and the Tropic of Capricorn. One aspect that is worth noting and that formed a contrast with the chart of Juan de la Cosa, published 30 years before, was the wealth of place names, not only in the Old World but also in America, particularly the eastern coast of the continent. Its ornate decoration is also worthy of attention. The depiction of many European ships on the length and breadth of the globe show the presence of Christians in nearly the whole world. The illustrations of astronomical and navigational instruments, such as the quadrant, astrolabe, and the same compass rose, ratified them as symbols of the authority and precision invested in the making of the map.

There is no better way to organize the portrayal of such vast stretches of land and so much geographical information than on a map. Maps are the most effective expressions of how to capture time and space in a simple and elegant flat representation. Thus, the great legacy of this geometrical conception of the world is the possibility of placing it on a sheet of paper. If we reflect on the ideas of historians of cartography such as J.B. Harley,[49]

49 Harley, John Brian., *La nueva naturaleza de los mapas. Ensayos sobre la historia de la cartografía*, Leticia García Cortes and Juan Carlos Rodríguez (trans.), Mexico, Fondo de Cultura Económica, 2005.

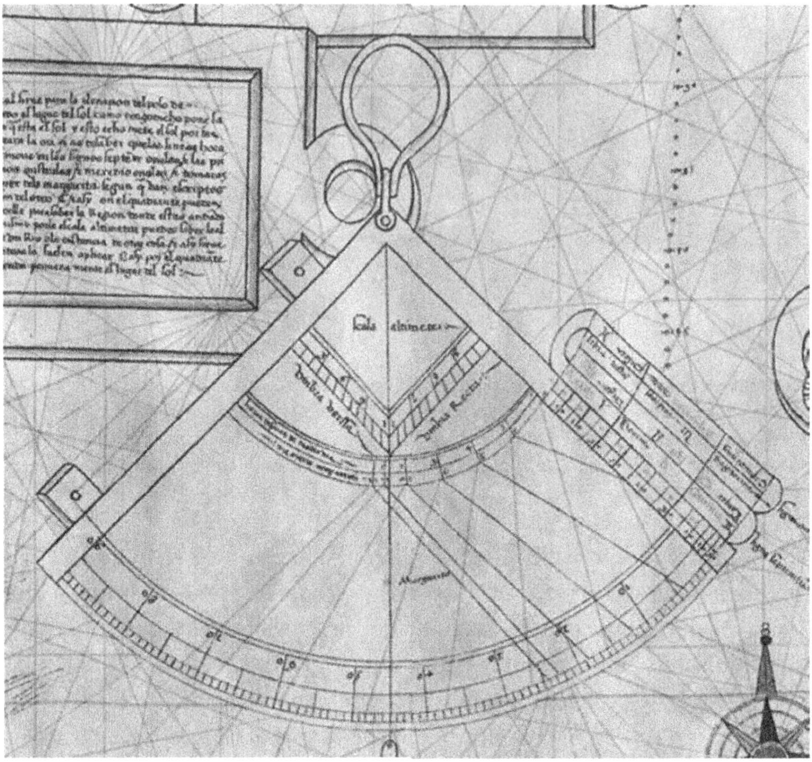

Illustration 5.11. Quadrant. Detail from the Carta Universal of Diego Ribero, 1529. Observational instruments were an important subject on the map. Here, we see a drawing that shows one how to use the quadrant.  Naval Museum, Madrid.

who understand maps not as simple representations of what there is but as ways to construct an order, we can better understand the political meaning of cartography in the imperial context. This possibility of visualizing large stretches of land and sea on a plane, or even the whole world, made the idea of global control plausible.

While it may be that most maps are copies of other maps, the information that allowed for the representation of the New World came from voyages and explorations, hence the concern for a suitable training of the pilots who brought useful information to Seville. This was a great technical challenge: to mobilize the information, draw, and control at a distance. The tension between pilots and cartographers and between sailors and cartographers on land revealed a very interesting epistemological problem. The solution was to create a form of observing and compiling information that could be summarized and incorporated in a single pattern. To achieve that, the instruments and units of measurement had to be standardized, the pilots

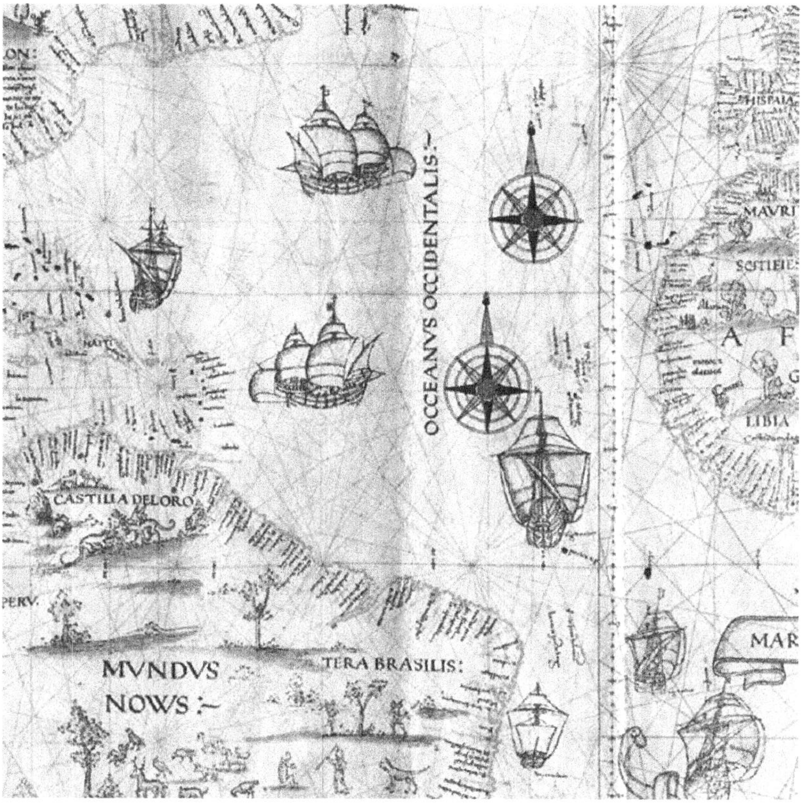


Illustration 5.12. Toponymy of the New World. Detail of the *Carta Universal* of Diego Ribero, 1529. Along the length and breadth of the map there are finely drawn figures of ships, as well as a host of place names drawn on the New World and Africa. All of it gives us an idea of the world conquered by the European explorers.  Naval Museum, Madrid.

and explorers disciplined and trained, and, of course, you had to make sure they came back from their travels.

The project of making a complete map of the world, the *Padrón Real*, implied a technical and scientific endeavor of enormous proportions in which many actors were added together and linked. Far from being an individual achievement, the creation of a map of the world was the work of powerful institutions that systematized the experience of many people within a common framework of reference. This is precisely one of the essential characteristics of modern science, and for that reason this episode of history should be included in any narrative that seeks to give an account of the birth of science.

# 6. The creatures of God never seen before: natural history

*'Yu-ana es una manera de sierpe de cuatro pies, muy espantosa de ver y muy buena de comer'.*[1]

**Abstract**

Chapter 6 journeys through the main problems of American natural history. Once Spanish travelers had overcome the great feat of crossing the Atlantic, they encountered an alien nature, which, through an arduous process of description and classification, they transformed to align with their domestic world. Novelties such as the armadillo, iguana, or the pineapple, among many others, challenged the authorities and obliged the sixteenth-century naturalists to rethink natural history.

**Key words**: Natural History, Local Knowledge, New World, Translation, Botany

## Nature in the New World

The conquest of the New World was not completed at sea. Arriving at the islands or coasts of the mainland of the Western Indies was the start of a new challenge. After the solitude of the sea and confinement of the ships, the travelers then had to confront lands of an undreamt expanse and an exuberant nature that was beautiful and hostile at the same time. As we have seen, the transatlantic route was a challenge full of obstacles, but once on land the Christians had to

1 'Yu-ana is a kind of serpent with four feet, horrible to look at but very good to eat'. Fernández de Oviedo, Gonzalo, *Sumario de la Natural Historia de las Indias*, [1526], Manuel Ballesteros Gaibrois (ed.), Madrid, Dastin, 2002, Chapter LVI, page 143.

Nieto Olarte, M., *Exploration, Religion, and Empire in the Sixteenth-century Ibero-Atlantic World: A New Perspective on the History of Modern Science*. Taylor & Francis Group, 2022
DOI 10.5117/9789463725316_CH06

deal with the resistance of the natives who occupied those territories and the conquest of America turned into a violent invasion. The penetration of the continent entailed sailing along unknown and complex rivers, long journeys of exploration in tropical climates, and interminable treks through jungles, deserts, and long mountain chains. Despite the obvious natural wealth of the New World, Europeans lacked the know how that was needed to navigate the torrential rivers of America, to find food, fight off animals, and protect themselves from tropical climates. On land, many explorers, obsessed with gold, silver, and other riches of the New World, lost their lives or their minds.

The explorer needed to acquire new kinds of knowledge in order to survive, but he had a mission that was even more difficult: to take the riches of the New World to Europe. In America, Christians came across a part of the Creation for which there were no testimonies and a nature that could be only dominated with very hard work. Controlling the New World from Europe thus required a systematic gathering and arrangement of information about coastal access routes and each object that made up the natural world: the rivers, mountains, settlements and their inhabitants, the sources of valuable minerals, and the plants and animals.

In his *Historia general y natural de las Indias*, Gonzalo Fernández de Oviedo[2] clearly described the challenge that the extensive and varied natural world of America implied for Europeans:

2 He grew up close to the courtly world, being a page of Prince Juan. In 1497, after the premature death of the Prince, he lost his place at court and was forced to travel to Italy, where he came into contact with the Italian Renaissance while he served as a scribe to figures such as Ludovico Sforza, the Duchess of Mantua, and Cardinal Cesare Borgia. Oviedo began his career in the New World in 1514, where he remained for the rest of his life, except for some short visits to Spain. Having been appointed the supervisor of gold smelting in Tierra Firme, he embarked from Spain on April 11, 1514, and, after some stops in the Canary Islands and Santo Domingo, reached Santa María de la Antigua on June 12 of the same year. His first activities on the Spanish Main had to do with capturing and branding *indios* for his friend Cochinillos and reading them the *Requerimiento* of Pope Alexander VI, which ordered them to turn into subjects of the King and convert to the Catholic religion. In 1526 King Charles commissioned him to write all that he knew of the Indies. As a result, he wrote his *De la historia natural de las Indias*, known as the *Sumario*, meant as a preliminary version of a much more complete account. In 1532, he was officially appointed to the post of royal chronicler, with an annual salary of 30,000 *maravedíes*. Since his assignment to this new post meant he had to write about everything that was happening in the Indies, Oviedo had the power to ask the colonial officials in the Indies to provide him with the available information on lands, animals, and events in the new continent. In 1535, he published, in Seville, the first part of his *Historia general y natural de las Indias Occidentales*, which comprised 19 books. This book was printed at the presses of Juan Cromberg, the former 'engraver' of the *Sumario*, and was dedicated to the then President of the Consejo de Indias, Archbishop don Fray García de Loaysa. In 1549, with the second and third part of the *Historia* in his hands, Oviedo returned to Santo Domingo on his last ocean voyage. He later wrote other

> What moral ingenuity will know how to comprehend such a diversity of lakes, habits, customs of the men of these Indies? Such a variety of animals, both domesticated and wild? Such an indescribable multitude of trees, full of diverse kinds of [edible] fruits, and others sterile, both those which the *indios* cultivate and those which Nature, on its own initiative, produces without the help of mortal hands? How many plants and herbs useful and advantageous for mankind? How many others, innumerable, which are not known to him, and with so many different kinds of roses and flowers and scented fragrances? Such a diversity of birds of prey and other sorts? How many very high and fertile mountains and other so different from each other and wild? How many meadows and grasslands apt for agriculture and with very suitable riversides? How many mountains which are more impressive and fearful than Etna or Mongibello, and Vulcano and Estrongol; and those and others under your monarchy?[3]

Attaining a knowledge of the different cultures, animals, and plants of the New World represented a major challenge for the Europeans of the 16$^{th}$ century. In the dedication to the King that opened his *Historia General de las Indias*, López de Gómara justified calling the Indias Occidentales the 'New World' by saying:

> They not only call it new for being newly found, insofar as it is very vast and almost as big as the old one, which contains Europe, Africa, and Asia. It may also be called new because all of its things are so very different from those of our own. The animals in general, while they are of few species, are of another kind; the fishes of the waters, the birds of the air, the trees, fruits, plants and grains of the earth, which is no small mark of esteem by the Creator, since the elements are a single thing both there and here.[4]

From a very early stage, the sixteenth-century accounts of the New World were already furnishing some of the emblematic features of the Christian view of the American continent, which was not only described as an

books, until his death on June 26, 1557, the year that also saw the publication of Book XX of the first part of the *Historia*. In theory, that should have been the first book of the second part of the *Historia* and should have been accompanied by a new edition of the *Sumario*. It was only three centuries later that a complete edition of the three parts of the *Historia general y natural de las Indias* was published, thanks to José Amador de los Ríos (1851–1855).

3 Fernández de Oviedo, Gonzalo, 'Historia general y natural de [...]', op. cit., page 8.

4 López de Gómara, Francisco. 'Historia General de las Indias', op. cit., 'To Don Carlos, Emperor of the Romans, King of Spain, Lord of the Indies and the New World', page 5.

inexhaustible source of wealth, but also a world inhabited by savage beings, cannibalistic and idolatrous, as well as all kinds of fantastic creatures.

Describing the islands of the Caribbean in his *Summa de Geographia*, Enciso wrote:

> They are all cannibals, who eat human flesh and go to the sea in canoes to make war in other places, some against others; and all they can capture they take to their lands, and if they are men, they eat them, and if they are women, they use them as slaves; and if some man of the islands is skinny, the fatten him, and when he is fat, they eat him and say that the most delicious parts are the fingers of the hands and the thin part of the flanks of a man.[5]

The riches of the Indias Occidentales were the subject of a constant and often exaggerated praise. There were frequent references to the ostensible abundance of gold to be found there. In his description of Hispaniola, for example, Enciso wrote: 'They find much gold there'.[6] But that wealth did not only consist of precious materials such as gold or pearls, its flora and fauna were also a kind of natural El Dorado and were often described with a paradisiacal flourish: 'It is a land of many kinds of meat and much fresh fish. Cattle multiply a great deal, because there is no animal which causes harm to any of them. The cows and mares give birth every year [...] the grass is always green and never dries; the foliage of the trees is constantly green, as happens here in May and June'.[7] The *Summa de Geographia* included descriptions of some animals, such as 'tigers and lions', 'lizards', and the 'iguana'.[8]

It is also interesting to note how the early narratives of the American cultures were already marked by a moral tension about the conquest. While the Americans were regarded as savage peoples, a condition that justified the imposition of the Christian doctrine, among other things, the problem that arose from the invasion of the land of others was also evident.

Enciso wrote an account of an incident when, following the royal instructions to read out the *Requerimiento*[9]—the 'Requirement', to be read aloud

5 Fernández de Enciso, Martín, 'Suma de Geographia que trata [...]', op. cit., 'cannibals', Folio LII.

6 Ibid.

7 Ibid.

8 Ibid., Folio LVII.

9 The *Requerimiento*, the Spanish 'Requirement' of 1513 (also known as the *Requerimiento de Palacio Rubio*), was a declaration by the Spanish Crown, authorized by the Laws of Burgos,

to the conquered Indians as a warning to accept Spanish rule and Christian preaching or risk subjugation, enslavement, and death—he explained to the *indios* of the Sinú that God had come to the world and left Saint Peter in his place; that the successor of the latter was the Holy Father, and that the Holy Father, Lord of the Universe, had granted the King of Castile the title to all the land of the Indies and of this peninsula. The *caciques* (chiefs) responded that the Holy Father had been very generous with the property of others and must have been drunk when he gave away what did not belong to him and that the King of Castile was impudent to threaten those he did not know.[10] If we put a special stress on the story told by Fernández de Enciso, in which the natives call the Pope and the King of Spain thieves and drunks, it is not so much for the veracity of this specific incident as the way it clearly revealed the profound contradictions of the Christian conquest—an aspect that the sixteenth-century chroniclers of the New World were not unaware of, of course.

In the middle of the 16th century, two new literary genres arose in Spain that were incorporated into its cosmographical and historical traditions: *natural history* and *moral history*. They were two sides of the same coin and had a single purpose, although they were usually divided into two sections. Natural history was devoted to geography, animals, plants, and minerals, while moral history dealt with humans and encounters between cultures. While for many of the writers about the New World, moral history and natural history were not two separate fields of knowledge, the following sections focus on the realm of nature and not so much on that of human beings.

Drawing on the works of Gonzalo Fernández de Oviedo, Francisco Hernández, José de Acosta, Bernardino de Sahagún, and Nicolás Monardes, possibly the most important chroniclers and naturalists of the 16th century, we will try to explain how describing, classifying, naming, and using a nature that was strange and at times monstrous enabled it to be domesticated. In short, the idea is to show how nature was turned into an instrument at the service of Europeans within the framework of an imperial, Christian order.

The lives and writings of these travelers and chroniclers had some common features that are worth recalling. The world into which the chroniclers

of Spain's divinely ordained right to conquer the lands of the New World and subjugate those indigenous peoples who refused to be evangelized. By means of this declaration, preached in Spanish in a standard form, the conquistadores had to inform the indigenous people that God, the Creator of the first men, had chosen Saint Peter and his successors in Rome as monarchs of the world, whose authority was higher than all the princes on Earth.

10 Fernández de Enciso, 'Suma de Geographia que trata [...]', op. cit., page 250.

were born rapidly expanded to the west Atlantic and the Orient, thanks to the expeditions of the Spanish and Portuguese. The discovery of Tierra Firme, Patagonia, the Strait of Magellan, the Amazon, and the Indian Ocean had forced Europe to understand the world in a different way and the chroniclers were witnesses of these discoveries. With a bit of imagination, one can recreate the scene the young Gonzalo Fernández de Oviedo witnessed in 1493, when Christopher Columbus led a triumphal procession along the streets of Barcelona after his first encounter with the New World. Oviedo, who was then twelve and, as a pageboy of Prince Juan, close to the world of the court, might well have had access to the information and rumors about the achievements of Columbus and his fantastic discoveries. When Columbus marched through Barcelona, he did so with the natives he had captured, parrots, and samples of gold, so that Oviedo, like the other spectators, must have felt an enormous curiosity about what they were seeing. A daring navigator, in the name of the Crown, had led three ships to places never before visited and had returned with the news and samples of an incredible world: human beings, animals, and riches beyond the grasp of the imagination.

The discovery of American nature forced Europeans to rethink the canons of natural history, medicine, military strategy, and trade, among other things, and the constant incorporation of the classics of the ancient world into the thought of the Italian Renaissance helped to renew humanism and encourage the appearance of new practices and bodies of knowledge, which were spread, as never before, by the printing press.

The books of the Iberian naturalists combined factors that are key for an understanding of the sixteenth-century Spaniard. On the one hand, their writings were inextricably linked to the Crown and formed part of the great imperial project of the Habsburg monarchy. On the other, the Iberian Catholics looked at the world from a religious standpoint and regarded themselves as the protagonists of a historic moment in the expansion of their faith.

Some of them belonged to religious communities and had a mission that was directly related to the evangelization and conversion of the natives to Christianity. But, even when they were not members of the clergy, the religious question was fundamental for them all. The Church, Christianity and the monarchy were part of a single source of authority that they mutually strengthened. Since trade was part of the imperial project, these books showed a clear interest in it as well. Finally, the exploitation of the resources of the New World was their justification for a detailed knowledge of its nature there.

These coincidences were also related to the kind of education those sixteenth-century travelers had received. Several of them were trained as doctors, naturalists, and explorers at the Universities of Alcalá de Henares and Salamanca, which not only taught the medical tradition of Galen, but, before they studied medicine or any other career, also required undergraduates to learn arts, philosophy, rhetoric, theology, and other bodies of knowledge of the humanist school. This was a crucial aspect, because it clearly motivated those travelers to undertake their work in the name of God, the King, and the Pope, which meant that they were aware of the need to strengthen the Catholic moral dogmas of the Empire. Alcalá de Henares was likewise an important center for the compilation and study of the works of Aristotle, Pliny, and Dioscorides, among other outstanding Greco-Roman students of nature. Those authors were the main referents for interpreting the American landscape and fitting it into the familiar schemes of the Old Word, or, in extreme cases, coming up with the idea of American savagery, as compared to the civilized world of Europe, a notion that began to emerge with the discovery of America.

This process of understanding the New World and translating it into European terms also required the publication of a large body of writing that had to be placed at the service of the king of Spain, and doubtless served as a justification for the delegation of the work of chroniclers and naturalists to many important political figures of the period. Their activities came to be part of the political structure of the Spanish monarchy, whether they served as doctors or informants of the king. The travelers of the 16$^{th}$ century, together with their assessments, catalogues, books, drawings, and cognitive schemes, were largely responsible for ensuring that America would be interpreted within frameworks of understanding that were familiar to the Old World. The final result of this whole process and the total of rhetorical and technical practices was that the new and the strange became domesticated and familiar, and all of those marvelous creatures were stripped of their enchantment.[11] Books like those of Oviedo, Acosta, Hernández, and many other chroniclers and naturalists in the Indies helped to build the idea of a universal empire.

To explain the work of the Christian naturalists of the 16$^{th}$ century, and understand this process of translating and comprehending the strange, we have to go back to the Euro-Christian frameworks of reference characteristic of the humanism of the 15$^{th}$ and 16$^{th}$ centuries. At the precise moment when the Iberians seemed to have the world at their feet, their knowledge and

11 Asúa, Miguel de and French, Roger, 'A New World of Animals [...]', op. cit., page 229.

beliefs were confronted by a world that was unknown, and that (though not for long) seemed to defy the common frameworks of reference they employed to understand it. Dominating nature is not a military problem, and therefore requires the use of instruments very different from metal swords or firearms. The naturalists of the 16th century formed part of a humanist tradition and their models of interpretation were closely related to ancient and medieval texts and the great classics of geography, medicine, and natural history, which, while they turned out to be insufficient for understanding the New World, in some manner were the starting point for the first attempts to recognize the strange.

## The classics and the Bible

The frameworks of reference that the earliest European travelers who went to the American continent relied on rigorously adhered to the ancient and medieval traditions of natural history. The work of Aristotle continued to be a key referent in cosmology and natural history, while the most direct referent for the naturalists of the 16th century was the Roman author Pliny the Younger's *Natural History*. Other frequent referents for the natural histories of the 16th century included the works of Dioscorides, Claudius Aelianius, and Theophrastus, along with texts, such as Isidore of Seville's *The Etymologies*, that were the foundation of the medieval bestiaries. The accounts of travelers to distant places were also widely read, such as the *Imago Mundi*, by Pierre d'Ailly; Marco Polo's *The Book of Wonders*; and John Mandeville's *The Book of Marvels and Travels*. Those treatises, published before the discovery of America, not only described creatures known to Europeans—African and Asian animals such as giraffes, rhinoceroses, lions, tigers, and elephants—but also fantastic creatures such as cyclopes, sirens, Amazons, and men with the head of a dog, among many others. When the Iberian exploration of the New World began at the end of the 15th century, these 'fantastic creatures' were believed to be real in the same way as other exotic animals such as the rhinoceros or the giraffe. The explorers of a new route to the Orient were prepared to come across lions, camels, giraffes, unicorns, and sirens, but they came up against something even more strange: animals that had never been seen before or described by any previous traveler or writer.

Thus, European travelers in unknown lands often wound up believing in the existence of and even 'seeing' creatures that were already familiar in Europe and in medieval literature. For example, in his *Diario de abordo*

(Captain's log), Columbus wrote: 'Yesterday, when the Admiral went to the Río del Oro, he said he saw three sirens which had come from far out at sea, but they were not as beautiful as they are depicted, that in some way their face had the shape of a man's'.[12]

Thus, in America, the first travelers 'saw' the exotic creatures that filled European literature. Another excerpt from the diaries of Columbus refers to the existence of cyclopes and cynocephali (dog-headed men) in the Indies: 'He also understood that far from there, there were men with one eye and others with the snouts of dogs who ate men and when they caught one, they cut his throat and drank his blood and cut off his genitals'.[13] Those kinds of descriptions show that the observer's own tradition and expectations determined what he might 'see'. Columbus came across a new type of marine animal, probably a manatee, and before considering the possibility that he might have been facing something unknown and strange, 'saw' a siren. Although its appearance was not as beautiful as he had been led to believe, the idea of encountering a sea nymph seemed more real and familiar than coming across an aquatic mammal for which he had no reference at all. While he admitted that he had not seen some things himself, he trusted in the accounts of others and was sure that there were cynocephali, cyclopes, cannibals, Amazons, and other creatures from medieval literature in the newly discovered lands.

The permanent references to the classics are a key to understanding this process of 'comprehension' since they were a necessary means for incorporating the unknown into familiar frameworks of reference. At the same time, however, there were often cases that contradicted the classical authorities or that clearly showed their limitations. The 'encounter' with the New World implied a clear break with classical geography, but America challenged the authority of the ancients not only in the field of geography, but natural history as well. Committed to an ambitious commercial and political project of world conquest and confronted with a natural setting and geography that were unknown to Europe, the Iberian naturalists and explorers never abandoned their Biblical and classical referents, despite their obvious limitations. The creatures of the New World forced Christian Europeans to create a new form of understanding nature. Its legitimacy was no longer exclusively based on the erudition of citing classical authors and writings, and it began to depend on a direct and personal experience of nature.

12 Colón, Cristóbal (Columbus, Christopher) 'Diario de a bordo' (Captain's Log), op. cit., January 9, 1493, page 206.

13 Ibid., November 4, 1492, page 132.

On several occasions, Oviedo stressed the absence of several New World animals from classical literature:

> The manatee is one of the most remarkable fishes, unheard of among the many I have read of or seen. Neither Pliny spoke of it, nor did Albertus Magnus write of it in his *Proprietatibus Rerum*, nor are there any in Spain. I have never heard a man of the sea say that he has seen or heard of them except in these islands and mainland of these Indies of Spain.[14]

The great treatises of natural history and encyclopedias of antiquity and the classical works written by Aristotle, Pliny, Dioscorides, and Ptolemy were the referents that were available for understanding the new but, in the face of what was being found in the New World, they proved to be insufficient. It was obvious that their universal nature was limited because they completely ignored a part of the world and its creatures.

The chroniclers' relation with the classics was not limited to making frequent references to specific animals or plants, since they continued to follow the classical thinkers in a deeper sense as they tried to become their successors. Oviedo insisted that his work sought to emulate Pliny: 'because, in some manner, I understand that I am following or imitating Pliny'.[15] We should not forget that Pliny dedicated his work to Titus, the future successor of his father the Emperor Vespasian, and he wrote it with the clear aim of serving the interests of the Emperor and celebrating the triumphs of the Roman Empire. Oviedo, in turn, dedicated his work to Charles V, whom he referred to as 'Cesar' and, as the other chroniclers did, rendered a great homage to the Empire of the Habsburgs. Celebrating the Christian empire was a common feature of all those who wrote about the New World. In the letter in which he presented his book, Oviedo remarked:

> From their fruits, it would seem that Our Lord has increased the victories of your Caesarian majesty and every day they grow more, and it seems that even the things of the Indies and their riches grow in your shade: the service of God flourishes in them; the republics of Christians there are ennobled; the realms of Spain are enriched, and everything goes from strength to strength.[16]

14 Fernández de Oviedo, Gonzalo, 'Historia general y natural de [...]', op. cit., vol. 2. Book XIII, Chapter IX, 'On the manatee and its big size and shape, and the way in which the Indians sometimes confused this big animal with a fish, and other peculiarities of it', page 63.

15 Ibid., 'General y Natural Historia de las Indias', page 11.

16 Ibid., page 5.

Oviedo used Pliny's *Natural History* in several ways. The Roman author adorned and enriched his work and granted him a certain erudition and authority to discuss the natural world, and, at times, Oviedo depended on him to describe some creatures and in that way, created a link between and showed a certain continuity in the animals and plants of the whole world.[17] However, Oviedo frequently paid special attention to those animals that do not appear in the writings of Pliny since he could thus point to the originality not only of the New World but also his own work.

Pliny was even more crucial for the work of Francisco Hernández,[18] who devoted part of his life to studying and translating the Roman author. The

17 Asúa, Miguel de and French, Roger, 'A New World of Animals [...]', op. cit., page 71.

18 Francisco Hernández (1515–1587) was born in Puebla de Montalbán, in the Kingdom of Toledo. He obtained a bachelor's degree in medicine from the Universidad de Alcalá de Henares in 1536 and a doctorate in 1539. Between 1570 and 1577 he traveled through the American continent, especially New Spain, where he not only worked in the field of medicine–public sanitation, but also as a cosmographer, a collector of plants and natural flora with medicinal properties, and a chronicler of the culture, religion, history, customs, and spiritual practices of the pre-Colombian societies of Mexico. After appointing him as chief physician of the Indies, King Philip II ordered him to sail for America in 1570, evidence not only of his commitment to medicine but also his closeness to the center of power and the political skills that had previously led to his appointment as 'Doctor to the Royal House of his Majesty'. Hernández wrote many works, including a Spanish translation of Pliny's *Natural History*. His most important book was the *Historia Natural de las Cosas de la Nueva España,* which he was not able to publish in full due to his premature death when he returned to Europe. He began to write this book during his stay in Mexico (between 1573 and 1576), which displayed ample proof of his knowledge of American plants, plant remedies, and antidotes and the diseases they served. His description of the natural and medicinal world of the Indies included a detailed analysis of the forms of animals and plants—the structure of their roots, branches, flowers, and fruits, among the most important parts—and he also spoke of how these findings were related to or independent of Pliny's *Natural History*. In general, his writings, which he wanted to be published in Latin, were made up of at least three thousand chapters, most of them devoted to the study of plants and animals, and, to a lesser extent, mining. The manuscripts of his works were sent to the library at El Escorial, where they remained unorganized, despite the fact that the Neapolitan doctor Nardo Antonio Recchi had been commissioned to put them into order in 1580, though it was largely the fault of Hernández's intention to follow the arrangement of the *Materia Medica* of Dioscorides. Thus, until 1671, when a large part of the unpublished documents was lost in a fire, there was not a single copy of a coherent, edited book. Even so, the information in the *Historia Natural de las Cosas de la Nueva España* was of great interest to foreign scholars and its influence was seen in a number of later studies of American natural life. One example was the *Historia Naturalis Brasiliae*, by Willem Piso, published in Amsterdam in 1648, which included descriptions of plants and animals similar to the ones Hernández found in Mexico. The same approach was seen in the studies done in the Dutch colonies of Brazil by the gardener Johan Maurits van Nassau (1604–1679), the doctor Willen Piso (1611–1678) and, especially, the naturalist Georg Marcgraf (1611–1643), who clearly owed much to the traditions and bodies of practical knowledge compiled by Hernández in New Spain. That shows how his work was

efforts of Hernández were complemented by those of Jerónimo de la Huerta, who, with the support of Philip II of Spain, published an annotated edition in Spanish of Pliny's *Natural History*. All the sixteenth-century naturalists made frequent references to that translation. Pliny continued to be widely admired and followed with respect, but it now became possible to criticize him.

Similarly, and even more worrying for the Christian World, the Holy Scriptures and the credos of Christianity were unable to explain the origin and meaning of a large part of creation. The classical authors, humanist erudition, theology, art, and literature had to confront a new reality about which the major sources of authority were silent.

Vespucci had already remarked on the difficulty of explaining the striking variety of creatures in the New World by referring to the Biblical story of Noah's Ark. In 1590, Father José de Acosta[19] clearly set forth the difficult question concerning the origin of American creatures. The title of Chapter XXXVI of the fourth book of his *Historia natural y moral de las Indias* was: 'How Is It Possible That the Indies Have Animals Which Are Not Found in Other Parts of the World'. He then addressed this alarming question in depth, trying to show that it was not only difficult to find an answer in the

acknowledged, influential, and accepted in places different from the Spanish colonies of the 16th century. In the end, however, the Spanish Crown was responsible for publishing the most complete edition of his *Natural History* when, in 1767, it commissioned Casimiro Gómez Ortega, director of the Botanical Garden of Madrid, to publish a compressed, three-volume edition of his medical, anatomical and botanical writings from manuscripts still conserved in the library of the *Colegio* run by the royal chronicler Juan Bautista Muñoz. Despite the efforts of Gómez Ortega, this edition was far from complete and lacked illustrations, which was later remedied by the compilation of a complete catalogue of the *Historia Natural de las Cosas de la Nueva España*.

19 Jose de Acosta (1540–1600) was a descendant of Jews and member of a merchant family. In 1551 he joined the Jesuit Order in his birthplace, Medina del Campo, and began a life of constant study, both as a student and as a professor. His religious vocation, which was followed by seven of his nine brothers, led him to make his novitiate in Salamanca in 1552 and take holy orders in 1557. Between 1568 and 1569, Acosta wrote to Father Borja, third General of the Company of Jesus, saying that he wished 'to be sent to the Indies among barbaric people to work without caring what people think or whether they honor me'. In response to his request, he was sent to the Vice-Royalty of Peru in 1571 in the armada of Pedro Méndez Avilés, on the Jesuits' third mission to the Indies. On September 1 of that year, he was appointed as rector of the Colegio de Lima and on January 1, 1573, Provincial Superior of the Company in Peru. After spending about a year in New Spain, he returned to Spain. By then, he had finished his Latin treatise *De Natura Novi Orbis* and had assembled the material for his new book on natural history. His most important book was published in 1590, under the title of the *Historia natural y moral de las Indias*.

naturalists of antiquity and the Renaissance, but it also raised problems about the principles of Biblical authority:

> It is very difficult to determine the origin of diverse animals which are found in the Indies and not found in our world. Because if the Creator created them there, there is no need to resort to Noah's Ark, not even if it were a matter of saving all of the species of birds and animals which existed then, if they later had to be raised anew; nor does it seem that with the creation of the world in six days, God would have left the world finished and perfect, if there was a lack of new species of animals to form, most of which are perfect animals and no less excellent than those which are known. But if we were to say that all of these species of animals which were conserved in Noah's Ark, living on like those other animals, went to the Indies from this world here, as well as these, which are not found in other parts of the world, and, that being so, I ask myself how was it that none of their species remained here? How is it that they are only found [in places] which are strange and foreign? The truth is that the question has perplexed me for a long time.[20]

While it is true that Oviedo, and to a certain extent, Hernández, emulated the work of Pliny and followed it as a model, José de Acosta, by contrast, based his studies of the New World on the work of Aristotle. His philosophical bent and inquiries into causes were clearly influenced by that Greek philosopher, even though, like Oviedo, he did not hesitate to criticize his maestro for his inability to explain American nature.[21]

The result of this challenge was the invention of a new natural history, one based on a different authority that would enable its followers to incorporate, transform, and explain that new world as part of the Christian universe, and which was not done without difficulties and paradoxes. As we shall see, the descriptions of the European naturalists revealed more about the dreams and fears of the Christian world and the culture of the Renaissance than the creatures of the New World.

In the following section, we will show how the strange and the unknown were incorporated into the classical frameworks of reference and explain the different strategies and forms of domesticating the natural world of America employed by the Christians of the 16th century.

20 Acosta, José de, *Historia natural y moral de las Indias,* [1590], José Alcina Franch (ed.), Madrid, Dastin Historia, 2003, Book IV, Chapter XXXVI, page 282.

21 Asúa, Miguel de and French, Roger, 'A New World of Animals […]', op. cit., pages 79–80.

## Monsters in paradise

To understand the sixteenth-century chroniclers and their comprehension of American nature, we must consider their astonishment, their proclamation of genuine discoveries, and the novelty and fear they felt in the face of a strange nature. Europe's fascination with the New World was nourished by marvelous accounts of fantastic creatures and terrifying beings, so that American nature became an inexhaustible source of extraordinary creatures. Largely as a result of the discovery of the Indias Occidentales, sixteenth-century Europe was marked by a fascination with the exotic and the new. In the face of the unknown, judgments of an aesthetic character[22] are a factor that must be taken into consideration. Just as something that is beautiful will always be something that is familiar and close, that which is unknown will always cause fear and be seen as monstrous. As we shall see, this monstrousness was a common denominator of descriptions of the new. The opossum, anteater, manatee, armadillo, and iguana, among other animals, were initially described as horrible creatures.

One of the animals which most struck the attention of travelers was the iguana. For a person who had never seen such an animal, which was true of the European reader of the 16th century, its description was not only incredible but terrifying. Amerigo Vespucci wrote of this fascinating creature as follows:

> where we saw that they were roasting a certain animal which resembled a serpent, except that it did not have wings and so ugly looking we were greatly amazed by its deformity. We thus walked by their houses or rather, huts and came across many of these serpents, who were alive and tied by the feet and had a string round the snout so that they could not open the mouth, as is done with bulldogs so they cannot bite you. They looked so fierce that none of us dared to touch them, thinking they were poisonous. They are the size of a little goat and a *braza* and a half long [a *braza* equals 167 centimeters]; they have long, thick feet, armed with strong nails. They have a hard skin and are of different colors. Their snout and face are those of a serpent and a crest issues from their nose which is like a saw, which

22 A suggestive analysis of the literary description of American Nature is found in: Cabarcas, Hernando, *Bestiario del Nuevo Reino de Granada. La imaginación animalística medieval y la descripción literaria de la naturaleza americana*, Bogotá, Instituto Caro y Cuervo / Colcultura, 1994.

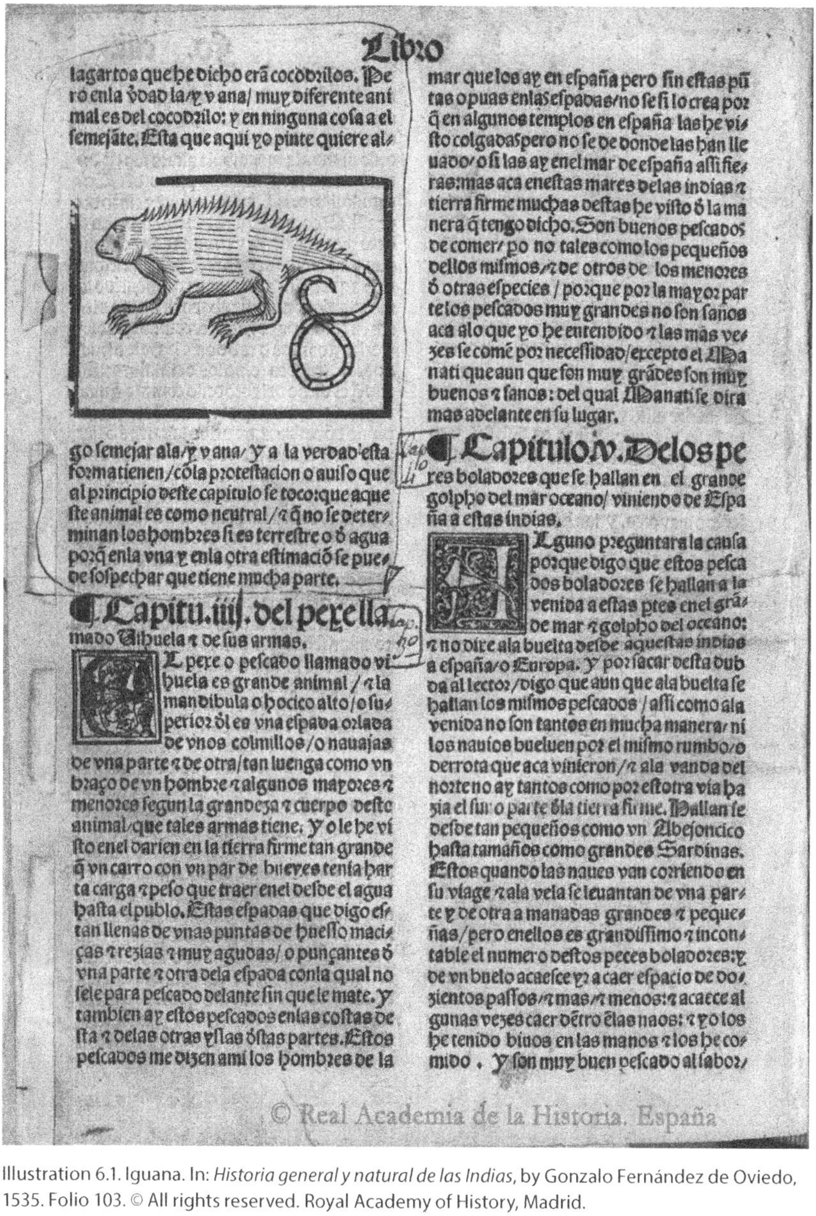

Libro

lagartos que he dicho erã cocodrilos. Pero enla vdad la yvana / muy diferente animal es del cocodrilo: y en ninguna cosa a el semejãte. Esta que aqui yo pinte quiere algo semejar ala yvana / y a la verdad esta forma tienen / cõla protestacion o auiso que al principio deste capitulo se toco: que aqueste animal es como neutral / ⁊ q̃ no se determinan los hombres si es terrestre o d' agua porq̃ enla vna y enla otra estimaciõ se puede sospechar que tiene mucha parte.

¶ Capitu. iiij. del pexe llamado Vihuela ⁊ de sus armas.

El pexe o pescado llamado vihuela es grande animal / ⁊ la mandibula o hocico alto / o superior d'l es vna espada orlada de vnos colmillos / o nauajas de vna parte ⁊ de otra / tan luenga como vn braço de vn hombre ⁊ algunos mayores ⁊ menores segun la grandeza ⁊ cuerpo deste animal / que tales armas tiene. Yo le he visto enel darien en la tierra firme tan grande q̃ vn carro con vn par de bueyes tenia harta carga ⁊ peso que traer enel desde el agua hasta el publo. Estas espadas que digo estan llenas de vnas puntas de huesso maciças ⁊ rezias ⁊ muy agudas / o punçantes d' vna parte ⁊ otra dela espada con la qual no se le para pescado delante sin que le mate. Y tambien ay estos pescados enlas costas desta ⁊ delas otras yslas d'stas partes. Estos pescados me dizen ami los hombres de la mar que los ay en españa pero sin estas pũtas o puas enlas espadas / no se si lo crea por q̃ en algunos templos en españa las he visto colgadas pero no se de donde las han lleuado / o si las ay enel mar de españa assi fieras: mas aca enestas mares delas indias ⁊ tierra firme muchas destas he visto d' la manera q̃ tengo dicho. Son buenos pescados de comer / yo no tales como los pequeños dellos mismos / ⁊ de otros de los menores d' otras especies / porque por la mayor parte los pescados muy grandes no son sanos aca alo que yo he entendido ⁊ las mas vezes se comẽ por necessidad / excepto el Manati que aun que son muy grãdes son muy buenos ⁊ sanos: del qual Manati se dira mas adelante en su lugar.

¶ Capitulo. v. Delos peces boladores que se hallan en el grande golpho del mar oceano / viniendo de España a estas indias.

Alguno preguntara la causa porque digo que estos pescados boladores se hallan a la venida a estas ptes enel grãde mar ⁊ golpho del oceano: ⁊ no dixe ala buelta desde aquestas indias a españa / o Europa. Y por sacar desta duda al lector / digo que aun que ala buelta se hallan los mismos pescados / assi como ala venida no son tantos en mucha manera / ni los nauios bueluen por el mismo rumbo / o derrota que aca vinieron / ⁊ ala vanda del norte no ay tantos como por estotra via hazia el sur o parte d'la tierra firme. Hallan se desde tan pequeños como vn Albejoncico hasta tamaños como grandes Sardinas. Estos quando las naues van corriendo en su viage ⁊ ala vela se leuantan de vna parte y de otra a manadas grandes ⁊ pequeñas / pero enellos es grandissimo ⁊ incontable el numero destos peces boladores: y de vn buelo acaesce yr a caer espacio de dozientos passos / ⁊ mas / ⁊ menos: ⁊ acaece algunas vezes caer dẽtro ẽlas naos: ⁊ yo los he tenido biuos en las manos ⁊ los he comido. Y son muy buen pescado al sabor /



Illustration 6.1. Iguana. In: *Historia general y natural de las Indias*, by Gonzalo Fernández de Oviedo, 1535. Folio 103.  Royal Academy of History, Madrid.

runs from the middle of the back to the end of the tail. In short, we judged that they were poisonous serpents, and they [the sailors] ate them.[23]

23 Vespucci, Amerigo, *Edición facsimilar de las Cartas de Vespucio*, Biblioteca Nacional de Colombia, Bogotá, 1942, page 41.

Similar and equally detailed descriptions were found in most of the chroniclers who followed Vespucci. Francisco Hernández wrote about the same animal:

> The creatures which the Haitians call iguanas and the Mexicans ACUECUETZPALLIN seem to belong to the species of *colotes*, but their shape and nature are very different. They reach a length of five palms, including the tail, which is three times the size of the rest of the body and thicker than a human arm. Their fingers are slender and very long in relation to their body. Their arms and legs are as thick as those of a newborn child or a little less. Their stomach is green with white and the rest of the body scaly and of a color between silver and green. Their nails are of medium size. Their tail is long and gradually tapers from the hind legs to its end. Some narrow fins [*Pinnae*], an inch long, rise from their back and stretch in a dense row toward the end of the tail, although they become smaller and smaller. The top of its head is deformed and horrible, very hard and as if studded, just like the top of the neck. They have black eyes. The lower jaw is blue and a membrane, four fingers wide, hangs down from it, circular, slender, scaly and like a jowl, studded or pinnate at the stretch where it starts, and of a green color shading into yellow. The same jaw is covered with large, round, blue plates and it mainly has one on each side which is the size of a *medio real* coin and with a color and gleam similar to the scale of a fish, and another, somewhat smaller and shinier, near the ears. The opening of its mouth is wide, and its teeth are small.[24]

Nevertheless, this new world and all of its creatures, strange and horrible as they may have seemed, also had to be part of the work of a benevolent God of limitless power. Thus, such unknown and monstrous animals had to be transformed into creatures of God at the service of mankind and into useful products for the Empire. All these rarities had to be regarded as examples of the infinite power, wisdom, and mercy of the Creator.

Hernández himself wrote the following about the same animal he initially described as somewhat fearful:

> It is a harmless animal which lives in the waters. It lays a large amount of eggs which taste good and are an excellent food, as is its meat as well

24 Hernández, Francisco, *Historia Natural de Nueva Espana*, [1571–1576], Mexico, Universidad Nacional Autónoma de México, 1959, Third Treatise, Chapter IV, 'Del ACUECUETZPALLIN o colote del agua', page 369.

> (we call it that because it seems more like meat than fish), which is not inferior to that of a chicken, neither in its flavor nor its nutritional quality. Practically no one who sees this animal for the first time will avoid being frightened or who once eats it will not be eager to try it again.[25]

The most convincing act of appropriating these 'horrible' creatures, or the most obvious and explicit form of mastering them, was to turn them into foodstuffs people would crave or, as we shall see below, by finding medicinal properties or other features in them that would be useful for mankind.

Some animals, such as reptiles and, especially, snakes, already belonged to a class of fearsome animals, and there were frequent references to them as evil creatures.[26] The New World did not seem to have an abundance of the large animals or mammals that were common in the Old World, but it did have a great variety of birds, insects, and reptiles that had never been seen in Europe.[27]

The monstrousness of the exotic turned out to be as striking as it was problematic for religious belief. A benevolent God could not have created horrible and malevolent creatures, and, for that reason, the idea that nature was made for the convenience of humans was always the dominant one; despite the dangers and fears they embodied, even serpents must be useful for mankind. Accordingly, Oviedo reasoned that the Creator had given a benevolent design to the rattlesnake by furnishing it with a warning signal for mankind:

> This serpent has five or six knots in its tail, separate and round, which seem to be strung together, and when this animal moves, they sound like real and veritable muffled bells, so that it would seem that benign nature (or rather, God), in his mercy, gave it that as a warning to human men, because, hearing those bells, they can protect themselves.[28]

The longing for a paradise on Earth contrasted with the fear of an unknown world. This tension between the fear of the unknown and the illusion of

25 Ibid.

26 Cabarcas, Hernando, 'Bestiario del Nuevo Reino de [...]', op. cit., page 101.

27 See, for example, Fernández de Oviedo, 'Historia general y natural de [...]', op. cit., vol. 2, Books XII–XV.

28 Ibid., vol. 1, Book VI, Chapter XXIII, 'which deals with a group or new kind of very poisonous snake which is found in the island of Margarita, which are called rattlesnakes, and other vipers or snakes which wish to imitate them with a rattle and are also very poisonous, in the province of los Alcázares', page 181.

an earthly paradise was seen in the description of the bird-of-paradise, which was beautiful and monstrous at the same time. Its plumage is of an extraordinary beauty: 'The Indians adorn their head with this dried bird, using it in the form of a crest, due to the beauty of its feathers, the variety of its colors and the rare and beautiful shape of the bird itself'.[29] However, aspects of it still struck Europeans as monstrous and strange. Writing about the same bird, Hernández remarked:

> Humanity finally found a true *'ápodos'* (legless animals) that is, birds which not only lack the use of their legs, as had been known for a long time in the world, but are absolutely devoid of the legs themselves and in place of them, furnished with some feathers which are tawny, bristly and thin, four and two quarter inches long, and which emerge near the middle of the body like very thick manes [of a horse]. They use them to hang from the trees (if they happen to stop flying), since they cannot perch on them, and to couple when the male and female are together. After laying her eggs in the hole on the back of the male, the female incubates, protects and embraces the chicks with her frontal cavity [...] Nature having provided that so that it can only live in the air, it does not need any food or it makes do with what it finds there.[30]

Despite the monstrous character that was attributed to the plants and animals of America in such detailed and meticulous descriptions, by giving them a name and acknowledging their usefulness, the Spanish chroniclers transformed the savage into the familiar and monsters into domesticated creatures that formed part of a Christian order that was both anthropocentric and imperial.

Possibly under the influence of Pliny, the naturalists of the 16th century thought of nature as something both capricious and creative. This playful character of nature, which was so important in Pliny's *Natural History*,[31] has been problematic for Christian writers at times. The fantastic aspects and rarity of nature had to be explained away, so that there would be no loose ends in a universal creation governed by an immaculate god, one who arranged the world to be at the service of mankind.

29 Hernández, Francisco, 'Historia Natural de Nueva España', op. cit., Second Treatise, Chapter CXXIX, 'On the Bird of Paradise', page 347.

30 Ibid., pages 347–350.

31 See: French, Roger Kenneth, *Ancient Natural History: Histories of Nature*, London, Routledge, 1994.

Partly because of the mysticism of the Hermetic tradition it drew on, the Renaissance gave new life to an idea found in many medieval scholars who wrote about the natural word, namely, that the word of God was not only expressed in the Holy Scriptures, but in nature as well; that is, the work of the Creator had to be understood as a display of its divine power and wisdom. For them, the natural order was a divine order and the most powerful evidence for the wisdom of the Creator lay in the perfection of his work. For the Christian writers of the 16th century and the chroniclers of the Spanish Crown, the task of describing nature had a double purpose: to display the sum of the empire's possessions in an orderly and intelligible manner and to celebrate the grandeur of the Creator.

José de Acosta insisted on the idea that the Lord should be praised for his marvelous works and clearly stated that the aim of all of his own efforts was a paean to the Creator: 'the purpose of this book is that, by bringing news of the natural works which the very wise Author of all of nature has made, it will praise and glorify God Most High, who is marvelous everywhere'.[32] It was in that context that the knowledge of the naturalists turned the monster into a familiar creature, which at times is docile and even beneficial for mankind.

## To describe, classify, and name

As we have frequently pointed out so far, the encounter of Europeans with the unknown nature of America implied new challenges for natural history. The animals of the New World did not fit into the personal knowledge of the travelers, nor had they been described by the authorities of antiquity, and, from the standpoint of Europe, they did not have names. The Bible taught that Adam had named all of the animals according to their kinds, and so these creatures, apparently without names, had to be baptized and incorporated into the Christian order of the Creation.

But how was this world to be brought to Spain herself? Or, rather, how to describe it in a convincing manner so that Europeans would have a faithful idea of a world that had never been seen before? One obvious path was a comparison with familiar referents. Hence, many of the descriptions of American creatures argued that they were hybrids or combinations of known animals or plants, like a jigsaw puzzle made up of pieces of familiar creatures.[33] The most basic way to establish links is to use analogy and

32 Acosta, José de, 'Historia natural y moral [...]', op. cit., 'Prologue for the reader', page 58.

33 Asúa, Miguel de and French, Roger, 'A New World of Animals [...]', op. cit., page 14.

comparison; thus, the unknown yields its strangeness and appears in a familiar framework of reference. In the opinion of Pedro Martir de Anglería, the opossum had the face of a fox, the tails and feet of a monkey, the ears of a bat, and the hands of a man. Alvar Núñez Cabeza de Vaca thought that peccaries were wild boars and the jaguar a tiger, while Pigafetta described the penguins of Patagonia as geese, and Urdaneta as 'wingless ducks'. Pedro Cieza de León explained that llamas looked like camels and were the same size as a small burro, and thus, one after another, names such as the *oso hormiguero* ('anteating bear') for an anteater and the *oso perezoso* ('lazy bear') for the sloth turned into familiar references for the naming of strange creatures.[34]

As follows, we will give some detailed examples of how strange objects were changed into something familiar. The descriptions that Gonzalo Fernández de Oviedo made of strange products that were unknown to Europeans illustrated this movement between the known and the unknown: the armadillo became the *encubertado* (a 'covered' or armored animal); the *zarigueya*, the Brazilian name for the opossum, became the *churcha*, its indigenous name on the Spanish Main; and the pineapple (*piña*), the *ananá* (from the Tupi word for the fruit), among many other examples. His chronicle revealed the novelty and distinctiveness of those products, but also enabled his readers to recognize their similarities to things with which they were already familiar.

Fernández de Oviedo's description of the armadillo began by highlighting its novelty and strangeness: 'The *encubertados* are animals it is very worth seeing and very strange in the sight of Christians and very different from all that have been spoken of or seen in Spain and other places'.[35] Oviedo then switched to the familiar references that would undo the amazement and distrust that such a strange creature inspired:

> These animals have four feet and a tail and all of it is the same skin, the skin being like the covering or hide of a lizard [...] its features and form are no more and no less than those of a covered horse [armored for combat] [...] in the same manner as the armored harness which was placed on warhorses in ancient times. It is the size of a little dog or common mongrel.[36]

34 Ibid., page 33.

35 Fernández de Oviedo, Gonzalo, 'Sumario de la Natural Historia de [...]', op. cit., Chapter XXII, '*Encubertados*' [Armadillos], page 114.

36 Ibid., pages 114–115.

This picture of a horse in armor was surely a familiar referent for Spaniards, who remembered the wars between Christians knights and Moors. The use of similar descriptive resources and techniques was also evident in his portrayal of the opossum: 'The *churcha* is a small animal, the size of a small rabbit and tawny-colored [...] and with a long tail in the manner of a mouse and the ears are very similar to it'.[37]

Many other references to American creatures were found in his *Historia general y natural de las Indias*, which used a familiar language and comparisons with domestic things to describe American plants and animals. The pineapple, for example, was depicted as a cross between the pine nut and the artichoke, while its properties were contrasted with those of fruits that were already known:

> The Christians gave them the name of *piñas* because, in a certain manner, they resemble our *piñones* (pine nuts), except that they are more beautiful and do not have the hardness of the cones of the pines of Castile, which are of wood or nearly so, and they can be cut with a knife, like a melon, or preferably into round slices, first removing the rind, which is made of up some scales in relief which make them look like pinecones. But you do not open and divide them at those joints of the scales, as you do pine nuts. Yet, in my opinion, a better name would be to call it an artichoke, in view of the thistles or thorns it grows out of, even though it looks more like a pinecone than an artichoke. The truth is that it does not wholly stop from being an artichoke [...] to be more specific, its specific flavor is that of the peach and it smells like both a peach and a quince, but the flavor of the pineapple is mixed with that of the muscatel [grape] and it therefore tastes better than peaches.[38]

A crucial aspect of the process of 'comprehension' is the act of naming in a way that eliminates ambiguity, variety, and the difference in local names. In the same way as geographical places and settlements, the natural history of the 16th century had to ensure that things only had a single name. As Oviedo remarked, 'and since the *indios* have many different languages, so

37 Ibid., Chapter XXVII, 'On the *Churcha*', page 120.

38 Fernández de Oviedo, Gonzalo, 'Historia general y natural de [...]', op. cit., vol. 1, Book VIII, Chapter XIV, pages 241–242. 'On pineapples [*piñas*], so-called by Christians because they resemble our pine nuts, which fruit is called *yayama* by the *indios*, and another type of the same fruit, *boniama*, and another, *yayagua*, as will be explained in this chapter even though it has other names in other places'.

with diverse names, they name them'.[39] While the name *piña* was not wholly suitable for the pineapple, it was preferable to the disorder and confusion caused by its many local names.

To that should be added that, like many of the plants and animals of the New World, pineapples were difficult to transport to Europe, hence the Empire's possession and control of them required more sophisticated techniques. Speaking of pineapples, Oviedo noted that:

> Some have been sent to Spain and very few arrive there. And when they arrive, they can no longer be perfect or good, because you have to harvest them when they are green and ripen them at sea and in that way they lose their appeal. I have tried to take them to Europe and as I was not certain of the route and the voyage took too long, all of them were lost or rotted and I tried to transport the shoots and they too were lost.[40]

The most effective way to possess and bring them to the notice of Europeans was with words and pictures. Oviedo wrote about the role of painting in natural history, as follows: 'our eyes provide us with a large part of the information about these things and since the same cannot be seen or touched, the plume which makes a picture of them is a great help'.[41]

As we mentioned when talking about the subject of cartography, one of the essential parts of the process of appropriation lies in the act of naming, which, like baptism itself, is a way to create a link between the familiar and the strange and incorporate the alien into a single, familiar framework of reference.[42] In order to possess the natural world, things must have a single name, so that the ambiguity and variety of local names are eliminated. European names such as *piña, armadillo, tigre, perico ligero* ('light parrot', for the three-toed sloth), *ave mosca, tominejo* (hummingbird), and *oso hormiguero* (anteater) comprised the unknown part of a familiar setting.

Closely related to descriptions and names is the classification and incorporation of each creature into a known group or class. The description and denomination of an object presupposes some form of cataloguing. Many of the American creatures did not fit into ordinary classifications, and their place in the categories established by Aristotle or implicit in Pliny is unclear.

39 Ibid.

40 Ibid.

41 Cited by José Rabasa, 'Inventing America: Spanish Historiography [...]', op. cit., page 146.

42 See: Chapter 5, page xx.

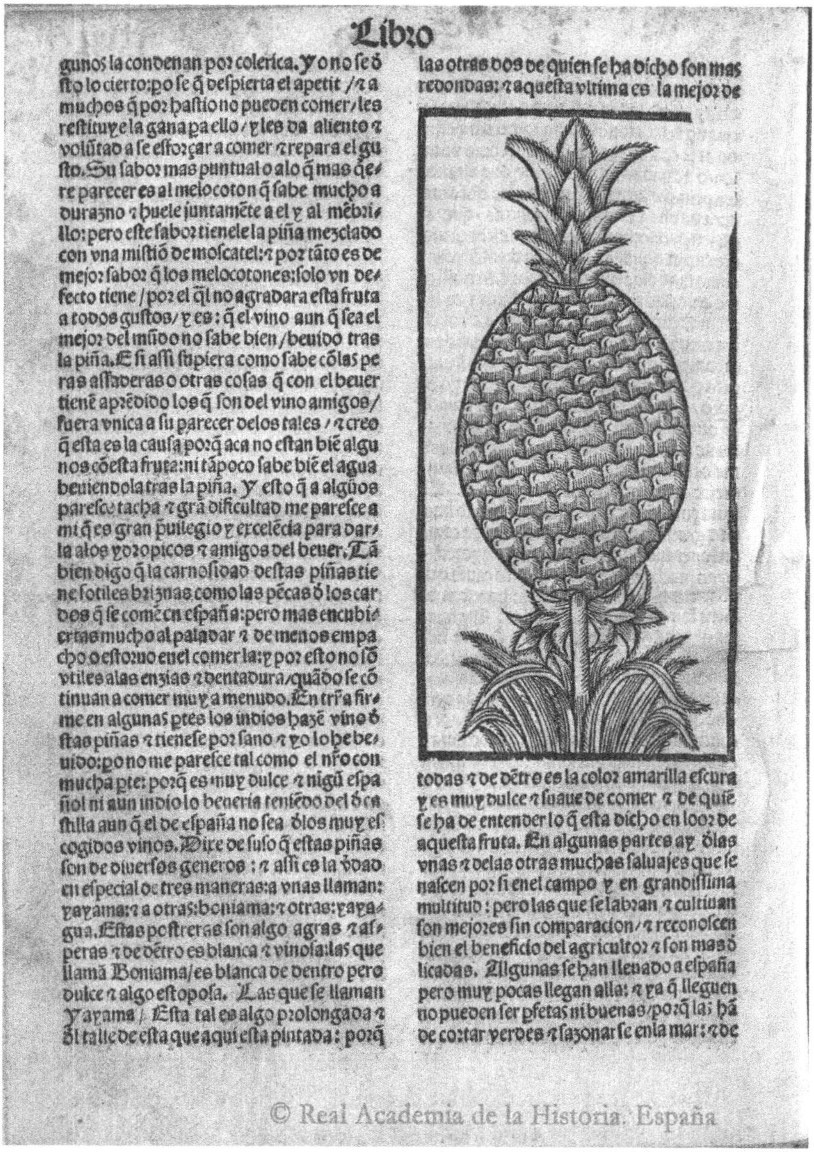
Libro

gunos la condenan por colerica. Yo no ſe ǒ ſto lo cierto: ꝑo ſe q̃ deſpierta el apetit / ⁊ a muchos q̃ por haſtio no pueden comer / les reſtituye la gana ꝑa ello / y les da aliento ⁊ volũtad a ſe eſforçar a comer ⁊ repara el guſto. Su ſabor mas puntual o a lo q̃ mas q̃e- re parecer es al melocoton q̃ ſabe mucho a durazno ⁊ huele juntamẽte a el y al mẽbri- llo: pero eſte ſabor tienele la piña mezclado con vna miſtiõ de moſcatel: ⁊ por tãto es de mejor ſabor q̃ los melocotones: ſolo vn de- fecto tiene / por el q̃l no agradara eſta fruta a todos guſtos / y es: q̃ el vino aun q̃ ſea el mejor del mũdo no ſabe bien / beuido tras la piña. E ſi aſſi ſupiera como ſabe cõ las peras aſſaderas o otras coſas q̃ con el beuer tienẽ aprẽdido los q̃ ſon del vino amigos / fuera vnica a ſu parecer delos tales / ⁊ creo q̃ eſta es la cauſa porq̃ aca no eſtan biẽ algunos cõ eſta fruta: ni tãpoco ſabe biẽ el agua beuiendola tras la piña. Y eſto q̃ a algũos pareſce tacha ⁊ grã dificultad me pareſce a mi q̃ es gran ꝑuilegio y excelẽcia para dar- la a los yoropicos ⁊ amigos del beuer. Tã bien digo q̃ la carnoſidad deſtas piñas tiene ſotiles briznas como las pẽcas ǒ los cardos q̃ ſe comẽ en eſpaña: pero mas encubi- ertas mucho al paladar ⁊ de menos empacho o eſtoruo enel comer la: y por eſto no ſõ vtiles a las enzias ⁊ dentadura / quãdo ſe cõtinuan a comer muy a menudo. En tr̃a fir- me en algunas ꝑtes los indios hazẽ vino ǒ ſtas piñas ⁊ tieneſe por ſano ⁊ yo lo he be- uido: ꝑo no me pareſce tal como el nr̃o con mucha ꝑte: porq̃ es muy dulce ⁊ nigũ eſpañol ni aun indio lo beueria teniẽdo del ǒ caſtilla aun q̃ el de eſpaña no ſea ǒ los muy eſcogidos vinos. Dixe de ſuſo q̃ eſtas piñas ſon de diuerſos generos: ⁊ aſſi es la ṽdad en eſpecial de tres maneras: a vnas llaman: yayama: ⁊ a otras: boniama: ⁊ otras: yayagua. Eſtas poſtreras ſon algo agras ⁊ aſ- peras ⁊ de dẽtro es blanca ⁊ vinoſa: las que llamã Boniama / es blanca de dentro pero dulce ⁊ algo eſtopoſa. Las que ſe llaman Yayama / Eſta tal es algo prolongada ⁊ ǒl talle de eſta que aqui eſta pintada: porq̃ las otras dos de quien ſe ha dicho ſon mas redondas: ⁊ aqueſta vltima es la mejor de todas ⁊ de dẽtro es la color amarilla eſcura y es muy dulce ⁊ ſuaue de comer ⁊ de quiẽ ſe ha de entender lo q̃ eſta dicho en loor de aqueſta fruta. En algunas partes ay ǒlas vnas ⁊ delas otras muchas ſaluajes que ſe naſcen por ſi enel campo y en grandiſſima multitud: pero las que ſe labran ⁊ cultiuan ſon mejores ſin comparacion / ⁊ reconoſcen bien el beneficio del agricultor ⁊ ſon mas ǒlicadas. Algunas ſe han lleuado a eſpaña pero muy pocas llegan alla: ⁊ ya q̃ lleguen no pueden ſer ꝑfetas ni buenas / porq̃ las hã de cortar verdes ⁊ ſazonar ſe en la mar: ⁊ de



Illustration 6.2. Pineapple. In: *Historia general y natural de las Indias*, by Gonzalo Fernández de Oviedo, 1535. Folio 77.  Royal Academy of History, Madrid.

The iguana, manatee, cactus, and American felines, among others, precisely showed some of these difficulties. In fact, the amount and variety of animals, plants, trees, reptiles, insects and serpents seemed to make naming and classifying an impossible task. Scientific nomenclature took a long time to reach the apparent clarity and universality of systems of classification, such as that of Carl Linnaeus, that came to be widely accepted in the European Enlightenment of the 18th and 19th centuries.

The felines of American were a very good example of the taxonomical difficulties caused by the creatures of the New World since it was unclear whether they were animals similar to those already known or different animals altogether:

> The first Spaniards who, in the Spanish Main, in the province of Cemaco or the Darien, saw those animals which the *indios* in that land call the *ochi*, called it the *tigre*, which are those like the one in the city of Toledo, in the abovementioned year, which was given to our lord, the Emperor, brought from New Spain.[43]

Part of the problem was that the name of the same animal varied in accordance with the place where it was named; thus, Oviedo explained about the tigers, 'in different provinces they name them in different ways'.[44] Those variables and the many local names were confusing, so that the easiest solution was to associate them with or give them the same name as known felines, such as the tigers of Asia. Nevertheless, a closer examination of the subject soon makes clear the problems of these simple assimilations. Oviedo resorted to the classics to explain the nature of such felines, but also to show the particularity and novelty of the American species:

> And the first which has to be said of the tiger is that, according to what the ancients wrote, it is the fastest of the land animals. In his *Ethimologias,* Isidore [of Seville] says: 'The tiger is thus named due to its speedy flight and the river Tigris was thus named for its rapidness and the Persians and Medes thus named the arrow. And that beast, distinguished by several spots and its talent and admirable velocity, is responsible for the name of the river Tigris, because it is the most rapid and torrential found in all the books' (Ethim; Book XII, Chapter II). According to Justinian, the river Tigris is born in Armenia, with little water, etc. (Just; book XLIII). Pliny (Plin., Book VIII; Chapter XVII) says that, due to the variety of their colors and diverse spots, the panther and the tiger are nearly different from all the other animals, because the other wild beasts have a single

43 Fernández de Oviedo, Gonzalo, 'Historia general y natural de [...]', op. cit., vol. 2, Book XII, Chapter X, 'On the animals in Tierra Firme which the Spaniards call tigers, or the *indios* have different names for, according to the language of those provinces where they are found' page 39.

44 Ibid.

> color, in accordance with their species [...] In my opinion, having said what I feel about whether these *ochis* are tigers or not, [I add that] whatever those in the world which are noted by their spotted skin may be, or if they happen to be another new animal which likewise has one and is not found in the accounts of those writings, because there are many animals on *Tierra Firme* and among those I mention here (or others among them), no writers among the ancients has left a record of them, perhaps because they are in a province they did not know, and neither the cosmography of Ptolemy nor other authors mentioned them, so we did not learn of them until Admiral don Christopher Columbus taught us.[45]

That problem led the Spanish naturalist to not only establish similitudes but to highlight differences as well and, even more interestingly, explain their causes in terms of the climate and environment in which the different animals and plants lived.

> [Speaking of] these *ochis* or tigers, or rather, panthers (because they lack the agility of the tiger, which is deduced from their habits and they do not have joints in the rear legs and run in leaps): there are many of them in the *Tierra Firme* and they eat many *indios*, and are very harmful. But, as I have said, I would not take them for tigers, seeing what is written about the agility of the tiger and what one sees of the slowness of those *ochis* which are called tigers in these Indies. It is true that, in accordance with the wonders of the world, and the extreme differences between creatures in one part of the world and those in another, and the diversity of the provinces and constellations in which they are reared, we see that the plants which are harmful in some parts are healthy and advantageous in others, like the *yuca* [cassava], which in these islands is lethal and is a good fruit in *Tierra Firme*. And that is why Saint Gregory says (Mor; c.6, job 38) that the herbs which sustain some animals kill others.[46]

The description and faithful recording of these creatures was done with meticulous attention to the details, and show us that Oviedo was an attentive observer:

> Its head is a wide as that of a lion or leopard, but thicker and it and the whole body and arms and legs are painted with black spots, some

45 Ibid., pages 39–40.

46 Ibid.

> joined to others, and outlined in a reddish color, which form a beautiful, harmonious painting: they are bigger on the back and along it and diminish toward the stomach and arms and head. The one they brought there was small and young, and in my view, might have been three years-old or less, but there are much bigger ones in *Tierra Firme*. I have seen ones that were taller than three handspans and others five ones long and their arms and legs are very bent over and strong, and they are strongly armed with teeth and fangs and nails, and in some manner fierce, which, in my view, does not apply to any real lion, however large, which is neither so fierce nor so strong. But I believe, instead, that lions are braver and more energetic.[47]

The case of the reptiles was even more complex than that of the felines. Although some of their characteristics were similar to those of European examples, the striking variety of reptiles in the New World obviously made them difficult to classify. For the diligent European observer, the varieties seemed 'numberless' and 'countless' and identifying each one an infinite task.

> There are innumerable lizards in this island of Hispaniola and in all of the islands of this gulf and the southern part of the Tierra Firme of these Indies, and there is so much to say about this that, particularly if I were to write it down, it would be a never-ending job. There are some that are green, and others brownish-gray, and others nearly black, and some more green than others, and some of a color which is nearly bright yellow and others of a tawny color. Just as their colors are different, so too are they different in size, and some larger or smaller than others, given that all are small. Some are spotted and others striped or striped with different patterns and colors; and of each kind, there are many. Others, when they stop to look at a man, release a round and colored crest or membrane from their dewlap and keep it extended, whether they are standing or still; and when they are excited, they take it in and cover it and let it out and cover it and put it back in the dewlap whenever they wish, or they flee. There are others which are somewhat larger than the common wall lizards of Spain, two or three times bigger, but not as big as the lizards of Castile. Let us now leave this subject of the wall lizards, which are very common things and innumerable and nearly infinite, and go on to speak of the serpents, which are the same as snakes. And it is

47 Ibid., page 39.

> not a brief subject, nor can I say enough to cover it. For one thing, they are innumerable, and for another, because neither I nor another has seen or can see all of them. There are many of them on this island, of many kinds, colors and sizes and the natives of these islands and even those Spaniards who have lived here for a long time commonly hold that they are not poisonous.[48]

The above examples of detailed description, as well as the naming and classification of strange plants and animals, reveal a process of domestication in which Christian naturalists transformed the unknown into something familiar and domestic; and natural history itself is a form of incorporating the alien into a framework of reference one knows, something that enables one to proclaim a dominion over nature.

## Medicine, botany, and the knowledge of the natives

Both for Aristotle and Pliny on the one hand, and for the Christians on the other, nature seemed to be governed by a design that was in accordance with the purposes of a clearly anthropocentric cosmology. Part of the natural order and of the religious and teleological meaning of nature, was the fact that it was useful to mankind. However, this did not mean that each and every one of the objects of nature had a beneficial character; in fact, many of these objects were described as being dangerous for man, but the knowledge of these objects evidenced a form of dominance. The biggest threat of nature lay in human ignorance.

To explore natural riches and transform the creatures of God into commercial goods were activities that fitted well into a religious order that sees man at the center of the Creation. Nature had to be turned into part of the Empire and the Empire into part of the Kingdom of God. In that context, understanding nature amounted to acknowledging that it was useful for man and the importance of his rule over it.

We have already discussed the different ways in which the New World was incorporated as a part of the Old World, but we have not yet dealt with one of the most powerful motivations for the study of the properties of American nature: medicine. While geography, botany, understanding other cultures, the study of animals, and mining were key fields of knowledge for

48 Ibid., vol. 2, Book XI, Chapter VIII, 'On the serpents or snakes and wall lizards and lizards of this island of Hispaniola and other places', page 35.

the enterprise of conquest, medicine seems to have played a leading role in that process. As early as the 16th century, Spain's interest in medicinal plants and the therapeutic uses of American nature was an important motive for the work of the Iberian naturalists of the era.

Philip II of Spain showed a special interest in the medicinal properties of American plants when he sent an expedition to New Spain, led by Francisco Hernández, with the specific aim of investigating the virtues of the plants and medicinal knowledge of the natives. An expert on and translator of the work of Pliny, Hernández wanted to emulate and update the *Natural History* of that Roman author who lived in the first century A.D. Like Pliny, this Spanish doctor also worked for an empire, hence his appointment to the post of *protomédico* (chief doctor) of the Indies, and, as such, he was ordered to regulate the practices of medicine in the New World in the name of Philip II, to whom he dedicated his book. From very early in the 16th century, the flora of America was regarded as an inexhaustible source of wealth, and part of its value lay in the possible medicinal virtues of the plants. With his marked interest in the pharmaceutical properties of plants, Hernández's *Historia Natural de Nueva España* described more than 3000 plants and more than 300 animals. American plants, such as maize, tomatoes, cacao, potatoes, pineapples, and plantains, among many others, had a strong impact on the culture of the West, and others, such as tobacco, sarsaparilla, coca, and quina bark, were important in the history of medicine.

The use of American medicinal plants by Europeans raises a question that deserves our attention: how was it possible for Europeans to learn of the properties of American plants? Recognizing the usefulness of the natural resources of America became possible thanks to the knowledge and traditions of the natives since, instead of referring to European discoveries during the scientific exploration of America, it would be better to speak of Europe's translation of local bodies of knowledge. The studies of Hernández, Sahagún, or Monardes evidenced the wealth both of nature in New Spain and the culture and knowledge of the natives. Even though it was done long before the upsurge of medicine during the Enlightenment, the work of these naturalists was the most important antecedent for the Royal Botanical Expeditions sponsored by the Spanish monarchs of the 18th century and was clearly a crucial referent for the projects of scientific exploration that defined the Enlightenment and Europe's quest to dominate every corner of the world and all the peoples and creatures on Earth.

Although a large part of Hernández's work remained as manuscripts in the library of the king at El Escorial and only became known to a wider public via an Italian version, it was the obligatory reference in a long Spanish tradition of the botanical exploration of America for pharmaceutical purposes.[49]

Though it is not our intention to explain the medical traditions of the American natives here, it is worth mentioning some examples of how the knowledge of the medicinal properties of New World plants and animals was closely related to the knowledge of the natives. As we have pointed out, in the process of comprehending the New World, a knowledge of the use of medicinal plants, was not only a mechanism for appropriating nature, but also an act of translating and appropriating the knowledge of the natives.

In his description of '*TLAOLLI* or maize and the drinks and kinds of cakes they usually prepare with it', Hernández wrote:

> I do not understand why the Spaniards, who are very diligent imitators of what is foreign and know how to exploit foreign inventions so well, have not yet adapted to its uses nor taken to their lands and cultivated this kind of grain, which, used in the correct manner, is exceedingly healthy both for the healthy and the ill, easy to cultivate, grows abundantly and very securely in almost any soil and is little vulnerable to drought and other hazards of the sky and earth and by means of which they might free themselves from hunger and the innumerable ills which derive from it.[50]

In the midst of America's enormous diversity, it was impossible to learn about the characteristics and possible virtues of its plants through direct experience. It is thus obvious that the plants that were most widely used and most important for the American cultures were those that struck the attention of Europeans. Coca, maize, and tobacco were notable examples. Hernández wrote the following about coca:

> They mix the chewed leaves with the powder of burnt clam [shells] and make this mixture into pills which they dry and then use. They say when these are put in the mouth and dissolved, they calm thirst, nourish the body in an extraordinary way, calm hunger when there is a scarcity of food and drink and do away with tiredness on long days of work.[51]

49 See: Nieto, Mauricio, *Remedios para el Imperio. Historia natural y la apropiación del Nuevo Mundo*, Bogotá, Instituto Colombiano de Antropología, 2000.

50 Hernández, Francisco, 'Historia Natural de Nueva España', op. cit., Book VI, Chapter XLIV, 'Account of the plants', 'On *TLAOLLI* or maize and the drinks and kinds of cakes they usually prepare with it', pages 288–292.

51 Hernández, Francisco, 'Historia Natural de Nueva España', op. cit., 'On the plant which is called Peruvian *COCA*', page 238.

Both Hernández and Monardes provided long and detailed descriptions of tobacco. The former's explanation of its properties was based both on his own experience and those of the American natives:

> Let us now discuss its properties, which are well known through daily experience: the leaves are set out to dry, later wrapped in the form of a tube and put in paper tubes or pipes, which are lit on one side, while the other is put in the mouth or the nose, and closing the mouth and nose, one inhales the smoke so that the vapor reaches the chest [lungs], which admirably causes expectoration, and miraculously relieves asthma, shortage of breath and related ailments. And inhaling this vapor in the indicated manner is not only a remedy for those ailments but also for problems of the uterus, and mainly the suffocations which usually arise from the rising of the same, since, when the medicament is applied, it returns to its proper site, the fainting stops, and it drives away the anguish and death which had already seized a large part of the body of the ill woman. It strengthens the head, produces sleep, calms pain, the stomach recovers its strength, it cures migraine, it dulls the feeling of sorrow and hardship, and completely instills the spirit with a restfulness which reaches to all its faculties (which might almost be called drunkenness). The same leaves, when they are green, softened and withered, as it were, with the hands smeared in oil, and then heated and applied to the pit of the stomach and the corresponding part of the back, but taking care that it does not reach the liver, which they might irritate or inflame, aids the digestion and cures indigestion in a remarkable way. When it is applied [to the body], it reduces inflammations of the spleen, calms the pains caused by cold, cleanses old and cancerous wounds, restores the flesh and aids cicatrization, crushing the leaves a little and putting some drops of their juice on the wound, and then sprinkling it with a powder made from the dry leaves and finally covering it with the same leaves from which the juice was extracted. It equally heals wounds on the head, always provided that the nerves and the bones are not damaged or disturbed. Heated, they soothe toothaches, wrapping them round the part which hurts or putting a paste of them in the cavities or holes in the teeth.[52]

Another figure responsible for an ambitious project to make a thorough compilation of the astronomical, historical, botanical, and religious knowledge of the indigenous people of Mexico in the 16th century, as well as a

52 Ibid., Book II, Chapter CIX, 'On PICIETL or the *Yetl* Herb [tobacco]', pages 80–82.

study of their political systems, was Friar Bernardino de Sahagún.[53] This author established a studio in which indigenous people made drawings that expressed their knowledge. Later, with the support of natives who had been taught Spanish and Latin,[54] Sahagún set out to transcribe the material and created a bilingual Spanish–Nahuatl book that conveyed the indigenous people's knowledge of geography, trade, religion, and politics. This difficult job of studying and interpreting the local cultures had one very clear aim: the evangelization of the indigenous people. However, what was intended to be a way to successfully convert them to Christianity turned into a sophisticated body of ethnographic knowledge and study of natural history that went far beyond its nominal purposes. Ultimately, it resulted in a number of large and complex treatises about American nature that served both as encyclopedias and aids to commerce.

The drawings and texts assembled by Sahagún differed from those of other sixteenth-century naturalists, such as Francisco Hernández or Fernández de Oviedo, who set out to describe nature in America without paying much attention to its local traditions. By contrast, the drawings of Sahagun's indigenous informants are the most authentic source we have of the vast knowledge of the natives and the expression of a mysterious and sophisticated culture. The meticulousness with which the plants and animals were described and the amount of information compiled at the studio of Sahagún teach us much about American nature and, even more important, the wealth of local knowledge.

53 Bernardino Ribeira, better known as Bernardino de Sahagún (from his birthplace, the villa de Sahagún in Spain), was born at the end of the 15th century and died on February 5, 1590. In 1529, after studying at the Universidad de Salamanca, he took holy orders at the convent of San Francisco. His most important book, *La Historia General de las cosas de la Nueva España* (General History/Account of the things of New Spain) was the result of his compilation and translation of local traditions and bodies of knowledge, and a pioneering contribution to his field of study. It is made up of twelve books in which the author deals with a number of religious, theological, and moral questions related to the cosmovision of time, the gods, the Aztec and Graeco-Latin male deities from higher to lower rank, the festivities, an adjustment of the Mexican calendar to the Christian one, the myths, fragments of ancient epics about the birth of Huitzilopochtli, the story of Quetzalcoatl and Huemac and the legendary adventures of the king of Tula, astrology, omens and fortune-telling, the forests, the gardens and the animals, and other aspects of the social and religious life of pre-Hispanic Mexico. Sahagún organized some of the information he compiled by following the order found in medieval encyclopedias and Pliny's *Natural History*: the latter seems to have been responsible for his description and arrangement of the taxonomical features of the animals and plants. It took some time for a coherent and complete edition of the *Historia General* to be published, perhaps around 1577, thanks to the valuable help Sahagún had from friar Rodrigo de Sequera, who won the support of the illustrious don Juan de Ovando, president of the Council of the Indies, for its translation into Romance language.

54 Capesciotti, Ilaria Palmeri, 'The fauna in Book XI of the *Códice Florentino* of friar Bernardino de Sahagún', in: *Estudios de Cultura Náhuatl*, vol. 32, 2001, page 199.

Illustration 6.3. *Tlazolteozacatl, Tlayapaloni, Axocotl, Chicomacatl.* In: *Librito de las hierbas medicinales de los indios*, by Martín de la Cruz, 1552. The text below the drawings explains: 'Remedy for a mistreated body. The body which is broken and mistreated should be anointed with a cataplasm made of *Tlazolteozacatl, Tlayapaloni, Axocotl, Chicomacatl*, with the moss which grows on trees, cypress cones, nettle seeds and the *ayauhcuahuitl* tree. He whose body is mistreated and broken should drink a juice well prepared from the stems of the *axocotl* flower, and *izquixoxhitl, tetlahuitl, eztetl, teamoxtli*, the liver of the *huexocanauhtli* aquatic bird and some leaves of *tlatlancuaye.* All of it should be ground in bitter water'.

Martín de la Cruz was another example of the vast botanical knowledge of the natives and the impact it had on Europe's understanding of the New World.[55] Martín de la Cruz was an indigenous doctor who was commissioned

55 Cruz, Martín de la, *Libellus de medicinalibus indorum herbis* [1552], Juan Badiano (translator), Mexico, Fondo de Cultura Económica, Instituto Mexicano del Seguro Social, 1991.

by Francisco de Mendoza, son of the first viceroy of New Spain, to write a book in the Nahuatl language that compiled the native knowledge of Mexican plants. It was translated into Latin by Juan de Badiano in 1552 so that it could be presented to Charles V. Beautifully illustrated, it displayed a great variety of plants that were used to cure different kinds of illnesses.

The works of Fray Bernardino de Sahagún and Martín de la Cruz thus enable us to see how the vast and detailed native knowledge of the natural history of America contrasted with Europeans' limited understanding of nature in the New World.

The books compiled by Sahagún and Martín de la Cruz were the result of an encounter between cultures, and formed bodies of knowledge that might be defined as *mestizos*.[56] They were not the pure expression of American bodies of knowledge, but neither were they pure expression of European ones. This phenomenon was already evident in the 16$^{th}$ century, one of a hybrid culture in which European traditions served as a framework of reference while being changed at the same time. It was precisely as a result of this encounter between different kinds of knowledge that the Spanish authors of the 16$^{th}$ century began to distance themselves from the classical medical traditions.

One of the best-known authors was Nicolás Monardes,[57] whose book on the plants and animals of America with medicinal properties was the

56 On the notion of 'mestizo thought', see: Gruzinzki, Serge, *El pensamiento mestizo. Cultura amerindia y civilización del Renacimiento*, Barcelona, Paidós, 1999.

57 Nicolás Bautista Monardes was born and died in Seville (1508–1588). Although he never traveled to America, Monardes was familiar with all the trading circuits between the port of Seville and those of Santo Domingo, Veracruz, and Cartagena de Indias, which were responsible for importing a wide variety of plants, animals and other natural products from America, such as parrots, monkeys, armadillos, plants, fruits, roots, precious stones, juices and gums, along with a number of other goods with alimentary or medicinal properties. He obtained one bachelor's degree in Arts and Philosophy (1530) and another in Medicine (1533) from the Universidad Complutense in Alcalá de Henares. In 1536, as an homage to the humanist school of Galen and Dioscorides, he published his first book in Seville: *Diálogo llamado pharmacodilisis o declaración medicinal*. It was printed by Juan Cromberger, who, a year before, had been ordered by the Spanish Crown to print the first part of the *Sumario Natural*, written by Gonzalo Fernández de Oviedo. The first part of his best-known book was published in 1568: *Historia Medicinal de las Cosas que se traen de nuestras Indias Occidentales que sirven en Medicina* (Medicinal Account of the Things which are brought from our Western Indies which serve in Medicine). In 1571, Alfonso Escribano published the second part, which bore a special dedication to Philip II. The third and final part was published in 1574. In the course of the 17$^{th}$ and 18$^{th}$ centuries, it was republished at least 14 times: seven editions in Italian, three in French, two in Latin, one in English, and one in German. Among those publications, there stood out a posthumous book by Nicolás Monardes, published by his friend, the printer Fernando Díaz. Among the foreign editions of his *Historia Medicinal* and several of his manuscripts, there stood out those published in Europe by authors such as: Jacques Gohory, who translated the chapter on the *mechoacán* root into French (1572); the pharmacist Antonine Colin,

Illustration 6.4. *Tezonpahtli, Huitzquilitl, Tecuammaitl*. In: *Librito de las hierbas medicinales de los indios*, by Martín de la Cruz, 1552. The text below the drawings explains: 'Scabies: the mangy head is washed in urines. Then, you put a mixture of the roots of *huitzquilitl, tezonpahtli, tetzmixochitl* on it, with the crushed barks of *copalcuahuitl* and *atoyaxocotl*'.

most widely read treatise of its kind both in and beyond Spain. Monardes never visited the New World, but he, unlike any other, knew how to take

who translated the *Historia medicinal* into French (1602 and 1619); John Frampton, who translated the same book into English (1577); and Charles de L'Écluse, who translated it into Latin (1576, 1579, 1582), as did a pharmacist from Seville, Castañeda (1605). Thus was immortalized the writings of a merchant and doctor who was one of the most important students of the plants of the New World.

Illustration 6.5. *Atzitzicaztli*. In: *Librito de las hierbas medicinales de los indios*, by Martín de la Cruz, 1552. The text explains: 'Herb to stem bleeding. When placed in the nose, the juice of nettles, ground with salt in urine and milk, stems nasal hemorrhages'.

advantage of the information which arrived from the Indies: '[Among] these riches from beyond the seas which are so great, from our Indias Occidentales they send us many trees, plants, herbs, roots, juices, gums, fruits, seeds, liquors, stones which have many great medicinal virtues and are responsible for very remarkable effects of great value and price'.[58]

58 Monardes, Nicolás. *La historia medicinal de las cosas que se traen de nuestras Indias Occidentales que sirven en Medicina* (1565–1574), Madrid, facsimile of the Seville edition (1580), Ministerio de Sanidad y Consumo, 1989, Folio I, page 92.

The place where he worked was the busy city of Seville, which was in a privileged position that offered obvious advantages to travelers and the explorers of the New World themselves: 'And since this city of Seville is a port and stage for all [voyages] to the Indias Occidentales, we know more than in any other place in the whole of Spain, because all things first arrive here, so that we are more knowledgeable and experienced'.[59] His book mentioned more than sixty plants and several animals with medicinal properties that came from places such as New Spain, the islands of the Caribbean, and Peru, although it is evident that he had also had access to natural products from the Orient. In fact, some of the useful plants that arrived from the Indias Occidentales were apparently identical to others that had been brought from the East. Hence, for example, the cinnamon and pepper from America could replace Oriental plants with the same or similar properties, while ginger, another species of Eastern origin, was successfully grown in America, as Monardes pointed out. This doctor from Seville found himself in precisely the right spot because it is unlikely that his *Historia medicinal de las cosas que se traen de nuestras Indias Occidentales*, known in English as *Joyful news out of the Newfound World*, could have been written in a place different from Seville.

Monardes was favored by the authority granted to him by a busy practice in a European city, so that we can assume that he often resorted to his own direct experience to certify the properties of some of his remedies. Nevertheless, the main source of information for the large majority of his medicaments was the original use of those plants or substances by the natives of the New World.

He wrote the following about a gum or resin which the *indios* called *Tacamahaca*: 'and it likewise removes any pain caused by cold or windy humors: the *indios* very frequently and commonly use it for that. And the Spaniards have used it for the same purposes'.[60] He often referred to classical notions of medicine, based on Galen, as with his theory of humors, but at the same time he resorted to the experience of the natives and Europeans to show their effectiveness. For example he described the specific practices deriving from the use of *Del Guayacan* and *Palo Santo* for the treatment of what was most likely syphilis:[61] 'Since a Spaniard was suffering great pains

59 Ibid., Folio II, page 93.

60 Ibid., Folio III, page 96.

61 *El mal de Bubas* (roughly, 'the disease of tumors'), also known in Spain as the 'French disease', was named after the sores caused by a sexually-transmitted disease, most likely syphilis, which Monardes thought had been brought to Europe from the Indies.

from *Bubas* [syphilis] which an *india* had given him, the *Indio* who was one of the doctors of that land gave him *Guayacan* water, which not only removed the pains he was suffering from but also cured the disease very well'.[62] He also added that many other Spaniards with the same disease were cured, which proved that 'it is the best and supreme remedy of those which have been found so far, and it heals and cures that illness with more certainty and firmness'.[63] The exotic aspect of his remedies was doubtless a reason why his book was attractive and successful. The novelty and strangeness of some of his prescriptions must have struck the attention of his readers. For example, Monardes devoted a section to describing the medicinal properties of the armadillo tail, the powder of which served to cure earache:

> From Tierra Firme, they likewise bring a bone which is the tail of a strange animal, which is covered all over with scales, down to the feet, like a horse [...] its active ingredient is only found in the tailbone, from which fine powders are made, and you take an amount the head of a thick pin, make a little ball of it and put it in the ear, where it marvelously gets rid of the pain there.[64]

Like Francisco Hernández, Monardes gave a detailed account of the medical virtues and native uses of tobacco and coca. In this, as in other cases, it is interesting to see that, while the native practices were acknowledged to be an original source of knowledge, Europeans did not seem to think they had the rigor of Western science and they also believed that their uses in the New World were always tinged by superstition. The sources of knowledge and the materials lay in the Indies, but the discoveries were proclaimed to be European. The English translation of his book evidenced this aspect of the unknown and the marvelous, *Joyful News out of the Newfound World*.

## The Empire and natural history

The job of sending the riches of the New World to Europe entailed shipping many products across the Atlantic. Some of them could be transported with relative ease, such as gold, silver, and other merchandises that did not alter or deteriorate when they were moved from their place of origin, but the

62 Monardes, Nicolás, 'La historia medicinal de las [...]', op. cit., Folio X, page 109.

63 Ibid.

64 Ibid., Folio LXVI, page 222.

appropriation of live beings—plants and animals—was a more difficult challenges, as we saw in the example of pineapples in Chapter 6, page xx.

In fact, with some exceptions, such plants and animals did not survive the transatlantic voyages, nor did they easily adapt to European climates, so that the mastery of them required complex forms of nomenclature, classification, and representation. It was by means of written and illustrated descriptions that Europeans proclaimed their dominion over the vast natural life of the world since, in contrast with live beings, they were easy to transport and store. Detailed descriptions that resorted to domestic referents and the use of familiar Christian names and illustrations were fundamental strategies for incorporating a savage and distant world into a Christian order. For these forms of long-distance control to be successful, links had to be established whose solidity would depend on the extent to which the representations were credible and faithful. A new code of legitimacy and truth was required, and that is why the problem of dominion was ultimately an epistemological one in large part, a subject we will return in the following chapter.

Building an empire demanded the existence of a scientific activity that made it possible to order nature and society in accordance with common codes. Therefore, cosmography, seamanship, and cartography, along with natural history and politics, must be regarded as expressions of a single purpose of control and rule. It may be useful to remember that the idea of linking natural history and political history may not be to the liking of some modern historians, but it should be stressed that this was not true of the 16th century. Most of those who first wrote about the New World, such as Oviedo or Acosta, regarded the description of nature and culture, natural history and moral history, God and the actions of man, and the natural world in general, as inseparable parts of the same story.

Perhaps it was the importance given to nature and morality as the main subjects for an understanding of the New World that was responsible for the dissemination of the writings of Gonzalo Fernández de Oviedo, Nicolás Monardes, Francisco Hernández, Bernardino de Sahagún, and José de Acosta: authors, at the service of the Spanish Empire, who left a large volumes of documentation about the animals, nature, and religious practices of the natives. These publications became so important that many of them were translated into different languages—Nahuatl, Latin, English, French, German, and Dutch—and were also used by countries other than Spain to lay out new geographical routes for exploring the New World between the 16th and 18th centuries.

Some of these strange creatures and objects were taken to Europe to be displayed as symbols of luxury and power in courts and palaces, while

Illustration 6.6. Museum of Ferrante Imperato in Naples, 1599.

others formed part of the collection of objects that, since they came from far places, turned into a sign of ruling and controlling the strange. The accumulation of possessions brought from remote places became fashionable among the richest merchants and princes of Renaissance Europe, and were held in private collections of exotic objects and the 'cabinets of curiosities' that made no distinction between natural and manmade objects.[65] All of these possessions were expressions of mankind's power over nature. In this process, where such artefacts were shipped from their natural setting and then classified and interpreted in a cabinet or catalogue, the savage was transformed into the domestic and nature became part of culture and property. Naturalists and collectors such as Athanasius Kircher, Conrad Gesser, Ulisse Aldrovandi, and Nicolò Serpetro, among others, were witnesses of the way in which the monstrous and savage features of the New World began to form part of European culture as early as the 17$^{th}$ century.[66]

65 See: Smith, Pamela and Findlen, Paula (eds.), 'Merchants and Marvels: Commerce [...]', op. cit.
66 Findlen, Paula, 'Inventing Nature. Commerce, Art, and Science in the Early Modern Cabinet of Curiosities', in: Smith, Pamela and Findlen, Paula (eds.), 'Merchants and Marvels: Commerce [...]', op. cit., pages 297–323.

# 7. The New World, global science, and Eurocentrism

*'The beginnings of global science occurred during the period of the rise of a global economy. Surely that was no coincidence'.*[1]

**Abstract**

The seventh and final chapter offers some reflections on the relations between the Spanish imperial endeavors of the 16$^{th}$ century and the history of Western science. The problem of knowledge, like that of control from a distance, is fundamentally a problem of communication, and, by the same token, of the processes of compiling, organizing, and systematizing information. The institutions and supply centers that were created in Spain in the course of the 16$^{th}$ century, such as the *Casa de Contratación* of Seville and the Consejo de Indies, had a definitive role in the construction of a new world order and a new technical and scientific horizon.

**Key words:** New World, Eurocentrism, Long Distance Control, Western Science, Standardization, 16$^{th}$ century

## *Plus ultra*

For the frontispiece of his *Instauratio Magna* (London, 1620), Francis Bacon chose the picture of a ship returning from the Atlantic and crossing the Pillars of Hercules at the Strait of Gibraltar, a symbolic representation of going beyond the limits of what medieval Europe regarded as the known world. At the bottom of the illustration, the phrase '*Multi pertransibunt*

1 Cook, Harold J., *Matters of Exchange: Commerce, Medicine, and Science in the Dutch Golden Age*, New Haven, Yale University Press, 2007, page 416.

Nieto Olarte, M., *Exploration, Religion, and Empire in the Sixteenth-century Ibero-Atlantic World: A New Perspective on the History of Modern Science*. Taylor & Francis Group, 2022
DOI 10.5117/9789463725316_CH07

*and augebitur Scientia'* stood out, a quotation from the Biblical Book of Daniel (12:4), which means 'many shall run to and for, and knowledge shall increase'. The choice of this saying from the Bible seemed highly appropriate as a motto for his great work of philosophy, which dealt with the foundations of a new and powerful science since, by means of science, it would be possible 'to extend human power over the universe'[2] in a new world in which 'nothing will be impossible for mankind'.[3] Those powerful galleons are the emblem of the power of human arts and that picture clearly represents the English philosopher's view of the sixteenth-century voyages of exploration and their relation to the history of modern science. The idea of associating the increase of knowledge with voyages and the exploration of the unknown perfectly complemented Bacon's philosophical approach to science because, from his standpoint, the advance of knowledge was the result of new experiences. This idea was likewise a clear expression of a moment in history when the technical achievements of sixteenth-century Europeans, in particular the explorers of new lands, marked the starting point of a new world order. The idea of a New World is a central feature of other books, Bacon wrote. His utopian novel, *New Atlantis*, was obviously inspired by the discovery and conquest of America. In it, Bacon spoke of the spirit of the Renaissance as a unique moment in human emancipation in which knowledge and the human arts are the means of building an ideal society where men have attained a complete control over nature and where 'man is a God for man'.[4] What the Spaniards of the 16th century wished to put into practice with the establishment of a complex institutional organization for the assembling of information and the production of knowledge at the service of the Empire seems to have been faithfully expressed in Solomon's House in Bacon's *New Atlantis* and the 17th century Royal Society of London.

No less interesting is to examine that prophecy in the light of the Biblical Apocalypse, since Daniel associated the end of times with the triumphs of a king whose power even defies the authority of God; a king with chariots, horsemen, and many vessels who invades foreign lands and spreads his power and possessions over all the countries on Earth, an act that seems analogous to that of the Catholic monarchs of the 16th century.

Daniel's prophecy has also been translated as '[m]any shall be running back and forth, and evil shall increase'. These Biblical references to the end of

2 Bacon, Francis, *Novum Organum*, [1620], Barcelona, Editorial Fontanella, 1979, page 39.

3 Ibid., page 38.

4 Ibid.

Illustration 7.1. Frontispiece of Francis Bacon's *Instauratio Magna*, 1620.

times do not seem to be a mere coincidence considering the sense of power of great Emperors such as Charles I and Philip II of Spain. Francis Bacon closely witnessed the birth of the great modern empires since Spain was the great military and religious rival of the English. England was the great enemy and, ultimately, the victor over Spain at sea, but it is also true that the English, like most of Europe, followed in the footsteps of the Spanish on the ocean. Sebastian Cabot was firstly pilot major of Spain before he was governor of an English royal company of expeditioners.[5] Likewise, Stephen Borough, the sixteenth-century English navigator who explored the Arctic and visited the navigational school in Seville, where he was impressed by its technical expertise, tried to convince Queen Elizabeth to establish a similar organization in England. Borough was not very successful in getting England to formalize a tradition of seamanship such as the one he had seen in the *Casa de Contratación*, but he did convince a group of London Merchants to translate and publish the *Arte de Navegar* of Martín Cortés in 1561, which came to be widely used by the British.[6] To claim that an institution in sixteenth-century Seville was a direct antecedent of the Royal Society of London might annoy a few historians of Western science, but leaving aside the marked nationalisms that still hinder a large part of modern history, the fundamental point is that the endeavor to encourage a science at the service of an empire seems to have been a common feature of both intitutions.

The illustration shown at the start of this chapter, an emblem of the start of modernity, becomes even more suggestive when we recall that the allegory of the Pillars of Hercules as the end of the known world also figured on the coat of arms of Charles I of Spain, where the columns (the Pillars) are enveloped in a ribbon that reads *'Plus Ultra'* ('Further Beyond'), a clear reference to his overseas possessions and limitless power and Spain's triumph over the unknown. In the abovementioned dedication of his book, López de Gómara wrote: 'God wished the Indies to be discovered in your time and by your vassals, so that they would be converted to His holy law, and take *Plus Ultra* for their motto, understood to be the rule over the New World'.[7] The motto 'further beyond' figured on the cover of several Spanish publications of the 16$^{th}$ century, such as the *Historia*

5 Sandman, A. and Ash, H., "Trading Expertise: Sebastian Cabot between Spain and England".

6 López Piñero, 'El arte de navegar en la [...]', op. cit., Chapter V; Taylor, Eva G. R., 'The Haven-Finding Art: A [...]', op. cit., page 196.

7 López de Gómara, Francisco, 'Historia General de las Indias', op. cit., 'To Don Carlos, Emperor of the Romans, King of Spain, Lord of the Indies and the New World', page 6.

*general and natural de las Indias* by Fernández de Oviedo (1535), and the *Milicia y Descripción de las Indias* (a manual of the military skills needed in America) by Bernardo de Vargas Machuca (1599). The same illustration of a vessel sailing through the Pillars of Hercules was used on the frontispiece of the *Regimiento de Navegación* by Andrés García de Céspedes, which was published in Madrid in 1606, more than a decade before the publication of Bacon's book.[8] Both for García de Céspedes and Bacon, this vessel, which made it possible to explore and conquer the unknown, was the symbol of the start of a new era in which Christian Europe proclaimed its rule over the world.[9]

The 'discovery' of the New World and the passage through the Pillars of Hercules symbolized the triumph of exploration and experience over the authorities of ancient times. The conviction that men had to go beyond the limits imposed by those authorities was not limited to geographical exploration. It was, in fact, the idea that defined the nature of the philosophy of modern science as an expression of mankind's emancipation, thanks to the confidence that humans could completely understand and control nature. The Pillars of Hercules represent the limits of knowledge and the arts, and, in the end, the limits of human power over nature. In his now classic study of Renaissance science, W.P.D. Wightman argued that if there is one characteristic that defined the Renaissance it was the change in the conception of man's relation to the cosmos.[10] Throughout his book, he gives many examples about developments in art, magic, and science where the guiding idea of the Renaissance appeared: 'men can do all things if they will'.[11]

Human emancipation had limits because it had always been subordinated to divine power and wisdom. Human pride was thus held up as a terrible sin in several passages of the Bible, but the very possibility of understanding the order of the Creation and defending the right to rule over it had a theological

8 On these two illustrations, see the commentaries of Cañizares-Esguerra, Jorge, 'The Colonial Iberian Roots of the Scientific Revolution', in: *Nature, Empire, and Nation: Explorations of the History of Science in the Iberian World*, Stanford, Stanford University Press, 2006, pages 14–45.

9 A thorough analysis of the imperial use of these images is found in: Vignolo, Paolo, 'Más allá de las columnas de Hércules: un emblema de la modernidad temprana', in: Hearing, Max S. and Pérez, Amanda (eds.), *Historia Cultural desde Colombia. Categorías y debates*, Bogotá, Universidad de los Andes, Universidad Nacional and Universidad Javeriana, 2012, pages 139–164.

10 Wightman, William P. D., *Science and the Renaissance*, 2 vols. Edinburgh, Oliver and Boyd / New York, Hafner Pub. Co., 1962, 'If there is any characteristic by which the Renaissance can be recognized it is [...] in the changing conception of Man's relation to the cosmos', page 16.

11 Ibid., page 18. Wightman recalls these words of Leon Battista Alberti.

Illustration 7.2. *Historia general y natural de las Indias*, Gonzalo Fernández de Oviedo, 1535.

Illustration 7.3. *Regimiento de Navegación*, by Andrés García de Céspedes, Folio I.

justification. López de Gómara clearly explained it in the first paragraph of his *Historia General de las Indias*:

> The world is so big and beautiful and has such a diversity of things which are so different from each other, that it causes admiration in he who thinks about and contemplates it. If they are no longer living like brutish animals, there are few men who have not paused to consider its marvels some time, because the wish to know is natural to everyone.[12]

Recalling the words of King Solomon, López de Gómara reminded his readers that God created the world for the benefit of mankind.[13]

The same idea was a very powerful notion in other contexts throughout the Renaissance. The hermetic traditions of magic, art, and science shared a basic premise: the world can be known because it arises from a rational creation. In that regard, God may be praised and known through his word, but also by his works.[14]

In the course of this book we have stressed the crucial role of religious feelings in the Christian conquest of the New Work; now we have to think again about the role of religion and Catholicism in the birth of modern science.

## Experience and authority

A lot of ink been spilled on the subject of the origins of modern science. The dominant approach in the literature on that subject, the result of a long historiographical tradition that goes back to the 17th century, agrees in attributing the origins of Western science to the Copernican revolution, the surpassing of Aristotle's physics and a geocentric cosmology, and the subsequent foundation of modern physics. In more general terms, the modern character of science is usually seen as a process in which the classical authors of antiquity were surpassed. In short, all of this seems to have become possible thanks to the triumph of reason and experience over dogma and faith. This idea has been inherited from a historiographical tradition

12 López de Gómara, Francisco, 'Historia General de las Indias', op. cit., 'First part of the General History of the Indies', page 7.

13 Ibid.

14 On the hermetic tradition of the Renaissance, see: Yates, Frances A., *Giordano Bruno and the Hermetic Tradition*, Chicago, The University of Chicago Press, 1964.

that pertains to the European Enlightenment, according to which nations such as Spain and Portugal seem to have made no contributions to science at all. On the contrary, the Catholic world has been regarded as an ambit that was clearly hostile to the progress of the sciences.

The idea that religion was an obstacle to science and philosophy, and that modern science represents the triumph of reason over faith, has already been questioned with sound arguments.[15] As a broad range of studies of the 'scientific revolution' have made clear, all of the great heroes of science—Copernicus, Kepler, Galileo, Boyle, Descartes, and Newton, among others—were deeply religious men. In addition, that same literature has plainly shown the crucial role of mystical traditions and magic in the history of science during the Renaissance.[16] What is rarely mentioned is the relation between the Catholic world and European modernity.[17]

As Serge Gruzinksi has rightly pointed out: 'It may seem paradoxical to associate modernity with the Iberian past. They are ideas which seem to be antagonistic, since in northern Europe we are used to associate the world of Portugal, Spain and Naples with the darkness of the archaic and the retrograde'.[18] Indeed, a quick look through some of the most influential accounts of the birth of modern science confirms the absence of the Iberian world from this history.[19] In anthologies such as the *Companion to the*

15 See, for example: Brooke, John Hedley, *Science and Religion: Some Historical Perspectives*, Cambridge, Cambridge University Press, 1991.

16 See, for example: Webster, Charles, *From Paracelsus to Newton: Magic and the Making of Modern Science*, Cambridge, Cambridge University Press, 1983; Rossi, Paolo, *Francis Bacon. De la magia a la ciencia*, Susana Gómez López (trans.), Madrid, Alianza, 1990.

17 A useful contribution to this subject is the study by Cañizares-Esguerra, Jorge, *Puritan Conquistadors: Iberianizing the Atlantic, 1550–1700*, Stanford, Stanford University Press, 2006.

18 Gruzinski, Serge, 'Passeurs y élites 'católicas' en las Cuatro Partes del Mundo. Los inicios ibéricos de la mundialización (1580–1640)', in: Godoy, Scarlett O'Phelan and Salazar-Soler, Carmen (eds.) *Passeurs, mediadores culturales y agentes de la primera globalización en el mundo Ibérico, siglos xvi–xix*, Lima, Pontificia Universidad Católica de Perú, Instituto Riva-Agüero, Instituto Francés de Estudios Andinos, 2005, page 13.

19 Burtt, Edwin Arthur, *The Metaphysical Foundations of Modern Physical Science: A Historical and Critical Essay*, 2nd ed., London, K. Paul, 1932; Butterfield, Herbert, *The Origins of Modern Science, 1300–1800*, New York, Macmillan Company, 1957; Cohen, I. Bernard, *The Birth of a New Physics*, New York, W. W. Norton & Co., 1985; Dijksterhuis, Eduard Jan, *The Mechanization of the World Picture: Pythagoras to Newton* [1950], Princeton, University Press, Princeton, NJ, 1986; Hall, Alfred Rupert, *The Scientific Revolution, 1500–1800: The Formation of the Modern Scientific Attitude*, London, Longmans and Green, 1954; Koestler, Arthur, *The Sleepwalkers: A History of Man's Changing Vision of the Universe*, London, Macmillan, 1959; Koyré, Alexander, *La Révolution astronomique. Copernic, Kepler, Borelli*, Paris, Hermann, 1961; Kuhn, Thomas S., *The Copernican Revolution: Planetary Astronomy in the Development of Western Thought*, Cambridge, Harvard University Press, 1957; Shapin, Steven, *La revolución científica. Una interpretación alternativa*,

*History of Modern Science*, published in 1990 with the aim of covering all of the subjects and major problems of modern science and its history, Spain and Portugal are conspicuous by their absence. In the same year, a book edited by David Lindberg and Robert Westman, titled *Reappraisals of the Scientific Revolution*, included a dozen articles by experts on the subject, none of which mentioned or discussed Spain or Portugal. In 1992, Roy Porter and Mikulas Teich published a volume titled *The Scientific Revolution in National Context*, wherein the British historian David Goodman *finally* deals with the subject in a chapter titled 'The Scientific Revolution in Spain and Portugal'.[20] However, despite showing the importance of Iberian science in the 16th century, Goodman concluded that the upsurge of science in that century was followed by stagnation, and that it is impossible to find any Iberian contribution to the scientific revolution in seventeenth-century Europe.

Several books recently published in English offer a different view of Iberian science and, with sound reasons, have shown that there is a need to examine the scientific practices of Portugal and Spain in a more rigorous manner, as well as their role in the discovery and conquest of the New World and the consolidation of scientific traditions in modern Europe. It is also important to highlight the contributions of books such as *Experiencing Nature: The Spanish American Empire and the Early Scientific Revolution*, by Antonio Barrera; *Secret Science: Spanish Cosmography and the New World*, by María Portuondo; and the essay 'The Colonial Iberian Roots of the Scientific Revolution', by Jorge Cañizares-Esguerra, along with Alison Sandman's valuable studies of cartography and the book by Asúa and French on the natural history of the New World.[21] Some other relevant works that have recently been released also

Jose Romo Feito (trans.), Barcelona, Paidós, 2000; Webster, Charles, 'From Paracelsus to Newton [...]', op. cit.; Westfall, Richard S., *The Construction of Modern Science: Mechanism and Mechanics*, 2nd ed., Cambridge, Cambridge University Press, 1977; and Wightman, William P. D., 'Science and the Renaissance', op. cit. These are all outstanding authors with original and significant contributions, yet none of them pays attention to the Iberian scientific activity of the 16th century.

20 Goodman, David, 'The Scientific Revolution in Spain and Portugal', in: Porter, Roy and Teich, Mikulas (eds.), *The Scientific Revolution in National Context*, Cambridge, Cambridge University Press, 1992, pages 158–177.

21 Barrera, Antonio, 'Experiencing Nature. The Spanish [...]', op. cit.; Portuondo, María M., 'Secret Science. Spanish cartography [...]', op. cit.; Cañizares-Esguerra, Jorge, 'The Colonial Iberian Roots of the Scientific Revolution', in: 'Nature, Empire, and Nation [...]', op. cit.; Sandman, Alison, 'Spanish Nautical Cartography [...]', op cit. and 'Mirroring the World [...]', op. cit.; Asúa, Miguel de and French, Roger, 'A New World of Animals [...]', op. cit. After the first Spanish edition of this book, the literature on Iberian science kept growing. For a complete review of recent literature, see Sánchez, Antonio, "The 'Empirical Turn' in the Historiography of the Iberian and Atlantic Science in the Early Modern World: From Cosmography and Navigation to Ethnography, Natural History, and Medicine", *Tapuya: Latin American Science, Technology, and Society*, 2019

carefully study these topics, such as the vast books *Imperio e Información: funciones del saber en el dominio colonial español*, by Arndt Brendecke, and *La Espada, la cruz y el Padrón; Soberanía, fe y representación cartográfica en el mundo ibérico bajo la Monarquía Hispánica, 1503–1598*, by Antonio Sánchez.[22]

All of these authors share the same concern: English-speaking historians of science have not given the Iberian exploration of the Atlantic the importance it deserves. The history of modern science cannot be limited to the Copernican revolution nor to the achievements of certain minds of genius in Western Europe. The idea that scientific modernity originated in crucial experiments or in the ideas of some philosophers who were miraculously modern and had a secular outlook is becoming is less and less convincing. By contrast, studies of more mundane cultural phenomena, such as the development of the modern printing press,[23] the formation of large empires, and the emergence of major commercial enterprises, seem to offer richer historical explanations of the foundations of modern culture.[24]

It has not been the intention of this book to engage in a debate about whether the sixteenth-century Iberians were 'modern'. That question would only make sense if there were an *a priori* definition of modernity or modern science. If we take science to mean the practices that seem scientific by contemporary standards, it may be anachronistic to speak of the 'science of the 16th century'. This book has been more interested in understanding the relations between knowledge, technology, religion, and empire than in finding a date and a place for the origin of modernity. The idea that a 'scientific revolution' gave rise to a new and unique form of doing science has already been sufficiently challenged.[25] Although we cannot ignore the magnitude of the political and cultural changes that occurred in the period traditionally known as the 'Renaissance', we also need to acknowledge the role that the sixteenth-century Iberian explorers played in the history of Europe and the Western world.

The Spanish ships that crossed the Atlantic in the 15th and 16 centuries did not get to India as they had hoped, but in failing to do so found a new world, one blessed with an exuberant and unknown nature. The exploration of those territories that were unknown in Europe and the endeavor to take

22 Brendecke, Arndt, *Imperio e Información* [...], op. cit.; Sánchez, Antonio, *La Espada, la cruz y el Padrón* [...], op. cit.

23 Einsenstein, Elizabeth, *The Printing Revolution in Early Modern Europe*, Cambridge University Press, 1983.

24 Harold J. Cook, 'Matters of Exchange: Commerce [...]', op. cit.

25 A fairly complete view of the problem is found in: Shapin, Steven, 'La revolución científica. Una interpretación [...]', op. cit.

possession of them had a notable impact on the development of Western science. This 'discovery' implied new challenges for European science. The explorers, cartographers, and naturalists of the 16th century had to establish transatlantic routes and means of communication, make maps of a new world, name and describe the unknown, and place new territories and creatures that had never been seen before in a familiar language and framework of reference.

When they confronted the challenges of incorporating knowledge of the New World into their maps and treatises, the Spanish cosmographers consolidated new forms of understanding the sciences of the Earth and, in that way, played a part in changing the way in which Europeans understood both the natural world and themselves.[26]

In the 16th century, chroniclers narrated and described 'incredible' things, but they did so with the intention of finding the truth and that is why it would be fair to say that they created a new standard of credibility. On the one hand, credibility had to be based on the incorporation of the new into familiar frameworks of reference, and, on the other, on the direct experience of impartial observers who shunned inventing or imagining things.

Efforts to compile encyclopedias and catalogues of the natural world have always been imperial endeavors. Natural and social order are inseparable, and all great empires have tried to order the world. The new order, the modern world, and Eurocentrism cannot be understood without explaining the cultural, technological, and discursive practices that, like cartography, history, and natural history, turned the unknown and alien into something that was familiar and one's own.

The new natural history and new geography written by the travelers and explorers in the territory of America incorporated two apparently conflicting features into their central arguments: the authority of the philosophical knowledge of the thinkers of antiquity and the undeniable value of the direct experience of qualified witnesses .

This idea of a direct and personal experience was crucial in the rhetoric of authors such as Oviedo, for whom 'all of these things and many others which might be said for this purpose are easy to prove and very worthy of belief by those who have read them or have traveled through the world, since what they have seen themselves will have taught them that what is said here is based on experience'.[27]

26 Portuondo, María M., 'Secret Science. Spanish cartography [...]', op. cit., pages 1–2.

27 Fernández de Oviedo, Gonzalo, 'Historia general y natural de [...]', op. cit., vol. 2, Book XII, Chapter X, page 39.

In that regard, both the meticulous description of the circumstances in which the observations were made and the realism with which the subject described what he saw were of central importance. That is why we often find circumstantial descriptions that explain who made the observations and when and how these observation took place:

> As I was going from the Tierra Firme to this island, in the year 1515, I passed the Neiva River in a balsa of reeds, near where that river which is very powerful and broad enters the sea, and ten or twelve *indios* were going there, swimming around a balsa of reeds, guiding it. I would like to explain how that happened, because it is well that the chroniclers who write about the things of the Indies from Spain should know how far they are from understanding them (or understanding themselves) when their eyes are distant from seeing the things [which occur] here.[28]

Detailed observation is not merely a contemplative experience. Very much the opposite, it presupposes the careful activity of an attentive observer who is skilled at establishing relations between his own experience and that of other authors, especially those who are authorities on the subjects in question. In some cases, these were methodical, almost experimental, observations in which conditions were set up to verify or refute some thesis. In some cases, there was even a resort to controlled experiences such as dissecting animals, as when Oviedo wrote of a snake: 'I ordered it to be opened up and thirty or so eggs were found there, like the yolks of a chicken's eggs'.[29]

The tension between the experience of the travelers and the authority of the classics, which we have seen in the treatises on navigation, cosmography, and natural history, led to the consolidation of new forms of scientific authority. The main paradigms of antiquity, the classical notions of geography and natural history and the premises about the peopling of the world, showed their limitations with the discovery of the New World. The Spanish cosmographers and naturalists were able to continue being faithful to the principles of classical scholars such as Pliny or Ptolemy, but those ancient authorities were not able to explain the new geographical or natural discoveries and, as a result, their disciples were forced to point out their errors or limitations.[30]

28 Ibid., Book XI, Chapter VIII, page 35.

29 Ibid., page 37.

30 For a complete analysis of the role of direct experience in the early modern Iberian science, see Antonio Barrera, *Experiencing nature* [...], op cit.

Oviedo wanted to emulate Pliny, but his authority lays in direct experience and not books: 'How much better do you understand the things which are seen than those which are read', he stated in his *Historia general y moral de las Indias*.[31]

The growing tendency since the end of the 20th century to call knowledge and reality a 'social construction' seems to lessen the authority of both the direct and bodily experience of the explorers and the objects of their study. This book has sought to show that the New World, its geography, and all the objects, animals, and plants that formed part of it were active and crucial actors in the shaping of a new world order and that they cannot be reduced to a mere 'construction' or 'invention' of Europe.

There are enough reasons to recognize key antecedents of what might be called a 'global science' in the Iberian World in the 16th century. Therefore, the meager attention that historians of science have paid to the Spanish and Portuguese scientific practices of that era is without justification. To acknowledge the importance of the Iberian continent in Western science has significant implications, because it leads us to carefully rethink the traditional history of European modernity. The names of the Iberian monarchs and merchants of the 16th century, with all of their pilots, cartographers, chroniclers and naturalists, should be added to the names of those who are usually regarded as the fathers of scientific modernity—people such as Francis Bacon, Rene Descartes, Robert Boyle, Galileo Galilei, and Isaac Newton. In addition, it would only be fair to pay more attention to the role of Arab bodies of knowledge and traditions in the history of modern science. In the age of Alfonso X of Castile, known as Alfonso the Wise (1252–1284), Spain was a center for the translation, reinterpretation, and conservation of Arabic texts, which means that the Iberian traditions of cosmography and navigation had their antecedents in Arabic astronomy translated into Latin in the then territory of Spain.[32]

The new experiences of the Iberian travelers and explorers clearly had a role in the endeavor of obtaining a commercial, political, and religious control of remote lands and peoples, and should be explained in the context of a cultural tradition in which Christian dogma, the Arab scientific legacy, and Renaissance humanism flourished in the courts and universities of Portugal and Spain. Of course, to explain the importance of that process

31 Ibid., vol. 5, Book I, Chapter XXVII, 'which deals with two hurricanes or storms which occurred in the island of Hispaniola and other islands close to it, and of certain shipwrecks caused by those storms in the months of August and September of 1545', page 407.

32 López Piñero, José María, 'El arte de navegar en la [...]', op. cit., Chapters I and II.

in terms of an 'early scientific revolution'[33] or see the 'Iberian Roots of the Scientific Revolution'[34] there implies upholding problematic theses that require a great deal of scrutiny and care. The idea of finding an alternative origin for the 'scientific revolution', now in the exploration of the New World, is clearly problematic. An epistemological shift of such magnitude—the notion of a great revolution on which modern science was founded and the attempt to define its essential characteristics in terms of a legitimacy based on personal experience—is full of difficulties.

Resting the authority of science on direct experience and trustworthy testimonies was not an invention of the travelers of 16th century Spain, nor was it exclusive to them. Aristotle, Pliny, Dioscorides, and Ptolemy tried to base their works on a rhetoric that was much the same, and in that regard, to believe that the classics of antiquity were less rigorous or 'scientific' than the work of the 16th century naturalists is simply anachronistic. Aristotle, Pliny, Fernández de Enciso, Pedro de Medina, Diego García, Alonso de Chaves, Juan Escalante de Mendoza, Gonzalo Fernández de Oviedo, and Nicolás Monardes—and even Robert Boyle or Galileo Galilei—had strong reasons to trust in their own observations, but also in the testimonies of others. In the case of all, their experiences or the witnesses they cited were legitimate sources of truth. Hence, it is not a matter of arguing here for the existence of a direct causal relation between the nautical treatises of 16th century Spain and the Copernican revolution, but it is important to point out that the 'crisis' of the great classical paradigms had clear antecedents in the field of geographical exploration, which as we have explained, was closely related to the history of modern astronomy.

The complexity and wealth of the knowledge about the American nature goes beyond the classical European traditions or the experience of the voyagers. One aspect, which has not received so much attention, is the role played by native traditions, which is more difficult for historians and much less visible in the Eurocentrism that characterizes Western historiography. As we saw in the sixth chapter of this book, the explorers not only encountered a strange nature, but that this encounter took place complex cultural contexts and the frameworks for studying American nature were always mediated by the practices and knowledge of the natives. Understanding the new natural history that arose from the exploration of the New World implies acknowledging a long process of interpreting the knowledge of others; that is, it requires us to understand that Europeans made use of the knowledge

33 Barrera, Antonio, 'Experiencing Nature. The Spanish [...]', op. cit.

34 Cañizares-Esguerra, Jorge, 'Nature, Empire, and Nation [...]', op. cit.

of the cultures they conquered and gradually incorporated those bodies of knowledge into their own frameworks of reference, even proclaiming themselves to be the legitimate spokesmen for the natural order.[35]

## The Empire and the challenge of standardization

The thing which turns out to be novel, special, and interesting about the Spanish treatises on seamanship, cosmography, and natural history, and the use they made of the classical traditions, was their wager on solving practical problems related to imperial expansion. The work of the Spanish cosmographers of the 16th century had a practical meaning and a clear aim to serve the Hapsburg Crown, whose legitimacy, we should recall, had a religious foundation.

The cosmography of the New World developed in Spain took shape in the midst of the academic requirements of the humanism of its universities and imperial needs linked to the experience of a new world. The humanist tradition, which María Portuondo calls 'Renaissance cosmographies', would be gradually replaced by a form of study that was mathematical, empirical, and pragmatic.[36]

Therefore, there was a close relation between the religious and imperial expansion of Spain and the development of a science that also had global pretensions. The history of the regulation of the colonial administration was inextricably linked with the history of science. Accordingly, the subjects of a bureaucratic, commercial, juridical, nautical, and cartographic nature—and moral and natural history—should be regarded as aspects of a single political and religious endeavor. The imperial Catholic State was a technical and scientific organization, and science and technology a matter of governance at the service of both the King and God.

The meticulous recording of data, based on the trustworthiness of the observed deed—a subject we already discussed in the chapters on navigation (Chapter 3), cartography (Chapter 5), and natural history (Chapter 6)—was not very different from what was done by an official of the Crown trained to record events or transactions in a legal language. Hence, the compilation of

35 Local and native traditions and knowledge have recently raised the interest of a number of authors. See, for instance, Barrera-Osorio, Antonio, and Nieto Olarte, Mauricio, "Ciencia, tecnología, saberes locales e imperio en el mundo atlántico del Siglo XVI al XIX", *Historia Crítica*, 73, 2019, pages 3–20.

36 Portuondo, María M., 'Secret Science. Spanish cartography [...]', op. cit., page 9 (and throughout the book).

scientific data was closely related to a legal rhetoric for recording evidence,[37] in accordance with the mission of the servants of the Crown, a rhetoric that was shared by those working in the fields of trade, geography, medicine, and natural history.

The machines of empire were also bureaucratic and administrative machines and that is why this book has concerned itself with the imperial institutions of the 16th century (Chapter 2). The *Real Consejo de Indias* and the *Casa de Contratación* were founded with the explicit purpose of setting those tools of conquest into motion: the latter served as a center for collecting, ordering, and disseminating an inexhaustible source of novel information: The New World. More than a storehouse, the *Casa de Contratación* was a place for recording data, a center for compiling and accumulating information. To keep a 'record' is a matter of writing things down, of producing texts, lists, tables, maps, or chronicles. For that reason, one notes a strong effort to formalize the practices of writing and standardize the visual representation of the natural world. Thus, the *Casa de Contratación* was the place where the knowledge and technical practices that were linked to the political interests of the Crown and merchants[38] were standardized and institutionalized.

In that way, *the Casa de Contratación* turned into a 'center of calculation',[39] which not only required the obtaining and building of devices but also the formalization of technical skills to ensure the optimization and sound administration of trade and imperial expansion.[40]

The project of collecting information about such vast territories ran into the enormous difficulties of cultural and geographical distances that were unprecedented in imperial history. In order for the information provided by the navigators and explorers to be useful to others, that is, to the imperial project, common game rules had to be laid down for compiling experiences and observations. Projects that were essential for the Empire, such as the making of maps, for example, would have been impossible without the standardization of experience, and it is clear that a heap of accounts of the individual experiences of explorers and navigators would have been useless without stable codes of observation.

37 Asúa, Miguel de and French, Roger, 'A New World of Animals [...]', op. cit., page 72.

38 Barrera, Antonio, 'Experiencing Nature. The Spanish [...]', op. cit., page 48.

39 The idea of a 'center of calculation' was introduced into the sociology of science by Bruno Latour and refers to places where information is collected and knowledge produced. See: Latour, *Ciencia en Acción: como seguir a los científicos e ingenieros a través de la sociedad*, Barcelona, Editorial Labor, 1992.

40 Antonio Barrera, op. cit.

The Iberian endeavor to catalogue, name, and describe the nature and geography of the New World was a colossal one, and the results were remarkable: ethnographic treatises, complex legal and moral debates, encyclopedias of natural history, treatises on medicinal botany, maps, and navigation manuals. The sum of those practices made sailing safer and trade more effective, and it also set into motion a huge scientific project that ultimately turned *Terra Incognita* into *Spanish America*. All of these scientific and technical achievements not only helped Spain's aim to control the New World, but they also circulated beyond its dominions and showed the world a different horizon.

Imperial control may be reduced to a fundamental problem of mobilizing persons, goods, and information. For that task of mobilizing, accumulating, and organizing information to be successful, the training and work of many persons and callings, linked by powerful institutions, was needed, but also, and no less important (as Chapter 4 explained in detail), many complex technological products as well, such as ships, navigation manuals, weapons, and instruments of observation. Human testimonies would not have been so trustworthy if they were not aided by calibrated instruments since they eliminated local and circumstantial factors. As we have explained, this depersonalization not only did away with the arbitrary nature of the senses, but it also forged a homogeneous discourse that made it easier to accumulate the experiences of different places and different persons. That required both standardized instruments and disciplined observers following common rules.

It is worth recalling that this need for standardization is crucial for an understanding of the nature of scientific knowledge. A datum or a measurement can only be regarded as precise insofar as it is useful to others. One of the fundamental characteristics of the idea of science, as we know it today, is the stability of and consensus on the units of measurement. That was a critical problem for shipbuilders, cartographers, and locating one's position at sea, as well as for the merchants and bureaucrats who oversaw trade. Only to the extent that there is a science with global parameters is it possible to proclaim the universality of knowledge.

The political dimension of scientific practices has been concealed, partly by the same historians of ideas, who have limited the problem of knowledge to a purely conceptual level and thus created an autonomous space for science, with some connection to society, nature, or artefacts. The power of knowledge cannot be explained by locating it in the 'world of ideas'. On the contrary, it is the result of the capacity to mobilize and transform the world in order to construct a new order that is both social and natural. For that reason, we cannot ignore the material components of science or relegate them to a secondary plane: institutions, people, machines, texts, maps, tables, pictures, periodical publications, and their publics, and, in general, the set of objects and persons that make an ordered and comprehensible world possible.

María Portuondo defines the mathematization of picturing the world, which is generally attributed to mechanical philosophy and the work of thinkers such as Rene Descartes, as a response to the need for control and order in the imperial administration. In his study of the role of Dutch global trade in the 17$^{th}$ and 18$^{th}$ centuries, Harold J. Cook argues that the careful description of natural products and their medicinal uses, the collection of facts and organized information about the natural world, and the accumulation and exchange of products from remote places required strong networks and practices that ultimately proved crucial for the consolidation of a new science. The economic changes that took place in the first age of global trade gave great importance to the careful accumulation of information and that practice shaped the way in which the natural world became known.[41] However, Cook's study does not deal with the Iberians, who initiated these practices in the 16$^{th}$ century.

Thus, the two pillars of what has traditionally been called 'modern science'—empiricism and mathematization of the natural world—were found in the Iberian endeavor of imperial control in the New World. Neither empiricism nor rationalism began with Bacon and Descartes; instead, these thinkers set out to explicitly formulate the foundations of a knowledge that had been put into practice long before and acquired a much higher level of formalization in the 17$^{th}$ century.

## Eurocentrism

We have already said that the problem of Eurocentrism has to be understood in terms of Europe's capacity to incorporate the strange within a single and familiar framework of reference. Now, this was only possible insofar as the information that came from the new experiences of many travelers in the New World could be added up, combined, and linked into a single and coherent representation of the world. Administering large territories at a distance was a less arduous job when one had a map or an orderly account

41 Cook, Harold J, 'Matters of Exchange. Commerce [...]', op. cit., pages 3–4.

of their resources; that is, it was a form of forging links, of domesticating the alien.

The set of practices that made up the Iberian exploration of the Atlantic—namely, navigation, cartography, natural history, medicine, and the translation of the knowledge of American cultures—converged in a greater historical problem: the consolidation of a Christian Empire with global pretensions and the forming of a new world order centered in Western Europe.

The powerful notion of a universal truth was an ideal. Both for philosophy and science, and religion; furthermore, as we have repeatedly noted, politics, religion, and knowledge were inseparable in the 16th century.

The Spanish Empire was therefore the result of a total of practices of imperial control that made it possible for the Christian monarchs to turn their wish to dominate the world into a reality. The 16th century was the scenario of the consolidation of the great empires of modern Europe; hence, the discovery and conquest of the New World was a process which is crucial for understanding modernity. Just as the great Iberian empires of the 16th century were major scientific endeavors, modern science can be also understood as a political and commercial endeavor in which the Iberian exploration of the Atlantic played a crucial role.

What had usually been an unknown world, different from the *Orbis Terrarum*, was integrated into the known world and adopted by Europe, which, in turn, went through a notable change as it happened. The world, as it was understood by Europeans, was subjected to considerable changes in what is known as 'the age of discoveries'. The *Orbis Terrarum*, the dwelling of the human species, was no longer the same when the New World was no longer spoken of as a new piece of the Earth and began to be understood as a new cosmology that also entailed the redefinition of the Old World.[42] This process of comprehension and integration required the building of links, of bridges, between the old and the new, in a process that was simultaneously one of construction, integration, and subordination. The depiction of the whole of the terrestrial sphere and the cataloguing of all of its creatures, just as it was wished to be done in the 16th century, was the powerful expression of a new relation between man and the cosmos.

As we have argued in the introduction of this book, the notion of a 'comprehension of the New World' may be useful, so long as one always takes into account that any act of comprehension presupposes the consolidation

42 This idea is one of the conclusions of the book by José Rabasa, 'Inventing America: Spanish Historiography [...]', op. cit.

of a new subject that is able to transform the unknown into the familiar. Such a transformation only becomes possible insofar as heterogeneous elements—human, religious, technical, and natural—interact to form powerful networks and a new natural and social order.

The comprehension of that New World of which America began to form a part is, at the same time, the mirror in which one can understand the formation of Europe itself and the framework in which the frontiers between the center and the periphery of the new world order are defined. This comprehension is the consolidation of Eurocentrism. The great religious mission of Christians and modern scientific practices turned Europe and the West into a universal culture and became an expansive and dominant form of thinking and acting.[43] It was in this New World that Europe held a central place as the cradle of a religion, a science and a civilization whose legitimacy and authority over others was conceived of in global terms.

43 Ibid., page 8.

# Bibliography

## Primary sources

### Manuals of seamanship


Cortés de Albacar, Martin. *Breve Compendio de la Sphera y de la arte de navegar. Con nuevos instrumentos y reglas ejemplificado con muy sutiles demostraciones*, Sevilla, 1551.

Chaves, Alonso de. *Quatri Partitu in Cosmographia Pratica y por otro nombre llamado Espejo de Navegantes*, [1518–1538], Castaneda, Paulino; Cuesta Domingo, Mariano, and Hernández Aparicio, Pilar (eds.), Madrid, Instituto de Historia and Cultura Naval, 1983. (Colección Clásicos Tavera. Fundación Mapfre).

Chaves, Jerónimo. *Repertorio de los tiempos*, Sevilla, 1548.

Escalante de Mendoza, Juan de. *Itinerario de Navegación de los mares y tierras occidentales*, Sevilla, 1575. (Colección Clásicos Tavera. Fundación Mapfre).

Falero, Francisco. *Tratado del Esphera y del arte de marear. Con el Regimiento de las alturas, con algunas reglas nuevamente escritas muy necesarias*, Sevilla, Cromberger, 1535.

Fernández de Enciso, Martin. *Suma de Geographia que trata de todas las partidas y provincias del mundo: en especial de las indias y trata largamente del arte de marear juntamente con la sphera en romance con el Regimiento del sol y del norte*, Sevilla, Casa de Andrés Burgos, 1546. (Colección Clásicos Tavera. Fundación Mapfre).

García de Céspedes, Andrés. *Regimiento de Navegación*, Madrid, Juan de la Cuesta, 1606.

García de Palacio, Diego. *Instrucción náutica, para el buen uso y regimiento de las naos, su traza y gobierno conforme a la altura de México*, [1587, Pedro Ocharte, Mexico], Madrid, Editorial Naval, 1993. (Colección Clásicos Tavera. Fundación Mapfre).

Medina, Pedro de. *Regimiento de Navegación. Contiene las cosas que los pilotos han de saber para bien navegar y los remedios y avisos que han de tener para los peligros que navegando les pueden suceder*, Sevilla, 1563. (Colección Clásicos Tavera. Fundación Mapfre).

—. *Arte de navegar en que se contienen todas las reglas, declaraciones, secretos y avisos que a la buena navegación son necesarios y se debe saber*, Valladolid, Francisco Fernández de Córdova, 1545. (Colección Clásicos Tavera. Fundación Mapfre).

Pigafetta, Antonio. *El primer viaje alrededor del mundo. Relato de la expedición de Magallanes y Elcano*, Leoncio Cabrero Fernández, (ed.), Madrid, Dastin, 1985.

Sacrobusto, Johannes de. *Tratado de la Sphera*, Jerónimo de Chaves (trans.), Sevilla, 1545.

Vargas Machuca, Bernardo. *Milicia y descripción de las Indias*, Madrid, 1599.

Vellerino de Villalobos, Baltasar. *Luz de navegantes donde se hallaran derrotas y señas de las partes marítimas de las indias, islas y tierra firme del mar océano*, [1592], Madrid, Museo Naval de Madrid / Universidad de Salamanca, 1984.

Vespucci, Amerigo. *Edición facsimilar de las Cartas de Vespucio*, Biblioteca Nacional de Colombia, Bogota, 1942.

Zamorano, Rodrigo. *Compendio de la Arte de Navegar*, Sevilla, Alonso de la Barrera, 1581.

## Books on natural history

Acosta, José de. *Historia natural y moral de las Indias* [1590], José Alcina Franch, (ed.), Madrid, Dastin Historia, 2003.

Colón, Cristobal. *Diario de a bordo*, Luis Arranz Márquez (ed.), Madrid, Dastin, 2000.

De la Cruz, Martín. *Libellus de medicinalibus indorum herbis* [1552], (que traduce *Libro medicinal de las hierbas de Indias*), Juan Badiano (trans.), Mexico, Fondo de Cultura Económica, Instituto Mexicano del Seguro Social, 1991.

Fernández de Oviedo, Gonzalo. *Historia general y natural de las Indias* [1535], 5 vols., Madrid, Atlas, 1959.

—. *Sumario de la Natural Historia de las Indias* [1526], Manuel Ballesteros Gaibrois (ed.), Madrid, Dastin, 2002.

Hernández, Francisco. *Historia Natural de Nueva Espana* [1571–1576], Mexico, Universidad Nacional Autónoma de Mexico, 1959.

López de Gomara, Francisco. *Historia general de las Indias* [1552], Guibelalde, Pilar y Aguilera, Emiliano (eds.), Barcelona, Iberia, 1965.

Mandeville, Sir John. *The Travels of Sir John Mandeville*, C. W. R. D. Moseley (trans.), Harmondsworth, Penguin Books, 1983.

Monardes, Nicolas. *La historia medicinal de las cosas que se traen de nuestras Indias Occidentales que sirven en Medicina (1565–1574)*, Madrid, facsimile of the edition of Seville (1580), Ministerio de Sanidad y Consumo, 1989.

Plinio Segundo, Cayo. *Historia natural*, Jerónimo de Huerta y Francisco Hernández (tras.), Madrid, Visor Libros / Universidad Nacional de Mexico, 1999, Book VIII.

Polo, Marco. *Libro de las maravillas*, Barcelona, Ediciones B / Grupo Zeta, 1997.

Sahagun, Bernardino de, *Historia general de las cosas de la Nueva Espana, I*, Juan Carlos Temprano (trans.), Madrid, Dastin Historia, 2001.

## Secondary sources

Acosta Rodríguez, Antonio, Gonzalez Rodríguez Adolfo and Vila Vilar, Enriqueta (coordinators.). *La Casa de la Contratacion y la navegación entre España y las Indias*, Sevilla, Universidad de Sevilla / CSIC / Escuela de Estudios Hispano-Americanos / Fundación el Monte, 2003.

Akerman, James R. (ed.). *Cartographies of Travel and Navigation*, Chicago, University of Chicago Press, 2006.

Almeida, Bruno. "Transmitting nautical and cosmographical knowledge in the 16th and 17th centuries: The case of Pedro Nunes". *Centaurus*, 60(3), 2018, pages 216–229.

Ash, Eric and Sandman, Alison. "Trading Expertise: Sebastian Cabot between Spain and England". *Renaissance Quarterly*, 57(3), 2004, pages 813–846.

Asua, Miguel de and French, Roger. *A New World of Animals: Early Modern Europeans on the Creatures of Iberian America*, Aldershot, England, Ashgate, 2005.

Bacon, Francis. *Novum organum* [1620], Barcelona, Editorial Fontanella, 1979.

Barrera, Antonio. *Experiencing Nature: The Spanish American Empire and the Early Scientific Revolution*, Austin, University of Texas, 2006.

Barrera-Osorio, Antonio and Nieto Olarte, Mauricio. "Ciencia, tecnología, saberes locales e imperio en el mundo atlántico del Siglo XVI al XIX", *Historia Crítica*, 73, 2019, pages 3–20.

Beck, Ulrich. *El Dios personal. La individualización de la religión y el "espíritu" del cosmopolitismo*, Rosa S. Carbo (trans.), Barcelona, Paidos, 2009.

Bleichmar, Daniela; De Vos, Paula; Huffine, Kristine and Sheehan, Kevin (eds.). *Science in the Spanish and Portuguese Empires, 1500–1800*, Stanford, Stanford University Press, 2009.

Bourguet, Marie-Noelle; Licoppe, Christian and Sibum, Otto H. (eds.). *Instruments, Travel and Science: Itineraries of Precision from the Seventeenth to the Twentieth Century*, London, Routledge, 2002.

Brannigan, Augustine. *The Social Basis of Scientific Discoveries*, Cambridge, Cambridge University Press, 1981.

Braudel, Fernand. *El Mediterraneo y el mundo mediterráneo en la época de Felipe II*, 4th ed., Mexico, Fondo de Cultura Económica, 1997.

Bray, Warwick (ed.). *The Meeting of Two Worlds: Europe and the Americas, 1492–1650*. New York, Oxford, Oxford University Press, 1993.

Briggs, Asa and Burke, Peter. *De Gutenberg a internet. Una historia social de los medios de comunicacion*, Marco Aurelio Galmarini (trans.), Madrid, Taurus, 2002.

Brooke, John Hedley. *Science and Religion: Some Historical Perspectives*, Cambridge, Cambridge University Press, 1991.

Brown, Lloyd A. *The Story of Maps*, New York, Dover Publications, 1977.

Burke, Peter. *Formas de historia cultural*, 2nd ed., Madrid, Alianza, 2006.

Burtt, Edwin Arthur. *The Metaphysical Foundations of Modern Physical Science: A Historical and Critical Essay*, 2nd ed., London, K. Paul, 1932.

Butterfield, Herbert. *The Origins of Modern Science, 1300–1800*, New York, Macmillan Company, 1957.

Cabarcas, Hernando. *Bestiario del Nuevo Reino de Granada. La imaginación animalística medieval y la descripción literaria de la naturaleza americana*, Bogotá, Instituto Caro y Cuervo / Colcultura, 1994.

Callon, Michel. "Some elements of a Sociology of Translation. Domestication of the Scallops and Fishermen of St Brieuc Bay". In, Law, John (ed.), *Power, Action, and Belief: A New Sociology of Knowledge?*, London, Routledge & Kegan Paul, 1986, pages 196–223.

Campanella, Tommaso. "La imaginaria ciudad del sol, idea de una república, filosófica", en Imaz Eugenio, *Utopias del Renacimiento*, 17th ed., Fondo de Cultura Económica, Mexico, 2009.

Cañizares-Esguerra, Jorge. *Nature, Empire, and Nation: Explorations of the History of Science in the Iberian World*, Stanford, Stanford University Press, 2006.

—. *Puritan Conquistadors: Iberianizing the Atlantic, 1550–1700*, Stanford, Stanford University Press, 2006.

Cañizares-Esguerra, Jorge and Seeman, Erik R. *The Atlantic in Global History, 1500–2000*, Upper Saddle River, NJ, Pearson Prentice Hall, 2007.

Capesciotti, Ilaria Palmeri. "La fauna del libro XI del *Codice florentino* de fray Bernardino de Sahagun. Dos sistemas taxonomicos frente a frente". In, *Estudios de Cultura Náhuatl*, vol. 32, 2001, pages 189–221.

Carmagnani, Marcello. *El otro occidente. América Latina desde la invasión europea hasta la globalización*, Jaime Riera Rehren (trans.), Mexico, Fondo de Cultura Económica, 2004.

Castro and Bravo, Federico de, *Las naos españolas en la carrera de las indias. Armadas y flotas en la segunda mitad del siglo XVI*, Madrid, Voluntad, 1927.

Cerezo Martínez, Ricardo. *La cartografía náutica española en los siglos XIV, XV y XVI*, Madrid, Consejo Superior de Investigaciones Científicas, 1994.

Chaunu, Pierre. *European Expansion in the Later Middle Ages*, Oxford, North Holland Pub. Co., 1979.

Cipolla, Carlo M. *Las maquinas del tiempo y de la guerra. Estudios sobre la génesis del capitalismo*, Barcelona, Critica, 1999.

Cohen, I. Bernard. *The Birth of a New Physics*, New York, W. W. Norton & Co., 1985

Collins, Edward. "Interactions of Portuguese artisanal culture in the maritime enterprise of 16th-century Seville". *Centaurus*, 60(3), 2018, pages 203–215.

Cook, Harold J. *Matters of Exchange: Commerce, Medicine, and Science in the Dutch Golden Age*, New Haven, Yale University Press, 2007.

Coroleu, Alejandro. "Humanismo en España". In, Kraye, Jill (ed.), *Introduccion al humanismo renacentista*, Cambridge, Cambridge University Press, 1998, pages 295–330.

Del Castillo, Fernando. *Las crónicas de Indias,* Madrid, Intervención Cultural, 2004.

Delbourgo, James and Dew, Nicholas (eds.), *Science and Empire in the Atlantic World,* London, Routledge, 2008.

Dickenson, Victoria. *Drawn from Life: Science and Art in the Portrayal of the New World,* Toronto, University of Toronto Press, 1998.

Dijksterhuis, Eduard Jan. *The Mechanization of World Picture: Pythagoras to Newton,* [1950 ed.], Princeton, Princeton University Press, 1986.

Dirlik, Arif. "History without a Center? Reflections on Eurocentrism". In, Fuchs, Eckhardt and Stuchtey, Benedikt (eds.). *Across Cultural Borders: Historiography in global perspective,* Boston, Lanham, Rowman & Littlefield, 2002, pages 247–284.

D'olwer, Luis Nicolau. *Cronistas de las Culturas Precolombinas,* Mexico, Fondo de Cultura Económica, 1963.

Dominguez Ortiz, Antonio. "Sevilla a comienzos del siglo XVI". In, Acosta, Antonio, González, Adolfo and Vila, Enriqueta (coordinators.), *La Casa de la Contratación y la navegación entre Espana y las Indias,* Sevilla, Universidad de Sevilla / csic / Escuela de Estudios Hispano-Americanos / Fundación el Monte, 2003.

Dussel, Enrique. "Europa, modernidad y eurocentrismo". In, Lander, Edgardo (coordinators.), *La colonialidad del saber. Eurocentrismo y ciencias sociales. Perspectivas latinoamericanas,* Buenos Aires, CLACSO, 2000, pages 41–53.

Eisenstein, Elizabeth. *The Printing Revolution in Early Modern Europe,* Cambridge, Cambridge University Press, 1983.

Elliot, John Huxtable. *Empires of the Atlantic World: Britain and Spain in America, 1492–1830,* New Haven / London, Yale University Press, 2006.

Fernández-Armesto, Felipe. *Colon,* Barcelona, Critica, 1992.

—. "Maps and Exploration in the Sixteenth and Early Seventeenth Centuries". In, Woodward, David, The *History of Cartography. Cartography in the European Renaissance,* vol. 3, Chicago / London, University of Chicago Press, 2007, pages 738–773.

—. *Pathfinders: A Global History of Exploration,* New York, W. W. Norton, 2006.

Findlen, Paula, "Inventing Nature. Commerce, Art, and Science in the Early Modern Cabinet of Curiosities", in Smith, Pamela and Findlen, Paula (eds.), *Merchants and Marvels: Commerce, Science, and Art in Early Modern Europe,* New York, Routledge, 2002, pages 297–323.

French, Roger Kenneth. *Ancient Natural History: Histories of Nature,* London, Routledge, 1994.

Gadamer, Hans-Georg. *Verdad y Método. Fundamentos de una hermeneutica filosofica,* Salamanca, Sígueme, 1992.

González Sánchez, Carlos Alberto. "La Casa de la Contratación y la historia cultural". In, Acosta, Antonio, González, Adolfo and Vila, Enriqueta (coordinators.), *La Casa de la Contratación y la navegación entre España y las Indias*, Sevilla, Universidad de Sevilla / CSIC / Escuela de Estudios Hispano-Americanos / Fundación el Monte, 2003.

Goodman, David C. "Cosmography and the Crown". In, *Power and Penury: Government, Technology, and Science in Philip II's Spain*, Cambridge, Cambridge University Press, 1988.

—. "The Scientific Revolution in Spain and Portugal". In, Porter, Roy and Teich, Mikulas (eds.), *The Scientific Revolution in National Context*, Cambridge, Cambridge University Press, 1992, pages 158–177.

Goody, Jack. *The Theft of History*, Cambridge, Cambridge University Press, 2006.

Grafton, Anthony. *New Worlds, Ancient Texts: The Power of Tradition and the Shock of Discovery*, Cambridge, The Belknap Press of Harvard University Press, 1995.

Gruzinski, Serge. *El pensamiento mestizo. Cultura amerindia y civilización del Renacimiento*, Barcelona, Paidos, 1999.

—. "Passeurs y elites 'católicas' en las Cuatro Partes del Mundo. Los inicios ibéricos de la mundializacion (1580–1640)". In, Godoy, Scarlett O'Phelan and Salazar-Soler, Carmen (eds.) *Passeurs, mediadores culturales y agentes de la primera globalización en el mundo Iberico, Siglos XVI–XIX*, Lima, Pontificia Universidad Católica de Perú / Instituto Riva-Aguero/ Instituto Frances de Estudios Andinos, 2005, pages 13–29.

Hall, Alfred Rupert. *The Scientific Revolution 1500–1800: The Formation of the Modern Scientific Attitude*, London, Longmans and Green, 1954.

Haring, Clarence H. *Trade and Navigation between Spain and the Indies in the Time of the Hapsburgs*, Cambridge, Harvard University Press, 1918.

Harley, John Brian. *La nueva naturaleza de los mapas. Ensayos sobre la historia de la cartografía*, Leticia García Cortes and Juan Carlos Rodríguez (trans.), Mexico, Fondo de Cultura Económica, 2005.

Headrick, Daniel R. *The Tools of Empire: Technology and European Imperialism in the Nineteenth Century*, Oxford, Oxford University Press, 1981.

—. *Power over Peoples: Technology, Environments, and Western Imperialism, 1400 to the Present*, Princeton, Princeton University Press, 2010.

Heers, Jacques. *Cristobal Colón*, José Esteban Calderón and Ortiz Monasterio (trans.), Mexico, Fondo de Cultura Económica, 1992.

Heidegger, Martin. *El ser y el tiempo*, José Gaos (trans.), Mexico, Fondo de Cultura Económica, 1983.

Hughes, Thomas. "The Evolution of Large Technological Systems". In, Bijker, Wiebe K., Hughes, Thomas and Pinch, Trevor (eds.), *The Social Construction of Technological Systems: New Directions in the Sociology and History of Technology*, Cambridge, MIT Press, 1989, pages 51–82.

Koestler, Arthur. *The Sleepwalkers: A History of Man's Changing Vision of the Universe*, London, Macmillan, 1959.

Koyre, Alexander. *La Revolution astronomique. Copernic, Kepler, Borelli*, Paris, Hermann, 1961.

Kraye, Jill (ed.). *Introducción al humanismo renacentista*, Cambridge, Cambridge University Press, 1998.

Kuhn, Thomas S. *The Copernican Revolution: Planetary Astronomy in the Development of Western Thought*, Cambridge, Harvard University Press, 1957.

Hutchins, Edwin. *Cognition in the Wild*, Cambridge / London, The MIT Press, 1995.

Latour, Bruno, "Drawing Things Together". In, Lynch, Michael and Woolgar, Steve (eds.), *Representation in Scientific Practice*, Cambridge, MIT Press, 1990, pages 19–68.

—. *Ciencia en acción: como seguir a los científicos e ingenieros a través de la sociedad*, Barcelona, Editorial Labor, 1992.

—. *La esperanza de Pandora. Ensayos sobre la realidad de los estudios de la ciencia*, Tomás Fernández Auz (trans.), Barcelona, Gedisa, 2001.

—. *Reassembling the Social: An Introduction to Actor-Network-Theory*, Oxford, Oxford University Press, 2005.

Law, John. "On Methods of Long-Distance Control: Vessels, Navigation, and the Portuguese Route to India". In, *Power, Action and Belief. A New Sociology of Knowledge?*, London, Routledge & Kegan Paul, 1986, pages 234–263.

—. *A Sociology of Monsters: Essays on Power, Technology, and Domination*, London / New York, Routledge, 1991.

Law, John and Hassard, John (eds.). *Actor Network Theory and After*, Oxford, Blackwell, 1999.

Leitâo, Henrique. "Ars e ratio: A náutica e a constituição da ciência moderna". In, Vicente Maroto, María Isabel and Piñeiro, Mariano Esteban (coordinators.), *La Ciencia y el Mar*, Valladolid, Server-Cuesta Impresores, 2006, pages 183–207.

—. "Instruments and Artisanal Practices in Long-distance Oceanic Voyages". *Centaurus*, 60(3), 2018, pages 189–202.

Lestringant, Frank. *L'Atelier du cosmographe ou L'Image du monde a la Renaissance*, Paris, Albin Michel, 1991.

—. *Mapping the Renaissance World: The Geographical Imagination in the Age of Discovery*, David Fausett (trans.), Berkeley, University of California Press, 1994.

Lester, Toby. *The Fourth Part of the World: The Race to the Ends of the Earth, and the Epic Story of the Map that Gave America its Name*, New York, Free Press, 2009.

Lindeberg, David. *Los inicios de la ciencia occidental. La tradición científica en el contexto filosófico, religioso e institucional (desde 600 a. c. hasta 1450)*, Barcelona, Paidos, 2002.

Linebaugh, Peter and Rediker, Marcus. *La Hidra de la Revolución. Marineros, esclavos y campesinos en la historia oculta del Atlántico*, Mercedes García Garmilla (trans.), Barcelona, Critica, 2005.

Lloyd, G. E. R. *Greek Science After Aristotle*, New York, W. W. Norton & Company, 1975.

López, Pinero José María, *Ciencia y técnica en la sociedad española de los siglos XVI y XVI*, Barcelona, Labor, 1979.

—. *El arte de navegar en la España del Renacimiento*, Barcelona, Editorial Labor, 1986. Lucena Salmoral, Manuel, *Piratas, corsarios, bucaneros y filibusteros*, Madrid, Síntesis, 2005.

Lynch, John (ed.). *Historia de España. Edad Moderna. El auge del imperio 1474–1598*, vol. 4., Barcelona, Critica, 2005.

Martin-Meras, María Luisa. *Cartografía marítima hispana. La imagen de América*, Madrid, Lunwerg, 1993.

—. "Las enseñanzas náuticas en la Casa de la Contratación". In, Acosta, Antonio, González, Adolfo and Vila, Enriqueta, (coords.), *La Casa de la Contratacion y la navegación entre España y las Indias,* University of Sevilla / CSIC/ Escuela de Estudios Hispano-Americanos / Fundación el Monte, 2003.

Martínez, José Luis. *Pasajeros de Indias. Viajes transatlánticos en el siglo XVI.* Mexico, Fondo de Cultura Económica, 1999.

—. *Cruzar el Atlántico*, Mexico, Fondo de Cultura Económica, 2004. Morison, Samuel Eliot, *El Almirante de la Mar Océano. Vida de Cristóbal Colon,* Mexico, Fondo de Cultura Económica, 1991.

Masters, Adrian. "A Thousand Invisible Architects: Vassals, the Petition and Response System, and the Creation of Spanish Imperial Caste Legislation". *Hispanic American Historical Review*, 98(3), 2018, pages 377–406.

Nieto, Mauricio. *Historia natural y política. Conocimientos y representaciones de la naturaleza americana*, Bogotá, Biblioteca Luis Ángel Arango / Universidad de los Andes / Universidad EAFIT, 2008.

—. *Orden Natural y Orden Social. Ciencia y política en el Semanario del Nuevo Reino de Granada*, Madrid, Consejo Superior de Investigaciones Científicas, 2007.

—. *Remedios para el Imperio. Historia natural y la apropiación del Nuevo Mundo*, Bogotá, Instituto Colombiano de Antropología, 2000.

O'Gorman, Edmundo. *La invención de América*, Mexico, Fondo de Cultura Económica, 1995.

O'Phelan Godoy, Scarlett and Salazar-Soler, Carmen (eds.). *Passeurs, mediadores culturales y agentes de la primera globalización en el mundo Ibérico, siglos XVI–XIX.* Lima, Pontificia Universidad Católica de Perú / Instituto Riva-Aguero / IFEA, 2005.

Pagden, Anthony. *Lords of all the World: Ideologies of Empire in Spain, Britain and France c. 1500–c. 1800*, New Haven, Yale University Press, 1995.

—. *Spanish Imperialism and the Political Imagination: Studies in European and Spanish-American Social and Political Theory*, New Haven, Yale University Press, 1990.

—. *The Fall of Natural Man: The American Indian and the Origins of Comparative Ethnology*, Cambridge, Cambridge University Press, 1987.

Pagel, Walter. *Science, Medicine, and Society in the Renaissance: Essays to Honor Walter Pagel*, Allen G. Debus (ed.), New York, Science History Publications, 1972.

Pérez-Mallaína, Pablo Emilio. *Spain's Men of the Sea: Daily Life on the Indies Fleets in the Sixteenth Century*. Baltimore, Johns Hopkins University Press, 1998.

Pestana Ramos, Fabio. *Por mares nunca dantes navegados. A aventura dos descobrimentos*, Sao Paulo, Contexto, 2008.

Porter, Roy and Teich, Mikulas (eds.). *The Scientific Revolution in National Context*, Cambridge, Cambridge University Press, 1992.

Portuondo, Maria M. *Secret Science: Spanish Cartography and the New World*, Chicago, University of Chicago Press, 2009.

Pulido Rubio, José. *El piloto mayor de la Casa de la Contratación de Sevilla, pilotos mayores del siglo xvi. (Datos biográficos)*, Sevilla, Tip. Zarzuela, 1923.

Pumfrey, Stephen. *Latitud. La verdadera historia del descubridor del magnetismo terrestre*, Barcelona, Editorial Juventud, 2008.

Rabasa, José. *Inventing America: Spanish Historiography and the Formation of Eurocentrism*, Norman, University of Oklahoma Press, 1993.

Rediker, Marcus. *The Slave Ship: A Human History*, New York, Penguin Books, 2008.

Romero, Luis Rene and Perez, Juan Felipe. *Naufragios y puertos marítimos en el Caribe colombiano del siglo XVI al XVII*, Mexico, Siglo XXI, 2005.

Rossi, Paolo. *Francis Bacon. De la magia a la ciencia*, Susana Gómez López (trans.), Madrid, Alianza, 1990.

Sánchez, Antonio. "The 'empirical turn' in the Historiography of the Iberian and Atlantic Science in the Early Modern World: From Cosmography and Navigation to Ethnography, Natural History, and Medicine", *Tapuya: Latin American Science, Technology, and Society*, 2019.

—— and Leitão, Henrique. "Artisanal Culture in Early Modern Iberian and Atlantic Worlds", *Centaurus*, 60(3), 2018, pages 135–140.

Sandman, Alison. "Mirroring the World: Sea Charts, Navigation, and Territorial Claims in Sixteenth-century Spain". In, Smith, Pamela and Findlen, Paula (eds.), *Merchants and Marvels: Commerce, Science, and Art in Early Modern Europe*, New York, Routledge, 2002, pages 83–108.

—. "Spanish Nautical Cartography in the Renaissance". In, Woodward, David, *History of Cartography: Cartography in the European Renaissance*, vol. 3, Chicago / London, University of Chicago Press, 2007, pages 1095–1142

Schäfer, Ernesto. *El Consejo Real y Supremo de las Indias. Historia y organización del Consejo y de la Casa de la Contratación de las Indias. La labor del Consejo de Indias en la administración colonial.* 2 vols. Madrid, Junta de Castilla and León / Marcial Pons Historia, 2003.

Selles, Manuel. *Instrumentos de navegación. Del Mediterráneo al Pacífico*, Madrid, Lunwerg, 1994.

Shapin, Steven. "Discipline and Bounding: The History and Sociology of Science as Seen through the Externalism–Internalism debate". In, *History of Science*, Vol. XXX, 1992, pages 333–369.

—. *La revolución científica. Una interpretación alternativa*, José Romo Feito (trans.), Barcelona, Paidos, 2000.

Smith, Pamela and Findlen, Paula (eds.). *Merchants and Marvels: Commerce, Science, and Art in Early Modern Europe*, New York, Routledge, 2002.

Sobel, Dava. *Longitude: The True Story of a Lone Genius Who Solved the Greatest Scientific Problem of his Time*, United States, Penguin Books, 1995.

Spivak, Gayatri Chakravorty. "Can the Subaltern Speak?". In, Nelson, Cary and Grossberg Larry (eds.), *Marxism and the Interpretation of Culture*, Chicago, University of Illinois Press, 1988, pages 271–313.

Taylor, Eva G. R. *The Haven-Finding Art: A History of Navigation from Odysseus to Captain Cook*, New York, Abelard-Schuman Limited, 1957.

Todorov, Tzvetan. *La conquista de América. El problema del Otro*, Flora Botton Burlá (trans.), Mexico, Siglo XXI, 1989.

Valverde, Nuria. *Actos de precisión. Instrumentos científicos, opinión pública y economía moral en la Ilustración Española*, Madrid, Consejo Superior de Investigaciones Científicas, 2007.

Vignolo, Paolo. "Más allá de las columnas de Hércules: un emblema de la modernidad temprana". In, Hearing Torres, Max S. and Perez, Amanda (eds.), *Historia Cultural desde Colombia. Categorías y debates*, Bogotá, Universidad de los Andes / Universidad Nacional / Universidad Javeriana, 2012, pages 139–164.

Vindel, Francisco (ed.). *Mapas de América en los libros españoles de los siglos XVI al XVII, (1503–1798)*, Madrid, Ministerio de asuntos exteriores, Dirección general de relaciones culturales y científicas, 1991.

Waldseemüller, Martin. *Cosmographiae Introductio* [1507] in Facsimile Followed by the Four Voyages of Amerigo Vespucci, Charles George Herbermann (ed.), New York, United States Catholic Historical Society, 1907.

Webster, Charles. *From Paracelsus to Newton: Magic and the Making of Modern Science*, Cambridge, Cambridge University Press, 1983.

Westfall, Richard S. *The Construction of Modern Science: Mechanism and Mechanics*, 2nd ed., Cambridge, Cambridge University Press, 1977.

Wightman, William P. D. *Science and the Renaissance*, 2 vols. Edinburgh, Oliver and Boyd / New York, Hafner Pub. Co., 1962.

Winner, Lagdon. "¿Tienen política los artefactos?". In, *La Ballena y el Reactor. Una búsqueda de los limites en la era de la alta tecnología*, Barcelona, Gedisa, 1987.

Yates, Frances A. *Giordano Bruno and the Hermetic Tradition*, Chicago, The University of Chicago Press, 1964.

Zamora, Margarita. *Reading Columbus*, Berkeley, University of California Press, 1993.

# About the Author

Mauricio Nieto Olarte has a Doctorate in the History of Science from London University. He is currently titular Professor at the Department of History and Geography as well as Dean of the Faculty of Social Sciences at the Universidad de los Andes, Bogota, Colombia. His research has focused on the relationship between science, technology, and politics in imperial and colonial contexts. He has worked on European expeditions to the New World, as well as on natural history, cartography, and navigation in order to explain the role of techno-scientific practices in the political and cultural history of the Hispano-American world from the 16$^{th}$ to the 19$^{th}$ centuries.

His most important publications include *Remedios para el imperio: historia natural y la apropiación del nuevo mundo*, Bogotá-ICANH, 2000, winner of the Silvio Zavala Award in Colonial History of America, (Mexico) 2001; *La Obra Cartográfica de Francisco José de Caldas*, Universidad de los Andes, 2006; *Orden Natural y Orden Social: ciencia y política en el Semanario del Nuevo Reyno de Granada*, Madrid-CSIC, 2007, which was awarded the Alejandro Ángel Escobar prize in Social Science (Bogotá) 2008; *Las máquinas del Imperio y el reino de dios. Reflexiones sobre ciencia, tecnología y religión en el mundo atlántico del siglo XVI*, Universidad de los Andes, 2013; and *Una historia de la verdad en Occidente*, Fondo de Cultura Económico-Universidad de los Andes, 2019.

# Index


Acosta, José de 249, 256, 257, 263, 282
Actor Network Theory 29, 30, 313
Accountant 62, 65, 66, 68
Admiral 8, 18, 56, 191
Aerodynamics 129
Africa 13, 18, 38, 39, 40, 41, 43, 46, 80, 81, 82, 106, 116, 130, 225, 232, 234, 237, 238, 243, 247
Agreement (s) 55, 185, 230, 232, 240
Alberto Magnus 110
Alcalá de Henares 61, 93, 251, 255, 277
Alcázar 61
Alexander VI 230, 246
Alfraganus 44
Alguacil 190
Almagest 92, 95
Álvarez Cabral, Pedro 18
Altitude 13, 104, 109, 118, 146, 149, 150, 160, 175, 193, 194, 232
Amazon 250
America 13, 18, 19, 20, 21, 22, 23, 24, 25, 31, 33, 34, 39, 44, 46, 51, 53, 64, 65, 91, 117, 118, 136, 139, 141, 144, 217, 220, 221, 222, 223, 234, 236, 237, 238, 240, 241, 246, 251, 252, 253, 255, 257, 262, 263, 266, 272, 273, 275, 277, 280, 286, 289, 296, 302, 304, 305, 309, 310, 311, 313, 314, 315, 316
American possessions 32, 55, 65
Ananá 264
Anchors 131, 135, 136, 139
Andalucía 97
Antarctic 99, 158
Antilles 136
Antwerp, Juan Lacio 221
Anteater 258, 264, 266
Apocalypse 286
Art of navigation 7, 75, 78, 79, 86, 88, 107, 110, 112, 178, 194
Artichoke 265
Arab 38, 44, 76, 92, 145, 298,
Arabs 92, 152
Aragon 65
Aristotle 28, 33, 34, 93, 94, 100, 101, 110, 124, 125, 251, 252, 254, 257, 266, 271, 299, 314
Armadillo 31, 34, 245, 258, 264, 266, 281
Armament 139, 198
Arrobas 51, 135, 204
Arte de navegar 11, 47, 76, 81, 88, 89, 90, 96, 129, 146, 151, 179, 214, 220, 221, 225, 288, 307, 308
Arctic 99, 288
Artillery 7, 80, 131, 135, 136, 139, 141, 142, 191, 198, 213
Artois 65
Archbishoprics 57
Asia 44, 82, 98, 106, 116, 232, 234, 237, 238, 239, 247, 268
Astrolabe 7, 12, 31, 77, 78, 79, 80, 81, 89, 96, 121, 122, 143, 145, 146, 147, 148, 149, 150, 151, 152, 153, 154, 166, 167, 170, 172, 175, 177, 179, 184, 187, 189, 192, 194, 196, 224, 226, 241
Astrology 72, 96, 110, 193, 194, 275
Astronomy 32, 33, 41, 42, 75, 80, 89, 91, 96, 101, 103, 109, 116, 145, 157, 160, 191, 193, 213, 293, 298, 299, 313
Astronomical table 8, 73, 89, 122, 145, 153, 170, 175
Atlantic Ocean 50, 75, 123
Audiencia of Mexico 57
Audencia of Santo Domingo 56, 67
Audiencias 56, 57, 67
Austria 65, 82
Axes 19, 141, 197
Axes 19, 141, 197
Azores Island 49, 230

Bacon, Francis 5, 13, 205, 285, 287, 288, 293, 298, 315
Bahamas 49, 50
Bahía del Caracol 228
Ballast 19, 131, 139, 234
Barber 8, 186, 187, 199
Barrels 19, 131, 198, 204, 205, 206
Barzán, Alonso 132
Barley grains 135, 232, 234
Battista Alberti, León 105, 289
Basque country 88
Basques 92
Bermuda Islands 49, 50, 113
Bible 8, 86, 93, 252, 263, 286, 289
Bilbao 127
Bird from paradise 262
Biscuit 40, 197, 204, 205, 206
Binnacle 131
Bilge pumps 131
Bilge 131, 197
Bishoprics 57, 68
Black Sea 37
Boat 127, 129, 140, 196
Boatswain 8, 186, 195, 200
Bohío 240
Borough, Stephen 288
Botero Benes, Juan 221
Boyle, Robert 298, 299
Botany 8, 245, 271, 302
Bow 129, 212
Brabant 65
Brasil 240
Brunelleschi, Filippo 105
Broad beans 205
Brasil wood 63

Cake 273
Cabin 131, 167, 174, 202
Cape Bojador 40
Cabo Lindo 228
Cape Verde 230
Cactus 267
Cádiz 49
Calafate 186, 190
Caulk 133, 134
Calatayud 135
Calms 48, 50, 70, 204, 274
Camels 252, 264
Canary Islands 49, 62, 117, 246
Cannibals 118, 248, 253
Canoes 71, 248
Cantabria 89
Cannons 28, 29, 33, 140, 141, 142
Captain 8, 57, 81, 88, 120, 122, 127, 131, 133, 142, 143, 186, 188, 190, 191, 195, 196, 197, 199, 200, 201, 202, 205, 253, 316
Capricorn 96, 99, 100, 106, 228, 241
Caravels 127, 129, 130, 143, 189
Carpentry 136
Carpenter (s) 8, 140, 186, 190, 197
Carrera de las Indias 48, 310
Cartagena 11, 49, 93, 118, 279
Cannibals 118
Cartography 7, 8, 24, 31, 34, 41, 42, 47, 48, 49, 55, 58, 69, 70, 73, 76, 77, 80, 82, 89, 90, 91, 94, 96, 98, 101, 103, 105, 106, 108, 109, 136, 160, 172, 175, 185, 217, 218, 219, 220, 223, 224, 228, 229, 230, 231, 234, 239, 240, 241, 242, 266, 282, 294, 296, 300, 304, 311, 315
Casa da Mina 41
Casa de Contratación 7, 29, 32, 35, 55, 56, 57, 58, 61, 62, 63, 64, 65, 66, 67, 68, 69, 72, 76, 77, 78, 81, 107, 113, 145, 184, 217, 218, 234, 240, 285, 288, 301
Casa de Guiné 41
Casa de Guiné e Mina 41
Casa de la Moneda 62
Castellán, Juan Antonio 221
Castilla de Oro 64
Castile 42, 43, 57, 64, 65, 67, 92, 235, 239, 249, 265, 270, 298
Catalans 92
Cathay 37, 38
Catholic Church 208
Canvas 119, 133, 135, 136, 186
Cabin boy 197, 200
Catholic kings 240
Caribbean 19, 34, 47, 111, 116, 117, 118, 140, 141, 248, 280
Celestial navigation 32, 75, 78, 80, 91, 172, 177, 185, 189, 193
Ceuta 40
Cédula real 78
16th century 17, 18, 19, 20, 25, 26, 29, 30, 31, 32, 33, 34, 35, 40, 41, 44, 45, 49, 55, 57, 59, 60, 62, 64, 71, 73, 75, 80, 81, 86, 87, 90, 92, 93, 94, 98, 100, 105, 106, 120, 122, 123, 130, 131, 136, 139, 141, 145, 151, 164, 166, 167, 172, 177, 178, 182, 184, 185, 188, 189, 194, 202, 204, 205, 206, 208, 211, 212, 218, 220, 223, 224, 229, 230, 236, 247, 249, 251, 252, 256, 257, 258, 262, 263, 265, 272, 274, 277, 282, 285, 286, 288, 294, 295, 296, 298, 298, 299, 300, 301, 303, 304, 310
Charles I 57, 288
Charles V 64, 65, 66, 86, 93, 254, 278
Chart 8, 12, 13, 31, 78, 79, 145, 158, 163, 170, 172, 173, 174, 175, 185, 192, 194, 196, 218, 219, 220, 222, 224, 225, 226, 227, 228, 234, 235, 240, 241
Chaves, Jerónimo 67, 78, 81, 307
Chaves, Alonso de 76, 78, 81, 87, 88, 92, 104, 113, 118, 133, 135, 141, 142, 143, 144, 147, 158, 161, 172, 173, 177, 178, 180, 190, 200, 202, 205, 209, 212, 218, 225, 226, 227, 307
Chile 67
China 38
Churcha 265
Chorography 70, 98, 116
Christianity 39, 53, 208, 212, 250, 256, 275
Christ 85, 210, 213, 227, 235
Chronicler 68, 211, 246, 255, 256
Christian faith 86
Church 57, 208, 212, 238, 250
Cinnamon 19, 39, 41, 51, 280
Cieza de León, Pedro 221, 264,
Cipango 18, 42, 43
Cloth 133, 199
Clergy 208, 250, 79, 87, 166, 168, 170, 196
Colón, Cristóbal 43, 50, 51, 52, 56, 253, 308, 312, 314
Colón, Diego 56
Compostela 57
Conquerors 23
Conquistadores 227, 249
Council of Castile 65
Consejo de Hacienda 68
Consejo Real y Supremo de las Indias 55, 56, 60, 64, 65, 68, 316
Consejo de las Indias 32, 35, 55, 56, 57, 58, 59, 60, 62, 64, 65, 66, 68, 70, 72, 91, 231, 246, 285, 301, 316
Consulate of Seville 68
Contramestre 190
Convoys 59, 142
Copernicus, Nicolaus 94, 293
Coca 272, 273, 281
Corsairs 39, 142, 229
Cortés, Hernán 65, 67, 136
Cortés, Martín 72, 81, 90, 146, 164, 221, 229, 288
Corvatones 137
Cosa, Juan de la 8, 18, 234, 235, 236, 237, 241
Cosmography 7, 25, 31, 32, 33, 44, 45, 67, 72, 73, 75, 78, 79, 80, 81, 86, 92, 94, 95, 98, 103, 105, 106, 109, 116, 117, 124, 125, 135, 172, 180, 193, 194, 230, 232, 234, 269, 282, 297, 298, 300

Cosmographer 67, 78, 90, 218, 240, 255
Cosmographer-chronicler 68, 69, 70, 72
Cosmographiae Introductio 236, 239, 316
Cosmology 72, 98, 100, 116, 124, 174, 178, 213, 252, 271, 292, 304
Copernican revolution 166, 292, 295, 299
Compass Rose 12, 13, 131, 160, 161, 162, 169, 226, 235, 236, 241
Council of the Indies 32, 55, 56, 65, 66, 275
Compass 7, 12, 13, 31, 41, 79, 87, 89, 131, 145, 156, 159, 160, 161, 164, 166, 175, 224, 228
Crossbows 140, 142
Cross-staff 155, 157
Cronberger, Jacobo 220
Cruz, Martín de la 276, 277, 278, 279
Crusaders 39
Crew 8, 33, 42, 45, 66, 80, 88, 89, 113, 119, 121, 122, 123, 124, 127, 129, 130, 131, 139, 167, 168, 185, 188, 189, 190, 191, 194, 196, 197, 198, 199, 200, 202, 204, 205
Customs house 56, 61
Cubit 137
Currents 7, 11, 32, 37, 40, 44, 45, 46, 48, 70, 80, 89, 109, 110, 113, 116, 124, 166, 192
Cuba 64, 78, 118
Cuesta, Juan de la 91, 222, 307

Da Vinci, Leonardo 105
Darts 198
D'Angelo, Jacopo 95
Daggers 142
Del Cano, Sebastián 18
De Triangulis 98
Declination 12, 42, 78, 80, 149, 150, 154, 163, 164, 176, 177, 178, 180, 192
Deck 131, 187, 197, 198, 202, 213
Devil 209, 212
Demand 130, 167, 240
Descartes 293, 298, 303
Despensero 186, 187, 190
Depth 31, 32, 70, 76, 104, 108, 113, 123, 130, 172, 217, 256
Degrees 13, 42, 43, 46, 95, 97, 118, 135, 145, 146, 148, 149, 151, 155, 156, 158, 162, 163, 164, 175, 177, 180, 184, 225, 230, 231, 232
Díaz, Bartolomé 40
Dioscorides 34, 251, 252, 254, 255, 277, 299
Discovery 17, 20, 21, 23, 24, 31, 44, 46, 50, 59, 64, 69, 80, 185, 219, 223, 230, 250, 251, 252, 258, 286, 289, 294, 296, 297, 304, 312, 313
Dominicans 57
Dog-headed men 253

Eanes, Gil 40
East 18, 19, 37, 40, 46, 49, 82, 98, 99, 109, 116, 120, 163, 168, 180, 189, 226, 228, 230, 280
Eclipses 69, 71, 231, 232
Ecliptic 177
Egypt 38, 119, 120, 239
El Escorial 72, 73, 105, 255, 272
El mal de Bubas 280, 281
Empiricism 303
Empire 1, 3, 7, 8, 9, 17, 21, 27, 29, 37, 55, 56, 58, 68, 72, 75, 79, 81, 82, 85, 86, 93, 105, 119, 186, 208, 217, 223, 230, 245, 251, 254, 260, 271, 281, 282, 285, 286, 289, 294, 299, 300, 301, 304, 309, 310, 311, 312, 314
England 26, 34, 38, 89, 90, 134, 189, 288, 309
Encubertado 264
Escalante de Mendoza, Juan 91, 97, 100, 101, 102, 103, 104, 128, 150, 160, 171, 182, 299
Etymologies 252
Ethnography 294, 315
Eurocentrism 7, 9, 17, 22, 23, 23, 26, 285, 296, 299, 303, 305, 311, 315
Europe 13, 18, 19, 20, 22, 23, 24, 25, 26, 27, 28, 30, 31, 34, 37, 40, 44, 45, 46, 51, 58, 59, 65, 67, 79, 80, 82, 90, 91, 93, 98, 106, 116, 120, 122, 127, 136, 141, 143, 225, 229, 232, 234, 237, 238, 239, 241, 246, 247, 250, 251, 252, 253, 255, 258, 261, 263, 266, 277, 280, 281, 282, 283, 285, 288, 289, 293, 294, 295, 298, 304, 305, 309, 311, 315, 316
Equatorial line 157
Equinoctial line 11, 100, 103, 149, 180
Experience 29, 30, 41, 70, 77, 79, 88, 92, 105, 106, 107, 108, 110, 115, 120, 134, 135, 160, 164, 173, 184, 189, 192, 193, 194, 195, 196, 197, 201, 202, 203, 210, 218, 229, 243, 253, 273, 274, 280, 289, 292, 296, 297, 298, 290, 299, 300, 301
Exploration 2, 17, 19, 20, 25, 26, 31, 32, 37, 38, 40, 41, 43, 45, 64, 75, 108, 123, 130, 136, 157, 172, 173, 188, 204, 207, 208, 219, 223, 224, 228, 229, 234, 235, 238, 246, 252, 272, 273, 286, 289, 295, 299, 304

Falero, Francisco 307
Fauna 34, 248, 275, 310
Felines 267, 268, 270
Fernández de Cordoba, Diego 221
Fernández de Cordoba, Francisco 76, 220, 307
Fernández de Enciso, Martín 11, 81, 82, 83
Fernández de Oviedo, Gonzalo 13, 119, 246, 249, 250, 259, 264, 267, 277, 282, 290, 299
Feet 135, 137, 234, 245, 251, 258, 264, 281
Fleur de lis 160, 226
Flora 34, 248, 255, 272
Flamenco, Juan 222
Flanders 65, 82, 127, 134, 189
Fleets 89, 315
Flat 34, 127, 156, 173, 173, 241
Food 8, 121, 131, 197, 202, 204, 205, 206, 207, 246, 260, 262, 273
Freight 38, 190
Frigate 133
France 26, 38, 134, 189, 314
Franciscans 57
Fray Antonio Guevara 188, 201, 206

Fray Bernandino de Sahagún 249, 275, 277, 282, 308, 310
Fray García de Loaysa 246
Friar Nicolás de Ovando y Cáceres 62
Fray Tomás de la Torre 203, 204, 205, 206, 210, 211, 213

Galleon 62, 127, 130
Galleys 123, 127, 130, 189
Galeza 132
Galileo Galilei 94, 298, 299
Galicians 92
García Céspedes, Andrés de 13, 73, 76, 81, 91, 178, 222, 289, 291
García de Palacio, Diego 81, 85, 88, 91, 111, 124, 136, 138, 140, 149, 153, 154, 157, 159, 165, 170, 172, 202
Gattinara, Mercurino 64
Genoa 37, 38, 60
Geography 23, 25, 30, 33, 35, 37, 41, 42, 44, 45, 58, 68, 70, 71, 72, 80, 82, 94, 95, 98, 99, 101, 106, 107, 116, 121, 145, 162, 217, 218, 225, 231, 239, 249, 252, 253, 271, 275, 296, 297, 298, 301, 302
Geometry 94, 95, 102, 103, 105, 135, 228
Ginger 39, 51, 280
Ginés de Sepúlveda, Juan 93
Girava, Jerónimo 221
Glass vials 168
Globe 18, 25, 41, 42, 43, 45, 76, 97, 98, 118, 120, 146, 167, 173, 175, 177, 180, 185, 226, 230, 231, 232, 238, 241
Gnomon 95
Gold 7, 19, 28, 50, 51, 58, 60, 61, 62, 63, 67, 86, 121, 192, 198, 227, 246, 248, 250, 281
God 8, 17, 33, 64, 82, 85, 86, 87, 113, 115, 122, 134, 200, 207, 208, 209, 210, 211, 212, 213, 215, 227, 245, 249, 251, 254, 257, 260, 261, 263, 271, 282, 286, 288, 292, 300
Greek 92, 94, 98, 121, 257, 314
Graduation 154, 226, 228
Guadalquivir river 49, 59, 120
Gulf of Mexico 19, 49
Guadalajara 67
Guadalupe 212
Guard 155, 168, 187, 191, 200, 215
Guardian 8, 186, 187, 195, 196
Guatemala 57, 67
Guayacan 280, 281
Gunpowder 139, 142, 198, 199

Hatches 131, 197
Hammock 204
Hammers 197
Health 8, 67, 189, 199, 204, 207,
Hernández, Francisco 249, 255, 260, 272, 275, 281, 282, 308
Herrera, Antonio de 222
Herrera, Juan de 72, 105
Hesiod 108
Hermetic tradition 87, 292
Hemp 127, 134, 135, 172
Historia general y natural de las Indias 13, 119, 246, 247, 259, 265, 267, 290, 308
Horizon 35, 42, 100, 146, 148, 151, 152, 157, 158, 177, 178, 210, 285, 302
Holy Spirit 227
Holy Trinity 227
Hourglasses 121, 143, 168, 196, 215
Huerta, Jerónimo de la 256
Humanism 7, 89, 92, 93, 94, 101, 105, 106, 250, 251, 298, 300
Hurricanes 50, 70, 109, 110, 141, 298
Hydrodynamics 129
Hydrography 70

Iberian Peninsula 7, 32, 37, 40, 58, 75, 92, 94, 119, 120, 189
Iguana 31, 34, 245, 248, 258, 267
Imago Mundi 252
India 18, 19, 29, 38, 39, 40, 41, 116, 164, 204, 240, 295, 313
Indias 13, 17, 18, 19, 32, 35, 46, 55, 56, 57, 58, 59, 60, 61, 62, 64, 65, 66, 68, 70, 71, 72, 78, 79, 91, 92, 117, 119, 131, 136, 141, 167, 184, 187, 188, 189, 190, 194, 199, 201, 202, 203, 206, 209, 210, 211, 213, 218, 231, 245, 246, 247, 248, 254, 256, 257, 258, 259, 265, 267, 277, 279, 280, 288, 289, 290, 292, 298, 301, 308, 309, 311, 312, 314, 316
Indian 18, 46, 60, 250, 315
Inquisition 66
Instauratio Magna 5, 13, 285, 287
Instrucción náutica 81, 85, 88, 91, 100, 110, 111, 124, 136, 137, 138, 139, 140, 141, 142, 145, 149, 153, 154, 155, 156, 157, 159, 163, 166, 168, 170, 172, 173, 178, 180, 186, 191, 194, 195, 234, 307
Instruments 7, 8, 29, 30, 31, 33, 34, 41, 42, 45, 59, 60, 64, 66, 73, 77, 78, 79, 80, 82, 88, 89, 91, 95, 103, 104, 108, 113, 119, 123, 124, 131, 135, 143, 144, 182, 184, 185, 186, 187, 191, 192, 195, 199, 218, 223, 224, 241, 242, 252, 302, 309, 313
Indian Ocean 18, 46, 250
Isla de Tortuga 228
Island of Gomera 49
Island of La Deseada 49
Islario 231
Italy 26, 37, 65, 92, 93, 119, 120, 246
Itineraries 38, 69, 76, 89, 116, 117, 229

Jaguar 264
Jamaica 64, 117, 118
Jews 66, 92, 256
Jesuits 57, 256
Juana 227
Juridical 55, 58, 65, 67, 73, 300
Junta de Guerra 68
Jupiter 99

Kamal 152
Kepler, Johannes 94, 166
King Solomon 86, 292
Kinetic energy 45
Kochab 169, 170

Lantern 131, 142
La Capitana 168
La Española 228
La Niebla 42
La Niña 130
La Pinta 130
La Santa María 130, 215, 227, 246
Latitude 41, 42, 68, 76, 80, 81, 97, 98, 99, 104, 117, 145, 146, 157, 174, 175, 177, 178, 180, 226, 230, 232
L'Ecluse, Charles de 72, 278
League 117, 135, 232, 234
Lizards 248, 270, 271
Light 115, 123, 127, 134, 266, 286
Lions 248, 252, 270
Libro de las Maravillas 121, 308
Libro de las Longitudes 231
Libya 98
Linen 119, 120, 211
Lima 57, 67, 256, 293, 312, 314
Lisbon 18, 19, 38, 40, 41, 81, 145
Liberal arts 103
Life on-board 119
Low tide 111
Load 130, 195
Lombards 142
Lombardero 198
López de Gómara, Francisco 17, 18, 247, 288, 292, 308
López de Legazpi, Miguel 18, 136
López de Velasco, Juan 70, 72
Local Knowledge 245, 275
Long Distance Control 20, 29, 30, 58, 282, 285, 313
Luxembourg 65
Lucitanian 39, 41

Mathematical Academy of the Court 72
Mast 124, 130, 131, 146, 210
Master 77, 108, 174, 200, 218
Maestre 8, 186, 187, 190, 195, 196
Magallanes 80, 143, 308
Magic 289, 292, 293
Maize 272, 273, 273
Mandeville, John 121, 308
Magnetic declination 163, 164
Manuel of Portugal 239
Manuals 7, 32, 37, 45, 75, 79, 80, 81, 82, 86, 87, 88, 89, 90, 91, 92, 96, 98, 105, 106, 107, 108, 109, 110, 113, 116, 117, 124, 125, 133, 135, 141, 144, 151, 158, 164, 170, 172, 177, 178, 179, 185, 186, 190, 191, 192, 194, 197, 198, 201, 204, 208, 214, 224, 225, 302
Maps 8, 20, 25, 34, 41, 48, 49, 66, 71, 76, 78, 80, 90, 95, 99, 108, 113, 118, 160, 161, 173, 174, 192, 217, 218, 219, 220, 222, 223, 224, 226, 227, 228, 229, 230, 231, 234, 235, 236, 239, 240, 242, 296, 301, 302, 303
Machines of the Empire 7, 119, 125
Máquinas del Imperio 4, 26, 28, 120, 141, 167, 310
Marco Polo 39, 51, 121
Marine 139, 253
Mars 99
Mártir de Anglería, Pedro 220, 264
Mast 124, 130, 131, 146, 210
Mathematization 80, 303
Mainland 77, 86, 92, 117, 118, 130, 219, 230, 245, 254
Manuals of Navigation 32, 75, 80, 214
Magnetic needle 31, 32, 161, 163, 164
Medieval bestiaries 252
Medicine 8, 207, 250, 251, 252, 255, 271, 272, 277, 280, 285, 294, 301, 304, 310, 315
Measurement 71, 95, 103, 104, 137, 144, 149, 150, 151, 153, 154, 163, 167, 175, 177, 184, 185, 230, 234, 242, 302
Measures 11, 63, 114, 135, 158, 185, 220, 232, 234
Medina, Pedro de 11, 47, 72, 78, 81, 84, 96, 108, 112, 114, 126, 129, 149, 151, 155, 156, 165, 174, 176, 178, 179, 181, 183, 192, 214, 220, 225, 233, 299
Mediterranean 19, 29, 32, 37, 39, 40, 41, 59, 75, 76, 123, 130, 143, 157, 159, 160, 161, 164
Mela, Pomponius 82, 93
Menéndez de Avilés, Pedro 132, 136
Merchandise 56, 58, 61, 64, 127, 132, 192, 195
Meteorology 7, 80, 89, 93, 108, 109, 110
Mexico 5, 19, 22, 23, 24, 43, 49, 50, 57, 67, 81, 85, 91, 111, 123, 136, 241, 255, 260, 274, 275, 276, 307, 308, 309, 310, 311, 312, 314, 315, 316
Midshipmen 8, 190, 195, 196, 205
Mizzen 130, 131
Mile 135, 234
Modern science 17, 18, 20, 31, 35, 41, 73, 80, 166, 243, 286, 289, 292, 293, 294, 295, 298, 299, 303, 304
Moral History 70
Moon 12, 79, 87, 94, 95, 99, 109, 110, 134, 167, 177, 180, 182, 192
Monardes, Nicolás 249, 277, 279, 281, 299, 308
Monarchy 7, 29, 64, 87, 209, 247, 250, 251
Monsters 8, 29, 121, 258, 262, 313
Monsoons 45, 46
Moors 66, 265
Modern Science 3, 17, 18, 20, 31, 35, 37, 41, 55, 73, 75, 80, 119, 166, 217, 243, 245, 285, 286, 289, 292, 293, 294, 295, 298, 299, 303, 304, 310, 316
Mundus Novus 240

Nautical needle 131, 161
Medicinal plants 71, 272, 273
Navigational charts 8, 34, 172, 173, 187, 217
Nail 149
Natural History 33, 70, 94, 119, 211, 245, 252, 255, 256, 262, 272, 275, 294, 311, 315
Natives 8, 21, 53, 65, 86, 116, 240, 246, 249, 250, 271, 272, 273, 274, 275, 276, 280, 282, 299
Náhuatl 275, 277, 282, 310
Nao 50, 113, 125, 131, 133, 134, 135, 136, 142, 172, 175, 180, 186, 195, 197, 200, 205, 209
Nao Victoria 120
Naos 85, 91, 125, 127, 130, 133, 134, 137, 217, 235, 307, 310
Naples 13, 65, 82, 283, 293
Navigation 7, 8, 13, 15, 29, 31, 32, 37, 40, 41, 43, 44, 45, 46, 47, 55, 58, 63, 67, 68, 72, 73, 75, 77, 78, 79, 80, 81, 82, 85, 86, 87, 88, 89, 90, 91, 101, 105, 106, 107, 108, 110, 111, 112, 113, 115, 116, 118, 119, 121, 123, 124, 125, 127, 129, 130, 133, 136, 141, 143, 144, 145, 154, 157, 158, 159, 160, 164, 167, 172, 173, 175, 177, 178, 180, 185, 186, 187, 189, 191, 192, 193, 194, 202, 205, 208, 209, 214, 218, 223, 224, 225, 227, 229, 234, 294, 297, 298, 300, 302, 304, 309, 312, 313, 315, 316
Nature 8, 15, 20, 23, 24, 25, 27, 28, 29, 30, 34, 45, 52, 79, 87, 98, 102, 105, 107, 108, 109, 110, 111, 112, 113, 118, 122, 124, 141, 144, 145, 151, 166, 188, 198, 201, 207, 208, 220, 224, 245, 246, 247, 249, 250, 251, 252, 253, 254, 257, 258, 260, 261, 262, 263, 268, 271, 272, 273, 275, 277, 282, 283, 286, 289, 294, 295, 297, 299, 300, 301, 302, 309, 310, 311
Naturalists 28, 30, 31, 34, 52, 62, 87, 245, 249, 250, 251, 252, 253, 256, 257, 262, 263, 271, 272, 275, 296, 297, 298, 299, 283
Nautical 8, 12, 33, 41, 60, 66, 67, 72, 77, 81, 85, 90, 91, 105, 113, 116, 131, 133, 135, 146, 152, 159, 161, 164, 172, 180, 185, 201, 208, 217, 219, 223, 224, 225, 229, 230, 294, 299, 300, 309, 315
Navegación de altura 32, 75
Navidad 227
Navíos 66
Navy 139, 239
Nebrija, Antonio de 93
Neoplatonism 102, 105
Neptune 194
Networks 19, 28, 30, 45, 123, 124, 188, 241, 303, 305
Newton, Isaac 105, 293, 294, 298, 311, 316
New Atlantis 286
North Pole 106, 154, 155, 157, 168
Noah 86, 116, 209
Notary 190, 198
Noah 86, 116, 209
Northwest 46, 49
North 12, 26, 39, 40, 42, 46, 49, 78, 79, 87, 95, 98, 106, 118, 123, 133, 146, 151, 154, 155, 156, 157, 158, 161, 163, 164, 165, 168, 169, 170, 171, 177, 178, 184, 192, 218, 226, 228, 235, 241, 310
Nueva España 57, 255, 256, 260, 262, 272, 273, 275, 308
Nueva Galicia 57
Nuevo Mundo 5, 220, 273, 314
Nutmeg 19, 39
Núñez Cabeza de Vaca, Alvar 264
Núñez de Guzmán, Hernán 93

Oak 133, 134
Oceanic exploration 32, 37, 75
Ocharte, Pedro de 85, 91, 307
Ochi 268
Ochoa de Isasaga, Pedro 63
Oil lamp 145
Old world 234
Onions 204
Opossum 258, 264, 265
Ordinances 61, 62, 67, 68, 69, 139, 186, 212
Ovando, Juan de 62, 68, 72, 275

Pacific 18, 46, 123, 136, 230
Palm 146, 232
Palo Brasil 51
Palo Santo 280
Palos and Moguer 42
Panama 67
Pao do mar 40
Parallel 49, 97, 120, 178, 180
Parla marinera 201
Passengers 19, 62, 63, 121, 127, 130, 131, 132, 187, 197, 198, 199, 201, 202, 204, 205
Pasamuros 142
Pataches 127
Patagonia 250, 264
Patronato Real 57
Pacific Ocean 20, 136
Pearls 51, 86, 248
Pepper 19, 38, 39, 41, 280
Penguins 264
Philip II 57, 67, 68, 69, 72, 86, 93, 209, 255, 256, 272, 277, 288
Philip III 72
Philippines 67
Pineapple 31, 34, 198, 245, 264, 265, 266
Pitch 140, 186
Pillars of Hercules 98, 285, 288, 289, 316
Pigafetta, Antonio 142, 143, 206, 207, 264, 308
Pilot 8, 11, 69, 70, 75, 78, 88, 90, 97, 107, 108, 109, 110, 113, 115, 117, 119, 122, 126, 127, 133, 145, 149, 153, 154, 166, 173, 174, 175, 177, 180, 184, 186, 187, 188, 190, 191, 192, 193, 194, 195, 197, 199, 201, 205, 218, 224, 228, 288
Piloto Mayor 7, 66, 76, 77, 218, 219, 315
Pinedo, Francisco 62, 63
Pingues 127
Pine trees 133, 134, 265
Pineapple 13, 31, 34, 198, 245, 264, 265, 266, 267
Pipes 274

Pirates 39, 59, 88, 141, 142, 207, 224, 229
Piracy 39
Pizarro, Francisco 22, 57
Planets 11, 94, 95, 99, 101, 104, 167
Plato 102, 105
Pliny 33, 34, 50, 51, 93, 94, 106, 108, 110, 119, 120, 251, 252, 254, 255, 256, 257, 262, 266, 268, 271, 272, 297, 298, 299
Plumb 8, 89, 113, 145, 172, 192, 194
Pole star 42, 78, 151, 152, 155, 160, 168, 169, 175, 184
Potatoes 272
Polo, Niccoló 37
Poles 11, 99, 100, 102, 121
Portugal 7, 19, 26, 32, 37, 38, 39, 40, 41, 60, 75, 80, 135, 141, 229, 230, 235, 239, 240, 241, 293, 294, 298, 312
Portuguese 15, 18, 19, 20, 29, 31, 38, 39, 40, 41, 42, 43, 46, 60, 63, 80, 89, 92, 123, 127, 130, 218, 240, 250, 298, 309, 310, 313
Poop deck 131
Portugal 7, 19, 26, 32, 37, 38, 39, 40, 41, 60, 75, 80, 135, 141, 229, 230, 235, 239, 240, 241, 293, 294, 298, 312
Prince Juan 246, 250
Prow 131, 168, 196, 203
Priest 8, 199
Provisions 40, 42, 62, 64, 76, 123, 131, 132, 139, 142, 190, 191, 197, 198, 202, 204, 205, 206
Ptolemy, Claudius 33, 42, 43, 94, 95, 98, 99, 101, 106, 110, 231, 232, 238, 254, 269, 297, 299
Puerto Bello 228

Quadrant 12, 77, 78, 79, 95, 145, 151, 154, 175, 178, 192, 194, 196, 241, 242
Quatri Partitu 78, 81, 87, 88, 92, 104, 113, 118, 133, 135, 141, 142, 144, 147, 158, 161, 168, 172, 173, 177, 178, 180, 190, 202, 205, 209, 212, 225, 226, 227, 307
Quito 67

Reefs 70
Real Consejo de Indias 32, 55, 56, 58, 301
Regimiento de Navegación 11, 13, 78, 79, 81, 84, 87, 88, 90, 91, 112, 113, 114, 126, 151, 156, 165, 166, 174, 176, 181, 183, 222, 224, 233, 289, 291, 307
Regiomontanus 98
Regulations 30, 63, 66, 122, 218
Religion 1, 3, 4, 17, 28, 29, 37, 55, 75, 86, 119, 208, 213, 217, 236, 245, 246, 255, 275, 285, 292, 293, 295, 304, 305, 309
Renaissance 18, 27, 28, 32, 34, 42, 77, 79, 87, 89, 93, 94, 95, 98, 101, 103, 105, 106, 218, 246, 250, 257, 263, 283, 286, 289, 292, 293, 294, 295, 298, 300, 309, 311, 313, 315, 317
Reptiles 261, 267, 270
Resin 136, 280
Reeds 297
Rhinoceros 252
Ribero, Diego 8, 13, 152, 234, 240, 241, 242, 243
River Tanais 98
Rio Tinto 130
Ribeira das Naus 40
Rigging 40, 61, 124, 125, 129, 131, 133, 136, 138, 196, 204, 206
Royal treasury 42, 56
Roman Empire 238, 254
Royal Society of London 286, 288
Robertis, Dominico 220
Romans 17, 247, 288
Rod 135, 151, 155
Roteiros 80
Routes 18, 19, 37, 38, 39, 40, 41, 46, 50, 69, 70, 76, 77, 80, 88, 90, 91, 92, 116, 127, 141, 163, 164, 189, 229, 241, 246, 282, 296
Rope 168
Royal treasury 42, 56
Rudder 113, 122, 124, 131, 134, 201

Sardinia 65
Sailors 8, 29, 31, 33, 45, 46, 49, 59, 60, 76, 78, 88, 89, 107, 108, 110, 113, 115, 122, 127, 130, 136, 149, 152, 153, 155, 164, 166, 168, 172, 184, 187, 188, 189, 190, 194, 196, 197, 198, 199, 200, 201, 204, 205, 207, 208, 209, 211, 213, 215, 223, 242, 259
Sacrobusto, Johannes de 81, 98, 308
Salamanca 24, 80, 88, 91, 93, 94, 105, 203, 204, 206, 210, 211, 213, 251, 256, 275, 308, 311
Salazar, Eugenio 184, 187, 194, 201, 203, 206, 201
San Francisco 275
San Lorenzo de El Escorial 72
Saint Paul 86
San Salvador 213, 227
Sancho de Matienzo 62
Sanlúcar 49, 67, 117, 131, 193
Sanlúcar de Barrameda 59, 67, 193
Santa Cruz 136, 231, 232, 240
Santa Cruz, Alonso 231
Santa Fe 43, 55, 57
Santa Fé de Bogotá 67
Santa María de la Concepción 227
Santa Marta 117, 118
Santángel, Luis de 43
Santo Domingo 49, 56, 67, 117, 246, 277
Saints 29, 200, 211, 212
Sardines 134, 205
Saturn 99
Sahagún, Bernandino de 249, 275, 277, 282, 308, 310
Sails 11, 28, 29, 33, 40, 45, 85, 91, 109, 119, 120, 122, 123, 124, 125, 127, 128, 129, 130, 131, 133, 134, 135, 136, 139, 142, 143, 186, 188, 192, 193, 195, 196, 197, 201, 211
Sand clock 168
Scythes 142

Science 2, 9, 15, 17, 19, 23, 25, 26, 27, 30, 31, 32, 33, 34, 35, 51, 72, 73, 75, 80, 87, 89, 90, 91, 92, 101, 103, 104, 105, 106, 107, 110, 157, 185, 186, 195, 223, 239, 243, 281, 285, 286, 288, 289, 292, 293, 294, 295, 296, 297, 298, 299, 300, 301, 302, 303, 304, 305
Scale 18, 19, 37, 45, 104, 120, 173, 174, 227, 228, 238, 260
Scurvy 207
Sevilla 59, 60, 61, 68, 70, 77, 78, 88, 217, 307, 308, 309, 311, 312, 314, 315
Seville 7, 11, 18, 19, 29, 32, 35, 41, 42, 49, 55, 58, 59, 60, 61, 63, 65, 67, 68, 69, 72, 76, 77, 78, 79, 80, 81, 82, 88, 90, 92, 120, 125, 131, 134, 135, 139, 144, 145, 177, 185, 186, 192, 218, 220, 221, 240, 241, 242, 246, 268, 277, 278, 279, 280, 285, 288, 308, 310
Sexquipies 135
Sea chart (s) 79, 159, 172, 173, 174, 191, 192, 222, 223, 224, 225, 226, 227, 228, 315
Seamen 8, 33, 89, 161, 166, 187, 188, 189, 190, 196, 201, 202, 205
Sea pilots 192
Shipyards 33, 61, 136
Shipyards of Seville 61
Sheet 146, 241
Shell 139
Ship (s) 7, 11, 12, 18, 19, 29, 30, 31, 32, 33, 38, 39, 40, 41, 43, 44, 45, 47, 48, 49, 50, 53, 59, 60, 61, 62, 63, 64, 66, 67, 75, 76, 77, 78, 79, 80, 86, 88, 89, 92, 107, 108, 112, 113, 114, 115, 119, 120, 121, 122, 123, 124, 125, 126, 127, 128, 129, 130, 131, 132, 133, 134, 135, 136, 137, 138, 140, 141, 142, 143, 147, 149, 151, 159, 163, 166, 167, 168, 172, 173, 174, 175, 182, 184, 185, 186, 187, 188, 189, 190, 191, 192, 193, 195, 196, 197, 198, 199, 200, 201, 202, 203, 204, 205, 206, 207, 208, 209, 210, 211, 212, 213, 215, 227, 228, 230, 234, 235, 241, 243, 245, 250, 285, 295, 302, 315
Shipbuilding 7, 39, 80, 91, 124, 132, 135, 136, 234
Shipmaster 8, 108, 113, 186, 190, 191, 194, 195, 196, 197, 199, 205
Shipwrecks 7, 59, 80, 109, 110, 111, 113, 128, 141, 205, 211, 298
Sinú 249
Siren 253
Simple steps 135, 234
Silver 7, 19, 50, 67, 86, 121, 192, 246, 260, 281
Silk 37, 38
Sicily 60, 65
Slaves 7, 50, 51, 60, 198, 248
Sloth 264, 266
Smuggling 63, 67
Snakes 261, 270, 271
Solomon's house 286
Solís, Juan 776, 219
South Pole 42, 106, 157
Southern Cross 12, 42, 158, 159
Storm 113, 210, 211
Spanish crown 17, 56, 57, 62, 64, 67, 248, 256, 263, 277
Spices 19, 37, 38, 39, 51, 52
Spice 39
Spain's Siglo de Oro 72
Spain 5, 7, 17, 19, 25, 26, 30, 32, 33, 34, 35, 37, 38, 49, 51, 52, 55, 56, 57, 59, 60, 61, 62, 63, 65, 68, 73, 75, 78, 79, 80, 81, 82, 84, 88, 89, 90, 91, 92, 98, 105, 116, 118, 120, 123, 127, 130, 134, 136, 141, 144, 189, 202, 205, 207, 208, 209, 218, 220, 221, 222, 227, 228, 229, 230, 240, 241, 246, 247, 249, 251, 254, 255, 256, 263, 264, 266, 268, 270, 272, 275, 277, 278, 280, 282, 285, 288, 293, 294, 297, 298, 299, 300, 309, 311, 312, 314, 315
Steward 8, 186, 187, 190, 197, 205
Storm (s) 45, 70, 108, 110, 111, 113, 115, 141, 204, 210, 211, 298
Strings 204
Stern 131, 137, 142, 143, 168, 210, 212, 213
Stella Maris 168
Standardization 8, 9, 67, 73, 182, 185, 201, 218, 230, 300, 301, 302
Strasbourg 167
Strait of Gibaltar 116, 285
Strait of Magellan 118, 250
Stove 131, 196, 205
Stem 137, 279
Supplies 40, 80, 89, 123, 132, 187, 197
Sugar 121
Surgeon 8, 187, 190, 199, 205, 207
Sun clock 95
Sun 11, 12, 42, 78, 79, 80, 86, 87, 94, 95, 99, 101, 103, 110, 120, 134, 142, 146, 148, 149, 150, 151, 153, 154, 158, 163, 166, 170, 174, 175, 177, 178, 179, 180, 184, 189, 192, 195, 203, 206
Suma de Geographia 11, 81, 82, 83, 99, 106, 116, 117, 178, 232, 248, 249, 307
Syria 38

Tar 134, 139, 186, 197, 203
Talavera, Jerónimo Hernando de 43
Tallow 40, 205
Tackle 138, 195, 196
Teredo 139
Terra de Santa Cruz 240
Terra incognita 240
Treasurer 61, 62, 68
Technology 2, 15, 25, 26, 27, 28, 29, 31, 44, 73, 112, 120, 122, 132, 188, 294, 295, 300, 312, 313, 315
The New World 5, 7, 8, 9, 12, 13, 17, 21, 23, 24, 25, 26, 27, 30, 31, 32, 33, 34, 35, 46, 50, 51, 53, 55, 56, 58, 59, 60, 62, 64, 65, 69, 75, 77, 80, 82, 84, 87, 88, 90, 91, 105, 106,107, 108, 116, 118, 136, 139, 141, 144, 192, 199, 204, 208, 217, 218, 219, 220, 221, 222, 223, 225, 228, 229, 230, 232, 234, 235, 237, 238, 240, 242, 243, 245, 246, 247, 249, 250, 251, 252, 253, 254, 255, 256, 257,

258, 261, 263, 266, 268, 270, 271, 272, 273, 276, 277, 278, 280, 281, 282, 283, 285, 285, 286, 288, 289, 294, 296, 297, 298, 299, 300, 301, 302, 303, 304, 305, 309, 311, 315
The Duke of Medinacelli 42
Theory 30, 77, 89, 105, 107, 108, 135, 230, 231, 247, 280
Tight 191, 198, 200, 203
Tide 109, 110, 180
Tides 11, 12, 45, 79, 108, 109, 110, 111, 180, 183, 192, 194
Tiger 264, 268, 269
Tools 19, 28, 30, 86, 94, 131, 136, 139, 144, 186, 187, 188, 197, 217, 301
Torrid zone 99, 106
Tobacco 121, 272, 273, 274, 281
Toledo 44, 92, 255, 268
Ton 91, 135, 202
Tonnage 66, 130, 139
Topography 70
Trade winds 45, 46, 49
Trees 7, 50, 51, 52, 91, 133, 134, 247, 248, 262, 267, 276, 279
Tratado del Esphera 81, 307
Tratado de Tordesillas 229, 230, 235, 240, 241
Tristan 97, 107, 109, 113, 115, 117, 153, 166, 193
Tropic 99, 106, 241
Tropic of Cancer 95, 99, 100, 106, 228, 241
Translation 4, 81, 90, 92, 95, 97, 228, 245, 255, 256, 272, 275, 281, 298, 304, 310
Trade winds 45, 46, 49
Tubal 116

Uncia 135
Unicorns 252
Urcas 123, 127

Vargas Machuca, Bernardo de 289
Vasco de Gama 18, 38, 187
Vessel (s) 11, 19, 29, 32, 33, 39, 41, 45, 49, 59, 66, 85, 86, 113, 114, 116, 117, 123, 125, 127, 129, 130, 131, 132, 133, 136, 140, 141, 142, 166, 167, 174, 187, 190, 191, 197, 198, 199, 200, 204, 207, 228, 286, 289, 313
Veedor 190
Vellerino de Villalobos, Baltasar 81, 88, 92, 93, 117, 308
Venus 99
Vera Cruz 240
Vespucio, Américo 218, 259, 308
Vinegar 197, 204
Virgin 13, 200, 212, 215, 227, 235, 236
Virgil 108, 110
Viceroyalties 57
Vegetables 197, 205
Viceroy 55, 56, 277
Visitors 66
Vizcaya 127
Voyages of exploration 25, 31, 45, 130, 207, 223, 235, 286

Waldseemüller, Martin 13, 239, 316
Western science 33, 35, 75, 92, 281, 285, 288, 292, 296, 298
Weather 11, 110, 112, 113, 143, 155, 180, 192, 205, 213
West Indies 19
Wind power 129
Winds 11, 29, 32, 37, 40, 41, 44, 45, 46, 47, 48, 49, 70, 76, 80, 85, 89, 99, 108, 109, 110, 113, 115, 117, 123, 124, 128, 129, 155, 161, 162, 163, 166, 173, 180, 192, 204, 226
Wine 19, 39, 131, 135, 197, 198, 204, 205, 206
Wicks 145
World economy 60
Wool yarn 197
Wood 122, 133, 134, 135, 139, 142, 168, 186, 205, 224, 265
Wool 197

Zamorano, Rodrigo 72, 76, 81
Zenith 148, 149
Zodiac 99, 100

Made in the USA
Monee, IL
13 September 2026

63176216R00181